URBANIZATION and COUNTER-URBANIZATION

Edited by

BRIAN J. L. BERRY

Volume 11, URBAN AFFAIRS ANNUAL REVIEWS

SAGE PUBLICATIONS / BEVERLY HILLS / LONDON

For information address:

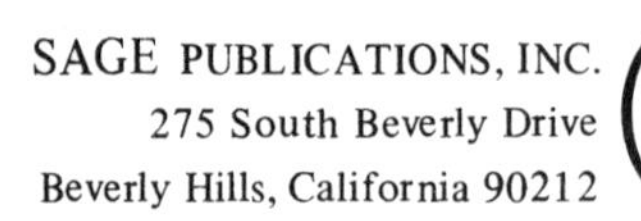

SAGE PUBLICATIONS, INC.
275 South Beverly Drive
Beverly Hills, California 90212

SAGE PUBLICATIONS LTD
St George's House / 44 Hatton Garden
London EC1N 8ER

Printed in the United States of America

International Standard Book Number 0-8039-0499-1 (cloth)
0-8039-0682-X (paper)

Library of Congress Catalog Card No. 76-15864

FIRST PRINTING

CONTENTS

Introduction:
On Urbanization and Counterurbanization

BRIAN J.L. BERRY

"From now on, I'll describe the cities to you," the Khan had said, "in your journeys you will see if they exist."

But the cities visited by Marco Polo were always different from those thought of by the emperor.

"And yet I have constructed in my mind a model city from which all possible cities can be deduced," Kublai said. "It contains everything corresponding to the norm. Since the cities that exist diverge in varying degree from the norm, I need only foresee the exceptions to the norm and calculate the most probable combinations."

"I have also thought of a model city from which I can deduce all the others," Marco answered. "It is a city made only of exceptions, exclusions, incongruities, contradictions. If such a city is the most improbable, by reducing the number of abnormal elements, we increase the probability that the city really exists. So I have only to subtract exceptions from my model, and in whatever direction I proceed, I will arrive at one of the cities which, always as an exception, exist. But I cannot force my operation beyond a certain limit: I would achieve cities too probable to be real."

Italo Calvina, *Invisible Cities*

□ HOW TIMES HAVE CHANGED. When the Social Science Research Council's Committee on Urbanization met in the early 1960s, the agenda was to review urban research in the social sciences and to identify the universals of structure, process, and stage that, transcending the idiosyncrasies of time and space, might provide the value-free basis for a general theory of urbanization. Without such a general theory, it was felt that the social sciences could advance no particular claims as sciences in the strict sense. The search was, however, short-lived, falling victim to radicalism at home and decolonization overseas. Thus, in the International Geographical Union's 1975 report *Essays on World Urbanization*, prepared by its Commission on the Processes and Patterns of Urbanization, the editor, Ronald Jones (1975:19) reported as follows:

> The phenomenon of urbanization, universal in the contemporary world, is being generated by so many different factors, operating with different emphases in each separate country, that it would be impossible, as well as unwise, to try to summarize the process in any meaningful way.

And one of Jones' collaborators, West Germany's P. Schöller, argued most strongly that urban research cannot be value free (quoted in Jones, 1975:38):

> In each country and in each urban society the assessment of problems created by urbanization inevitably includes some political and ideological premises. . . . Within the perspective of liberal capitalism, the forces of economic concentration and agglomerative urbanization are seen as natural ones. Problems and difficulties are seen to be unavoidable but temporary: governmental planning policies have to alleviate social problems but must not hinder the effectiveness of free-enterprise operations.
>
> Marxism, the counteraction to capitalism, takes economic forces also as the leading agent of history, and urbanization as the social symptom of industrialization. But the aim is not concentrated agglomeration. The overcoming of the traditional urban-rural antagonism by overall urbanization and through the destruction of traditional peasant life is seen as progress on the way to Communism. But in China, the peasant communism of Mao Tse Tung has found a different solution to Marxist ideas. Not only industrial urbanization but also industrial ruralization are held out as a means of eliminating the antagonism between city and country and of leading to new forms of collective life. . . .
>
> Anti-urban effects and sentiments are still prevailing in most conservative ideologies. European fascism and especially the Nazi ideology, with Rosenberg's 'Blut-und-Boden Mythos,' are extreme cases: here it is the imbalance between nature and human civilization that are seen as the major problem, whilst the problems of urbanization are held to be exaggerated and, if anything, not entirely relevant.

The SSRC committee sought value-free universal theory, a search deemed futile on idiosyncratic and ideological grounds by the IGU commission. Are we thus doomed to particularism? I think not. There is a middle ground that may be sensed both in Schöller's remarks and in the following quotations, derived from chapters that appear in this volume, *Urbanization and Counterurbanization:*

- A turning point has been reached in the American urban experience. Counterurbanization has replaced urbanization as the dominant force shaping the nation's settlement patterns. . . . What has finally been achieved in the 1970s is not something new, but something old, the reassertion of fundamental predispositions of the American culture that . . . are antithetical to urban concentration.
- In the Soviet Union a positive attitude toward the city and city life has prevailed. The city is in fact seen as the expression of the ideals of

socialist society, in sharp contrast to rural living which is to be gradually transformed by the infusion of urban life styles. . . . The central articulation of planning objectives, combined with the central allocation of resources, clearly has a powerful influence on all aspects of urban growth in the Soviet Union.

- [In Britain] the dominant trend is accelerating decentralization. . . . By the 1960s this relative decline [of the central cities] had been turned into an absolute loss of population. . . . Natural change in the system has, however, been dramatically affected by the New and Expanded Town legislation.
- There are differences between the West and the East as far as city sizes are concerned. In the West the small towns have been the most successful ones, especially in the German-speaking countries. . . . Ironically, it has been the East that has been most successful to date in maintaining the traditional European concept of the city core as the center of urban life and economic activity.
- Urbanization in Australia and Canada appears to have entered a new period. Growth rates are beginning to decline, and their distribution across the two urban systems is changing. Internal migration streams have shifted away from the two dominant major metropolitan areas in each country toward medium-sized cities and to small centers just outside the metropolitan regions. . . . However, a national consensus on urban goals, in pluralistic federated states such as Australia and Canada, is not likely to be obtainable in the foreseeable future.
- It is against the general background of a largely unconstrained urbanization in independent black Africa, and the generally ineffective attempts to constrain it, . . . that the more deliberate, politically motivated policies of the white South African government with regard to the cityward flow of blacks [must be] examined. . . . The racialized society is maintained by coercive organization.
- The evolution is very much related to the whole process of Brazilian development: the concentration of economic power in an important urban-industrial focus, . . . Sao Paulo, and spilling out for about 200 kilometers from it. . . . [There has been] mushrooming of the biggest places, . . . emergence of medium-sized towns, . . . asymmetry in growth between the South and the Northeast, . . . but accelerating diffusion.
- The basic outlines of present-day urban patterns [in North Africa] were established during the colonial period. Many of the contemporary problems and issues of change and policy derive from this "false start," and each of the countries must, to some extent, reorder its urban system and its cities to reflect the new goals of independence, distributive justice, and cultural integrity. Policies designed to alter [the present] realities . . . are at the core of governmental deliberations everywhere.
- West African urbanization presents an interesting problem of diversity and contrasts, deriving from the rich variety of indigenous cultures and

the different impact of French and British colonial policies. Yet permeating this pattern of diversity are several distinct and significant similarities which stem from the basic process responsible for the growth of West African urban centers—colonialism.

- Urbanization in East Africa is almost entirely a 20th-century phenomenon and quintessentially the product of European colonialism and economic exploitation. . . . [It] cannot be understood apart from the more pervasive process of underdevelopment, initiated during the period of European colonial contact.
- The center-periphery forces so evident in Central America today were initially set in motion by the Spanish invaders.
- The pattern of contemporary urban development [in Southeast Asia] dates from European colonialization of the region, beginning in the 16th century [and includes a] dualistic urban structure.
- The continuous Chinese experience with urbanization is perhaps longer than that of any other society. . . . The Maoist design is not being imposed upon a *tabula rasa;* the Chinese past helps explain why current development has been so successful. . . . The two major revolutionary goals, as identified by Chairman Mao, are the elimination of the twin and closely interrelated distinctions between mental and manual labor and between city and countryside.

Underlying all of these quotations is a common response to the laissez-faire industrial urbanization that produced the big cities of the 19th century—large, dense agglomerations of socially heterogenous immigrants. In the cities, economic opportunities multiplied, and traditional peasants were transformed into modern men. But the cities were poorly built and inadequately serviced, suffering from epidemic disease, poverty and inequality, anomie, alienation, and social disorganization. There was a shared revulsion, but there have been many attempted solutions—by free-enterprise private builders providing a suburban haven to which the upwardly mobile could escape and by various social reformers seeking more radical planned solutions for those who could not. From these reactions have emerged three types of counterurbanization, each an expression of contemporary urbanization policy: the individualistic decentralization that has culminated in the decreasing size, decreasing density, and increasing homogeneity of cities and the more ruralized life styles of the more liberal capitalist states; planned new towns as counterpoints to speculative private interest in the welfare states of Western Europe; and the Marxist search for a new settlement pattern for mankind, the city of socialist man in which traditional antagonisms between city and country are no more.

These forms of counterurbanization have been joined with attempts at decolonization in the Third and Fourth Worlds. As the new industrial powers grew in size and influence, they reached overseas for guaranteed supplies of raw materials and monopoly markets, producing direct political control of much of the rest of the globe. Colonial development was centered in primate cities and resulted in "patterned socioeconomic inequality traceable to the exploitative practices through which national and international institutions are linked in the interests of surplus extraction and capital accumulation" (Walton, 1975:34-35). Even after political independence was achieved, most Third and Fourth World countries were places with inherited neocolonial institutions and sociospatial inequalities; and, learning from both Eastern and Western examples, many of these countries have proposed counterurbanization policies of deliberate decentralization or explicit ruralization as solutions to their ills.

The results are as described by Lloyd Rodwin in 1970. "Before World War II," he said, "almost no one wanted the central government to determine how cities should grow." But, he continued, "radical changes in technology and in analytical and planning methods now make significant changes in the urban system not only feasible, but to some extent manipulable," and, because of this, "today, national governments throughout the world are adopting or are being implored to adopt urban growth strategies."

Rodwin put his finger on two important bases of today's near-universal desire in some manner to counter urbanization, the planning method, and the political process, linked as they are both ideologically in terms of world view (which involves the perception of urban problems and the specification of goals to be achieved) and practically in terms of the power to implement the means thought likely to achieve the ends. Societies differ in their planning capabilities, and this introduces one important element of differentiation. But, more importantly, cultural differences have produced divergent goals, divergent planning methods, divergent plans, and now divergent paths being followed by urban and counterurban development in the world today. As I concluded in 1973 in my book *The Human Consequences of Urbanisation:*

> The diverse forms that public intervention is taking, the variety of goals being sought, and the differences in manipulation and manipulability from one society to another are combining to produce increasingly divergent paths. . . . This makes it all the more important to understand the relations between socio-political forms and urbanisation, because the socio-polity determines the public planning style. Urbanisation, in this sense, can only be understood within the broad spectrum of closely interrelated cultural processes; among these processes, planning increases rather than decreases the range of social choice as modernisation takes place, while simultaneously restricting the range of individual choice to conform to the social path selected. . . . Images of the desirable future are becoming the major determinants of that future in societies that are able to achieve closure between means and ends. Political power is thus becoming a major element of the urbanization process. Combined with the will to plan and an image of what might be, it can be directed to produce new social forms and outcomes, making it possible for a society to create what it believes should be rather than extending what is or what has been into the future.

This book, *Urbanization and Counterurbanization,* is about contrasts. Part I focuses upon the industrial heartlands of the First and Second Worlds, where the opposing liberal and Marxist ends of the ideological scale are producing contrasting results: liberal capitalism, with its acceptance of big cities and growth as the inevitable accompaniment of success, is seeing its cities disintegrate under pluralized individualistic choices, while Marxist societies, committed to the emergence of a new settlement pattern for mankind, are preserving cities and centrality in a traditional sense. Part II then turns to the peripheries of the West, and to its interface with the Third and Fourth Worlds; and Part III treats urbanization and counterurbanization in the contexts of the Third and Fourth Worlds, where, again, the ideological contrasts inherited from East and West are becoming increasingly sharp. It is hoped that what the essays reveal is not Jones's idiosyncratic denial of urban theory, but the basis of a restructuring of that theory about decision processes and planning strategies.

An urbanization strategy is, axiomatically, concerned with the future. It involves goals; it involves motivated and informed decision-makers; it involves the will to act and the power to achieve. A society with an urbanization strategy is, necessarily, a planning society.

But the nature of planning varies with sociopolitical structure, as does the nature and degree of future-orientation and the capability to achieve consensus on goals. In consequence it is possible to identify a sequence of planning styles—and, by extension, of urbanization strategies and of determinants of the future—roughly paralleling the broad continuum from free-enterprise conditions to the conditions of the directed state.

The simplest strategy is simply *ameliorative problem-solving*—the natural tendency to do nothing until problems arise or undesirable dysfunctions are perceived to exist in sufficient amounts to demand corrective or ameliorative action. Such "reactive" or "curative" planning proceeds by studying "problems," setting standards for acceptable levels of tolerance of the dysfunctions, and devising means for scaling the problems back down to acceptable proportions. The focus is upon present problems, which implies continually reacting to processes that have already worked themselves out in the past. In a processual sense, then, such planning is past-oriented. And the implied goal is the preservation of the "mainstream" values of the past by smoothing out the problems that arise along the way.

A second style of planning is *trend-modifying problem avoidance.* This is the future-oriented version of reactive problem-solving. Present trends are projected into the future, and likely problems are forecast. The planning procedure involves devising regulatory mechanisms to modify the trends in ways that preserve existing values into the future, while avoiding the predicted future problems. Such is Keynesian economic planning, highway building designed to accommodate predicted future travel demands, or Master Planning using the public counterpoint of zoning ordinance and building regulations.

The third planning style is explicitly *speculative profit-seeking.* Analysis is performed not to identify future problems, but to seek out new growth opportunities. The actions that follow pursue those opportunities most favorably ranked in terms of returns arrayed against feasibility and risk. Such is the entrepreneurial world of corporate planning, the real-estate developer, the industrialist, the private risk-taker—and also of the public entrepreneur acting at the behest of private interests, or the national leader concerned with exercising developmental leadership, as Ataturk did in building Ankara, or as the Brazilians are doing in developing Amazonia today. It is in this latter context in already-developed situations that the concept of strategy planning was developed.

Finally, the fourth mode of planning involves explicitly *normative goal-orientation.* Goals are set, based upon images of the desired future, and policies are designed and plans implemented to guide the system toward the goals or to change the existing system if it cannot achieve the goals. This style of planning involves the cybernetic world of the systems analyst, and it is only possible when a society can achieve closure of means and ends—i.e., acquire sufficient control and coercive power to ensure that inputs will produce desired outputs.

The four different planning styles have significantly different long-range results, ranging from haphazard modifications of the future produced by reactive problem-solving, through gentle modification of trends by regulatory procedures to enhance existing values, to significant unbalancing changes introduced by entrepreneurial profit-seeking, to creation of a desired future specified ex ante. Clearly, in any country there is bound to be some mixture of all styles present, but, equally, predominant value systems so determine the preferred policy-making and planning style that significantly different processes assume key roles in determining the future in different societies.

The publicly supported private developmental style that characterizes the American scene, incorporating bargaining among major interest groups, serves mainly to protect developmental interests by reactive or regulatory actions, ensuring that the American urban future will be a continuation of present trends, only changing as a result of the impact of change produced by the exploitive opportunity-seeking planning of American corporations.

On the other hand, hierarchical social and political systems, where the governing class is accustomed to govern, where other classes are accustomed to acquiesce, and where private interests have relatively less power, can more readily evolve urban and regional growth policies at the national level than systems under the sway of the market, local political jurisdictions, or egalitarian political processes. This is one reason why urban growth policies burgeoned earlier in Britain than in the United States. Controls are of several kinds. Most basically, use of the land is effectively regulated in conformity to a plan that codifies some public concept of the desirable future and that welcomes private profit-seeking development only to the extent that it conforms to the public plan. Such is the underpinning of urban development in Britain, Sweden, France, and the Netherlands, in Israel's limited privately owned segments, and in the designated white areas of South Africa. Such a situation also obtains, it might be added, in the planning of Australia's new capital, Canberra. To understand the developmental outcome in these circumstances, one must understand the aspirations of private developers or of public agencies involved in the development process on the one hand and the images of the planners built into the Master Plan on the other. It is the resolution of the two forces that ultimately shapes the urban scene. In much of Europe, the planners' images of the desirable future have been essentially conservative, aiming to project into the future a belief that centrality is an immutable necessity for urban order, leading to the preservation of urban forms that are fast vanishing in North America. Thus, the utopian image that becomes embedded in the specific plan, and the efficacy with which the public counterpoint functions to constrain private interests, are the key elements.

Nowhere has the imagery of the social reformers been more apparent than in Soviet planning for the "city of socialist man." Reflecting the reactions against the human consequences of 19th-century industrial urbanization, the public counterpoint of the "mixed" economies has been the realization that such sought-after futures can be made to come true. Yet, ironically, Marxist counterurbanization has produced a new kind of city and has preserved and enhanced centrality, whereas the laissez-faire counterurbanization of many millions of Americans has produced a new form of dispersed urban region in which the 19th-century antithesis of town and country has disappeared.

REFERENCES

BERRY, B.J.L. (1973). The human consequences of urbanisation. Basingstoke, Eng.: Macmillan.

JONES, R. (ed., 1975). Essays on world urbanisation. London: George Philip.

RODWIN, L. (1970). Nations and cities. Boston: Houghton Mifflin.

WALTON, J. (1975). "Internal colonialism: Problems of definition and measurement." Pp. 29-50 in W.A. Cornelius and F.M. Trueblood (eds.), Urbanization and inequality: The political economy of urban and rural development in Latin America (Latin American Urban Research, vol. 5). Beverly Hills, Calif.: Sage.

Part I

EMERGING CONTRASTS: EAST AND WEST

1

The Counterurbanization Process: Urban America Since 1970

BRIAN J.L. BERRY

☐ A TURNING POINT has been reached in the American urban experience. Counterurbanization has replaced urbanization as the dominant force shaping the nation's settlement patterns. A similar tendency has been noted in other Western nations (Alexandersson and Falk, 1974). This paper lays out the facts of the change, and speculates about the nature of the process.

To those who wrote about 19th- and early 20th-century industrial urbanization, the essence was size, density, and heterogeneity. "Urbanization is a process of population concentration," wrote Hope Tisdale in 1942. "It implies a movement from a state of less concentration to a state of more concentration." But since 1970 American metropolitcan regions have grown less rapidly than the nation and have actually lost population to nonmetropolitan territory–1.8 million persons between March 1970 and March 1974 according to the estimates of the U.S. Bureau of the Census. Because migration has been selective of particular social and economic groups, very specific subgroups have been left behind.

The process of counterurbanization therefore has as its essence decreasing size, decreasing density, and decreasing heterogeneity. To mimic Tisdale: *counterurbanization is a process of population deconcentration; it implies a movement from a state of more concentration to a state of less concentration.*

RECENT POPULATION CHANGES: THE FACTS

Many of the facts of the recent changes have been spelled out in an excellent report by Forstall (1975), from whose materials Table 1 and the following summary have been assembled:

TABLE 1

POPULATION, CHANGE, AND COMPONENTS OF CHANGE FOR VARIOUS GROUPS OF METROPOLITAN AND NONMETROPOLITAN COUNTIES: 1960 TO 1970 AND 1970 TO 1973

(Numbers in thousands. Minus sign (–) denotes decrease)

Residence Category	Population			Population Change				Natural Increase				Net Migration			
	July 1, 1973 (provisional)	April 1, 1970 (census)[4]	April 1, 1960 (census)	1970-1973		1960-1970		1970-1973		1960-1970		1970-1973		1960-1970	
				Number	Percent	Number	Percent	Number	Percent	Number	Percent	Number	Percent	Number	Percent
UNITED STATES	209,851	203,300	179,323	6,551	3.2	23,977	13.4	4,917	2.4	20,841	11.6	1,634	0.8	3,135	1.7
Inside SMSAs[1]	153,350	149,093	127,348	4,258	2.9	21,744	17.1	3,768	2.5	15,637	12.3	489	0.3	6,107	4.8
Outside SMSAs	56,501	54,207	51,975	2,293	4.2	2,232	4.3	1,149	2.1	5,204	10.0	1,144	2.1	–2,972	–5.7
Metropolitan Areas over 3,000,000[2]	56,189	55,635	47,763	554	1.0	7,872	16.5	1,218	2.2	5,464	11.4	–664	–1.2	2,408	5.0
New York Area	16,657	16,701	15,126	–45	–0.3	1,576	10.4					–305	–1.8	218	1.4
Los Angeles Area	10,131	9,983	7,752	147	1.5	2,231	28.8					–119	–1.2	1,164	15.0
Chicago Area	7,689	7,611	6,794	78	1.0	817	12.0					–124	–1.6	–17	–0.2
Philadelphia Area	5,653	5,628	5,024	25	0.4	604	12.0					–75	–1.3	91	1.8
Detroit Area	4,691	4,669	4,122	22	0.5	547	13.3					–114	–2.4	9	0.2
San Francisco Area	4,544	4,423	3,492	121	2.7	932	26.7					23	0.5	485	13.9
Boston Area	3,783	3,710	3,358	73	2.0	352	10.5					15	0.4	32	0.9
Washington SMSA	3,042	2,910	2,097	132	4.5	813	38.8					34	1.2	426	20.3
Metropolitan Areas of 1-3,000,000 by Region	35,705	34,448	28,497	1,257	3.6	5,951	20.9	861	2.5	3,510	12.3	396	1.1	2,441	8.6
Northeast	3,720	3,751	3,712	–30	–0.8	38	1.0	42	1.1	289	7.8	–73	–1.9	–251	–6.8
Pittsburgh SMSA	2,367	2,401	2,405	–35	–1.4	–4	–0.2					–56	–2.3	–167	–6.9
Buffalo SMSA	1,353	1,349	1,307	4	0.3	42	3.2					–16	–1.2	–84	–6.4
North Central	12,427	12,381	10,868	46	0.4	1,513	13.9	318	2.6	1,369	12.6	–272	–2.2	144	1.3
St. Louis SMSA	2,388	2,411	2,144	–23	–0.9	266	12.4					–78	–3.2	21	1.0
Cleveland SMSA	1,997	2,064	1,909	–67	–3.2	154	8.1					–109	–5.3	–45	–2.4
Minneapolis-St. Paul SMSA	1,994	1,965	1,598	28	1.4	368	23.0					–30	–1.5	119	7.4
Milwaukee SMSA	1,432	1,404	1,279	28	2.0	125	9.8					–4	–0.3	–38	–3.0
Cincinnati SMSA (part)[3]	1,126	1,134	1,039	–8	–0.7	95	9.2					–36	–3.2	–31	–3.0
Kansas City SMSA	1,295	1,274	1,109	21	1.6	165	14.9					–14	–1.1	30	2.7
Indianapolis SMSA	1,139	1,111	944	28	2.5	167	17.7					–8	–0.7	37	3.9
Columbus SMSA	1,057	1,018	845	39	3.8	173	20.4					+6	0.6	52	6.1
Florida	3,376	2,976	2,078	400	13.4	898	43.2	21	0.7	135	6.5	379	12.7	764	36.7
Miami Area[2]	2,106	1,888	1,269	218	11.5	619	48.8					196	10.4	511	40.2
Tampa-St. Petersburg SMSA	1,271	1,089	809	182	16.7	279	34.5					182	16.7	253	31.3

Other South Atlantic	3,845	3,667	2,973	178	4.9	694	23.3	102	2.8	411	13.8	76	2.1	282	9.5
Baltimore SMSA	2,117	2,117	2,071	45	2.2	267	14.8					5	0.3	53	2.9
Atlanta SMSA	1,728	1,596	1,169	133	8.3	426	36.5					71	4.4	230	19.7
South Central[5]	5,930	5,675	4,305	255	4.5	1,370	31.8	213	3.8	693	16.1	42	0.7	677	15.7
Dallas-Fort Worth SMSA	2,442	2,378	1,738	63	2.7	640	36.8					–24	–1.0	362	20.8
Houston SMSA	2,138	1,999	1,430	139	7.0	569	39.8					52	2.6	311	21.8
Cincinnati SMSA (part)[3]	257	251	229	6	2.4	22	9.4					(Z)	0.1	–5	–2.1
New Orleans SMSA	1,093	1,046	907	47	4.4	139	15.4					14	1.4	8	0.9
West	6,406	5,998	4,561	408	6.8	1,438	31.5	163	2.7	613	13.4	244	4.1	825	18.1
San Diego SMSA	1,470	1,358	1,033	112	8.2	325	31.4					72	5.3	169	16.4
Seattle SMSA	1,385	1,425	1,107	–40	–2.8	317	28.7					–68	–4.8	187	16.9
Denver SMSA	1,366	1,239	935	127	10.2	305	32.6					86	7.0	163	17.4
Phoenix SMSA	1,119	969	664	150	15.5	306	46.1					114	11.7	190	28.6
Portland SMSA	1,066	1,007	822	59	5.9	185	22.5					40	4.0	117	14.2
Other SMSA Territory by Region	61,456	59,009	51,088	2,447	4.1	7,921	15.5	1,690	2.9	6,662	13.0	757	1.3	1,259	2.5
Northeast	13,517	13,225	11,828	292	2.2	1,397	11.8					60	0.5	312	2.6
North Central	14,761	14,447	12,820	313	2.2	1,627	12.7					–88	–0.6	6	(Z)
Florida	3,072	2,735	2,015	338	12.3	720	35.7					271	9.9	443	22.0
Other South Atlantic	7,556	7,317	6,285	238	3.3	1,032	16.4					6	0.1	126	2.0
South Central[5]	14,458	13,753	12,178	705	5.1	1,575	12.9					218	1.6	–248	–2.0
West	8,093	7,532	5,962	561	7.4	1,570	26.3					290	3.8	620	10.4
Counties with 20% or more commuters to SMSAs	4,099	3,848	3,474	251	6.5	373	10.7	74	1.9	315	9.1	177	4.6	58	1.7
Northeast (11 counties)	1,047	970	794	77	8.0	176	22.2					63	6.4	111	14.0
North Central (51 counties)	1,212	1,154	1,077	58	5.0	78	7.2					37	3.2	–5	–0.5
Florida (6 counties)	78	67	48	11	16.7	19	39.6					11	16.7	17	34.9
Other South Atlantic (41 counties)	730	697	659	33	4.7	38	5.8					17	2.4	–37	–5.6
South Central[5] (57 counties)	963	898	848	65	7.2	50	5.9					45	5.0	–36	–4.2
West (5 counties)	68	61	49	6	10.6	12	25.4					5	7.8	8	16.5
Counties with 10-19% commuters to SMSAs	9,683	9,269	8,636	414	4.5	633	7.3	182	2.0	792	9.2	232	2.5	–159	–1.8
Northeast (27 counties)	1,933	1,843	1,703	90	4.9	140	8.2					64	3.5	23	1.3
North Central (107 counties)	3,327	3,228	3,019	99	3.1	209	6.9					38	1.2	–51	–1.7
Florida (8 counties)	283	246	193	37	15.2	53	27.3					35	14.3	37	19.2
Other South Atlantic (61 counties)	1,569	1,511	1,468	58	3.8	43	2.9					20	1.3	–124	–8.4
South Central[5] (96 counties)	2,174	2,083	1,952	91	4.4	131	6.7					46	2.2	–67	–3.4
West (16 counties)	396	357	300	39	10.8	57	19.0					30	8.3	22	7.4

TABLE 1 (continued)

	Population			Population Change				Natural Increase				Net Migration			
				1970-1973		1960-1970		1970-1973		1960-1970		1970-1973		1960-1970	
Residence Category	July 1, 1973 (provisional)	April 1, 1970 (census)[4]	April 1, 1960 (census)	Number	Percent	Number	Percent	Number	Percent	Number	Percent	Number	Percent	Number	Percent
Peripheral Counties by Region	42,719	41,091	39,865	1,628	4.0	1,226	3.1	893	2.2	4,097	10.3	735	1.8	−2,871	−7.2
Northeast	3,823	3,673	3,490	150	4.1	183	5.2					84	2.3	−119	−3.4
North Central	13,493	13,101	12,919	392	3.0	182	1.4					201	1.5	−823	−6.4
Florida	868	767	617	100	13.1	150	24.4					80	10.5	67	10.9
Other South Atlantic	7,585	7,347	7,183	239	3.2	164	2.3					44	0.6	−694	−9.7
South Central[5]	10,021	9,723	9,718	298	3.1	5	0.1					76	0.8	−1,061	−10.9
West	6,929	6,481	5,938	449	6.9	542	9.1					250	3.9	−243	−4.1

(Z) Less than 500 or 0.05%.

1. SMSAs as defined by OMB as of December 31, 1974, except in New England, where definitions in terms of entire counties have been substituted.
2. Reflects certain combinations of SMSAs, as specified below. Population size groups are as of 1973.
3. Boone, Campbell, and Kenton Counties, Ky., are in the South Central Divisions; the remainder of the Cincinnati SMSA is in the North Central Region.
4. Includes corrections in local and national totals determined after 1970 census complete-count tabulations were made.
5. Comprises East South Central and West South Central Divisions.

SMSA combinations are as follows: New York Area comprises New York SMSA, Jersey City SMSA, Long Branch-Asbury Park SMSA, Nassau-Suffolk SMSA, New Brunswick-Perth Amboy-Sayreville SMSA, Newark SMSA, and Paterson-Clifton-Passaic SMSA; *Philadelphia Area* comprises Philadelphia SMSA, Trenton SMSA, and Wilmington SMSA; *Boston Area* comprises Essex, Middlesex, Norfolk, Plymouth, and Suffolk Counties, Mass.; *Chicago Area* comprises Chicago and Gary-Hammond-East Chicago SMSAs; *Detroit Area* comprises Detroit and Ann Arbor SMSAs; *Miami Area* comprises Miami and Fort Lauderdale-Hollywood SMSAs; *Los Angeles Area* comprises Los Angeles-Long Beach, Anaheim-Santa Ana-Garden Grove, Oxnard-Simi Valley-Ventura, and Riverside-San Bernardino-Ontario SMSAs; *San Francisco Area* comprises San Francisco-Oakland, San Jose, and Vallejo-Napa SMSAs.

SOURCES: Richard L. Forstall, "Trends in Metropolitan and Nonmetropolitan Population Growth Since 1970" (Washington, D.C.: Population Division, U.S. Bureau of the Census, Rev. May 20, 1975), summarizing the following—1960 population and 1960-1970 natural increase from U.S. Bureau of the Census, *Current Population Reports,* Series P-25, No. 461, "Components of Population Change by County: 1960 to 1970"; and *1970 Census of Population and Housing,* PHC(2)-1, "General Demographic Trends for Metropolitan Areas, 1960 to 1970." 1970 and 1973 populations and 1970-1973 natural increase and net migration from U.S. Bureau of the Census, *Current Population Reports,* Series P-25, Nos. 527, 530-532, and 535, "Estimates of the Population of [State] Counties and Metropolitan Areas: July 1, 1972 and 1973," for New York, Maryland, Alaska, California, and Texas, respectively; *Current Population Reports,* Series P-26, Nos. 49-93, "Estimates of the Population of [State] Counties and Metropolitan Areas: July 1, 1972 and 1973," for the other 45 states. The 1970 populations in these reports include corrections in local and national totals determined after 1970 census complete-count tabulations were made. 1960-1970 population change computed from 1960 and 1970 populations; 1960-1970 net migration computed by subtracting 1960-1970 natural increase from 1960-1970 population change; these data may differ from those in Series P-25, No. 461 and Series PHC(2), No. 1 as a result of reflecting corrections in 1970 local and national totals.

(1) Since 1970, U.S. metropolitan areas have grown more slowly than the nation as a whole, and substantially less rapidly than nonmetropolitan America, a development that stands in contrast to all preceding decades back to the early nineteenth century.

(2) On a net basis, metropolitan areas are now losing migrants to nonmetropolitan territory, although they still show a slight total increase in immigration because of recent immigrants from abroad.

(3) The decline in metropolitan growth is largely accounted for by the largest metropolitan areas, particularly those located in the Northeast and North Central regions. The eight metropolitan areas exceeding three million population have lost two-thirds of a million net migrants since 1970, and their central counties have declined in population absolutely by more than a quarter of a million. Altogether the central cities of the nation's SMSAs [Standard Metropolitan Statistical Areas] grew at an average annual rate of 0.6 percent between 1960 and 1970, but declined at an average annual rate of –0.4 percent after 1970 (annexations excluded). Much of the decline is attributable to a post-1970 decline of central city white populations at a rate of 1 percent per annum (1960-1970 the white population remained stable in the aggregate). Meanwhile Black and other minority populations have continued to decline in nonmetropolitan America since 1970, and the farm population has stabilized at approximately 9.5 million persons.

(4) Rapid growth has taken place in smaller metropolitan areas, particularly in Florida, the South, and the West; in exurban counties located outside SMSAs as currently defined, but with substantial daily commuting to metropolitan areas; and in peripheral counties not tied into metropolitan labor markets.

(5) Particularly impressive are the reversals in migration trends in the largest metropolitan areas and in the furthermost peripheral counties: the metropolitan regions with populations exceeding 3 million gained migrants between 1960 and 1970 but have lost since 1970; the nation's peripheral nonmetropolitan counties lost migrants between 1960 and 1970 but have gained since 1970. The balance of migration flows has been reversed.

Some sense of the accompanying restructuring of older metropolitan regions can be obtained by examining Figures 1-2, drawn from a nationwide study by the author (Berry and Gillard, 1976). The data relate to the 1960-1970 decade. Figure 1 shows the commuting field of the city of Cleveland in 1970; the longest one-way daily journey to work in the central city exceeded 70 miles in that year, revealing that many workers in Cleveland's factories, offices, and shops selected places of residence not simply outside the central city or Cuyahoga County, but beyond the Cleveland SMSA as currently defined.

Figure 2 shows the decadal changes in Cleveland's daily urban system, as indexed by changes in the percentages of workers commuting to the central city.

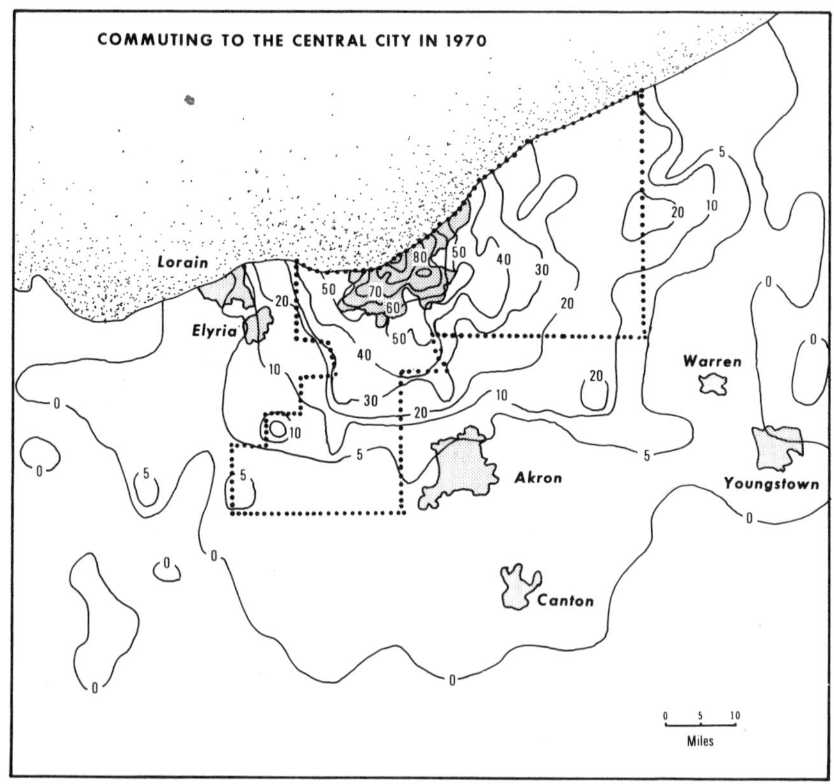

Figure 1: COMMUTING FIELD OF CLEVELAND, OHIO IN 1970

The contours show the percentage of workers commuting to the central city of Cleveland in 1970, based on census tract data. The outer limit is some 70 miles from the city center, and the daily urban system is revealed to extend far beyond the limits of the Cleveland SMSA, shown by dotted lines.

Crosshatched zones are those areas into which Cleveland's workers extended their places of residence. The thrust was outwards, in association with newly constructed expressways. Dashed areas to the east and south of the city reveal zones in which there were dramatic decreases in the volume of daily commuting to the city. These were the zones of active suburban and exurban development, where new residential and employment complexes enabled people to seek out new life-styles and to cut their ties to the older central city.

Similar illustrations for Akron reveal that city's commuting field to be substantially the same as Cleveland's, changing during the decade in the same ways. Some of Akron's workers moved much further afield, stretching their choice of place of residence outwards along new interstate highways. But the dependence of many areas on Akron's jobs declined drastically, particularly in that interurban belt between Akron and Cleveland that had also freed itself of dependency on Cleveland jobs during the decade. Thus, what had previously

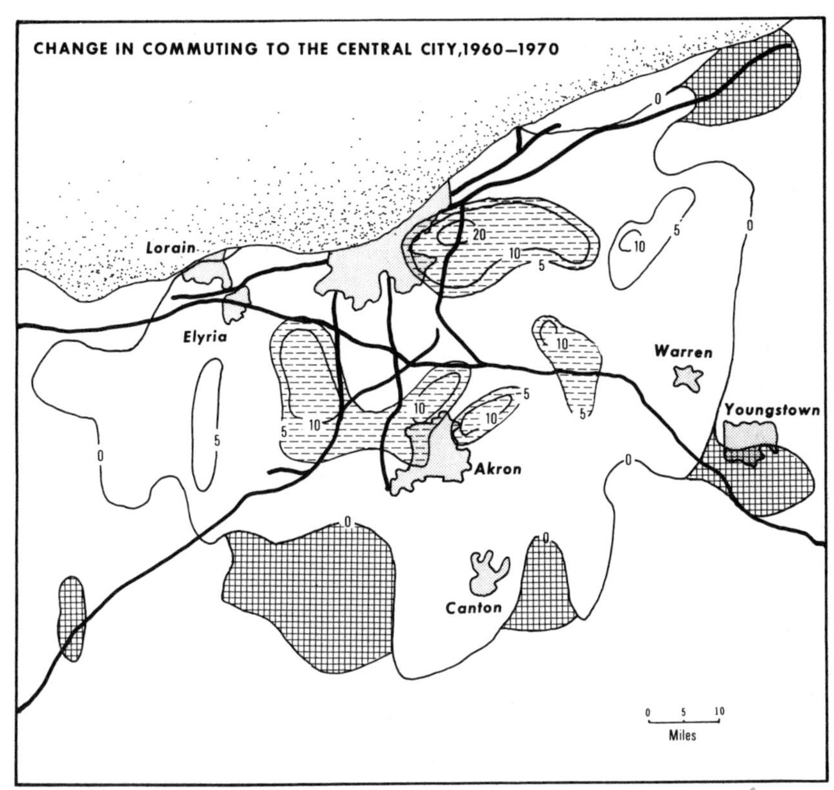

Figure 2: CHANGES IN THE CLEVELAND COMMUTING FIELD, 1960-1970

The contours show the change in percentage of workers commuting to the central city of Cleveland between 1960 and 1970. Dashed shades pick out the zones in which the commuting percentage dropped by 5 points or more—east and south of Cleveland within the SMSA, and outside the SMSA north of Akron. These were the zones of rapid suburban residential and industrial growth in the decade. Crosshatched areas pick out the zones into which the commuting field extended during the decade, generally associated with expressway extensions (solid black lines).

been the intermetropolitan periphery was now displaying newly found independence as one of the region's new growth centers—not in the form of a traditional concentrated industrial-urban node, but rather in a low-slung and far-flung form, like the new metropolitan regions of the South and West (Holleb, 1975). Northeastern Ohio had always been a multicentered urban region. During the sixties it became more thoroughly dispersed as the older central cities declined, decentralization proceeded apace, metropolitan regions were restructured internally, and more amenity-rich outlying areas were brought into daily interaction with other parts of metropolitan America by expressway-related accessibility changes (Lamb, 1975; Berry and Gillard, 1976).

TEMPORARY PERTURBATION, LONG-TERM TREND, OR CULTURAL PREDISPOSITION?

To some, the census figures summarized in Table 1 are but a temporary perturbation, an anomaly caused by the recession that will vanish when the health of the economy improves. But such an attitude is hardly credible; 20th-century trends have all pointed in the same direction—creation of nothing less than "an urban civilization without cities," at least in the classical sense (Kristol, 1972). As early as 1902, H.G. Wells wrote that the "railway-begotten giant cities" he knew were

> in all probability . . . destined to such a process of dissection and diffusion as to amount almost to obliteration . . . within a measurable further space of years. These coming cities . . . will present a new and entirely different phase of human distribution. . . . The city will diffuse itself until it has taken up considerable areas and many of the characteristics of what is now country. . . . The country will take itself many of the qualities of the city. The old antithesis . . . will cease, the boundary lines will altogether disappear.

Similarly, Adna Weber suggested in his remarkable 1899 study that

> the most encouraging feature of the whole situation is the tendency . . . towards the development of suburban towns [which] denotes a diminution in the *intensity* of concentration. . . . The rise of the suburbs it is, which furnishes the solid basis of hope that the evils of city life, so far as they result from overcrowding, may in large part be removed. If concentration of population seems desired to continue, it will be a modified concentration which offers the advantages of both city and country life.

Later Frank Lloyd Wright argued that "Broadacre City" was the most desirable settlement pattern for mankind, and Lewis Mumford called for a new reintegration of men and nature in dispersed urban regions, to cite but a few cases.

Throughout the 20th century all trends have pointed in the directions suggested by these writers (Berry, 1970; Lamb, 1975). Every public opinion survey has indicated that popular preferences are for smaller places and lower densities, with richer environmental amenities (Sundquist, 1975). The trend has been one leading unremittingly toward the reversal of the processes of population concentration unleashed by technologies of the Industrial Revolution, a reversal finally achieved after 1970.

Viewed more generally, though, what has finally been achieved in the 1970s is not something new, but something old—the reassertion of fundamental predispositions of the American culture that, because they are antithetical to urban concentration, have resulted in many of the contradictions and conflicts of recent decades.

We should go back 200 years, to Hector de Crèvecoeur's *Letters from an American Farmer.* "Who, then, is this new man, the American?" he asked, and his answer was a description of basic American culture traits. Foremost among these was *a love of newness.* Second was the overwhelming desire to be *near to nature. Freedom to move* was essential if goals were to be realized, and *individualism* was basic to the self-made man's pursuit of his goals, yet *violence* was the accompaniment if not the condition of success—the competitive urge, the struggle to succeed, the fight to win. Finally, Crèvecoeur (though he did not use the later, now-familiar phrases) perceived a great *melting pot* of peoples and a sense of *manifest destiny* (Watson, 1970).

THE LOVE OF NEWNESS

There has been no more evocative description of the consequences of the love of newness for American metropolitan structure than Homer Hoyt's discussion of *The Structure and Growth of Residential Neighborhoods in American Cities,* published in 1939. Hoyt said:

> The erection of new dwellings on the periphery . . . sets in motion forces tending to draw population from older houses and to cause all groups to move up a step, leaving the oldest and cheapest houses to be occupied by the poorest families or to be vacated. The constant competition of new areas is itself a cause of neighborhood shifts. Every building boom, with its crop of structures equipped with the latest modern devices, pushes all existing structures a notch down in the scale of desirability. . . . The high grade areas tend to preempt the most desirable residential land. . . . Intermediate rental groups tend to occupy the sectors in each city that are adjacent to the high rent area. . . . Occupants of houses in the low rent categories tend to move out in bands from the center of the city by filtering up. . . . There is a constant outward movement of neighborhoods because as neighborhoods become older they tend to be less desirable. A neighborhood composed of new houses in the latest modern style . . . is at its apex. . . . Physical deterioration of structures and the aging of families . . . constantly lessen the vital powers of the neighborhood. . . . The steady process of deterioration is hastened by obsolescence; a new and more modern type of structure relegates all existing structures to lower ranks of desirability.

Hoyt's perceptions cut right to the core of much of that which has transpired, for the accompaniment of the process of counterurbanization is urban decay and the abandonment of the nonachieving social underclass (Berry, 1975); ghetto growth is a product of the white exodus (Long, 1975).

NEARNESS TO NATURE

The love of newness joins with the desire to be near nature. H.G. Wells' 1902 forecasts should be recalled:

> Many of our railway-begotten giant cities are destined to such a process of dissection and diffusion as to amount almost to obliteration . . . within a measurable further space of years. . . . These coming cities . . . will present a new and entirely different phase of human distribution. . . . The social history of the middle and later thirds of the nineteenth century . . . all over the civilized world has been the history of a gigantic rush of population into the magic radius of—for most people—four miles, to suffer there physical and moral disaster . . . far more appalling than any famine or pestilence that ever swept the world. . . . But new forces . . . bring with them . . . the distinct promise of a centrifugal application that may finally be equal to the complete reduction of all our present congestions. . . . What will be the forces acting upon the prosperous household? The passion for nature . . . and that craving for a little private *imperium* are the chief centrifugal inducements. . . . The city will diffuse itself until it has taken upon considerable areas and many of the characteristics of what is now country. . . . We may call . . . these coming town provinces 'urban regions.'

Almost as an echo of Wells comes that sociological essay of the 1950s that proclaimed in its title that "the suburbs are the frontier," and Lamb's (1970) analyses of the role of amenities in peripheral expansion during the 1960s. But again antithetically, the greater the numbers trying to get near to nature, the more that which is sought is degraded.

FREEDOM TO MOVE

To occupy this new frontier, close to nature, and to keep on adjusting to succeeding waves of growth have demanded freedom to move. Americans are the world's most mobile people. Forty million Americans change residence each year; Americans change residence an average of 14 times in a lifetime. As Peter Morrison has remarked recently (1974):

> The typical American's life might be characterized as a prolonged odyssey. Marriage, childbearing, military service, higher education, changes from one employer to another or shifts from one plant or office location to another with the same employer, divorce, retirement—all may bring a change in residence and locale, not to speak of upward social mobility which may impel people to move for other reasons as well. . . . Now as in the past Americans continue to migrate for reasons that are connected to the working of national economic and social systems. . . . The quick exploitation of new resources or knowledge requires the abandonment of old enterprises along with the development of the new, . . . and migration is also an *assortive* mechanism, filtering and sifting the population as its members undergo social mobility.

Yet again, there is an antiphonal note. Filtering in housing markets, for example, is a process that has positive welfare consequences if new construction exceeds the rate necessary to normal growth of housing and produces an excess housing supply at the point where the filtering originates; if such new construction exerts a downward pressure on the rents and prices of existing housing, permitting lower income families to obtain better housing bargains relative to their existing housing quarters; if the upward mobility is apart from any changes caused by rising incomes and/or declining rent-income ratios; and if a decline in quality is not necessarily forced by reductions in maintenance and repair to the extent that rents and prices are forced down; and finally if a mechanism exists to remove the worst housing from the market without adversely affecting rents and prices of housing at the lowest level. Part of the reason for urban decay is that the last two conditions have not been met: deterioration has accelerated in many older neighborhoods, and abandonment has become contagious, frequently adversely affecting access by low-income residents to the better quality housing available locally.

INDIVIDUALISM

Contrary to the views of most radicals, however, urban expansion and urban decay are not caused by a single-minded conspiracy among large-scale institutions and investors. They result instead from myriad decisions made individually, within a tradition of privatism. This tradition has been called by Sam Bass Warner (1968)

> the most important element of American culture for understanding the development of cities. It has meant that the cities of the United States depended for their wages, employment, and general prosperity on the aggregate successes and failures of thousands of individual enterprises, not upon community action. It has also meant that the physical forms of American cities, their lots, houses, factories and streets have been the outcome of a real estate market of profit-seeking builders, land speculators, and large investors. And it has meant that the local politics of American cities have depended for their actors, and for a good deal of their subject matter, on the changing focus of men's private economic activities.

Privatism has prevailed throughout America's history, and a consequence is a preference for governmental fragmentation and for interest-group politics under presumed conditions of democratic pluralism. Antithetically, it has also meant that American city planning has been curative rather than future-oriented, reactive rather than going somewhere.

VIOLENCE

Although achievement in the mainstream has involved an individual fight to succeed, violence also is a pervasive underpinning of American life if only because fights have to have more than one participant. Acrimonious confrontations mark the fights to control turf within cities, while, for the underclass abandoned in deteriorating ghettos, crime and violence are a way of life. President Johnson's Commission on Crimes of Violence reported that if present trends continue,

> we can expect further social fragmentation of the urban environment, greater segregation of different racial groups and economic classes . . . and the polarization of attitudes on a variety of issues. It is logical to expect the establishment of the "defensive city" consisting of an economically declining central business district in the inner city protected by people shopping or working in buildings during daylight hours and "sealed off" by police during nighttime hours. Highrise apartments and residential "compounds" will be fortified cells for upper-, middle-, and high-income populations living in prime locations. . . . Suburban neighborhoods, geographically removed from the central city, will be "safe areas," protected mainly by racial and economic homogeneity.

THE MELTING POT(?)

In the expanding frontiers of suburban America, upwardly mobile individuals from a variety of backgrounds have been readily integrated into the achievement-oriented mainstream of society. When the heterogeneity of American cities was caused primarily by the influx of successive immigrant waves, the policy of encouraging such assimilation was taken for granted ideologically. But even in the suburbs, what poured out of the melting pot rapidly crystallized into a complex mosaic of sharply differentiated communities of achievers, counterposed against those who have been unable or unwilling to move out of the cities. Thus, the national ideal of integration remains inaccessible for many—in particular, for the unassimilable blacks, browns, and reds, for whom segregation within the central cities remains the rule as battle lines are drawn along neighborhood boundaries and at the gates of the schools. For others, integration is perceived as undesirable: unassimilated ethnics regard it as destructive of self-identity; members of new communities avoid it as they seek a return to simpler ways, sometimes in rural communes. When, then, we might ask, all the surprise at the discovery that forced racial integration of the schools accelerates white emigration from the central cities? It's the American way—the way in which the individuality of the homogeneous subgroup is maintained.

MANIFEST DESTINY

Yet alongside individualized withdrawal to the periphery, there remains the continuing feeling that Americans should no longer be willing simply to *react* to problems, or to *act* in a privatized mode, but should—collectively—*go somewhere,* to achieve goals, to "win" the "wars" on poverty, underprivilege, and urban decay—to perfect America and Americans. One version is a favorite of planning professionals. How many recall the call of the National Resources Committee in 1937, repeated many times since?:

> If the city fails, America fails. The Nation cannot flourish without its urban-industrial centers. . . . City planning, county planning, rural planning, State planning, regional planning, must be linked together in the strategy of American national planning and policy, to the end that our national and local resources may best be conserved and developed for our human use. . . . The Committee is of the opinion that the realistic answer to the question of a desirable urban environment lies . . . in the judicious reshaping of the urban community and region by systematic development and redevelopment in accordance with forward looking and intelligent plans.

And how many public programs pointed in such directions have been failures? For the other version *is* the collective decision of individual Americans. Planners' predilections notwithstanding, the American mainstream is one of counter-urbanization processes that are being stemmed by planning—federal, state, and local—as effectively as King Canute stopped the tide.

REFERENCES

ALEXANDERSSON, G., and FALK, T. (1974). "Changes in the urban pattern of Sweden, 1960-1970: The beginning of a return to small urban places?" Geoforum, 8:87-92.

BERRY, B.J.L. (1970). "The geography of the United States in the year 2000." Transactions of the Institute of British Geographers, 51:21-53.

--- (1973). The human consequences of urbanisation. Basingstoke, Eng.: Macmillan.

--- (1975). "The decline of the aging metropolis: Cultural bases and social process." Paper prepared for a conference at the Center for Urban Policy Research, Rutgers University.

BERRY, B.J.L., and GILLARD, Q. (1976). The changing shape of metropolitan America: Commuting patterns, urban fields and decentralization processes, 1960-1970. Cambridge, Mass.: Ballinger.

CREVECOEUR, J.H. St. John de (1782). Letters from an American farmer. London: Thomas Davies.

FORSTALL, R.L. (1975). "Trends in metropolitan and nonmetropolitan population growth since 1970." Washington, D.C.: Population Division, U.S. Bureau of the Census.

HOLLEB, D.B. (1975). "Moving towards megacity: Urbanization and population trends," Chicago: Center for Urban Studies, University of Chicago.

HOYT, H. (1939). The structure and growth of residential neighborhoods in American cities. Washington, D.C.: Federal Housing Administration.

KRISTOL, I. (1972). "An urban civilization without cities." Washington Post Outlook, Sunday, December 3, p. B1.

LAMB, R. (1975). Metropolitan impacts on rural America. Department of Geography Research Paper No. 162, University of Chicago.

LONG, L.H. (1975). "How the racial composition of cities changes." Land Economics, 51:258-267.

MORRISON, P. (1974). "Toward a policy planner's view of urban settlement systems." Unpublished manuscript. Santa Monica, Calif.: Rand Corporation.

––– (1975). "The current demographic context of national growth and development." Unpublished manuscript. Santa Monica, Calif.: Rand Corporation.

SUNDQUIST, J.L. (1975). Dispersing population. Washington, D.C.: Brookings Institution.

TISDALE, H. (1942). "The process of urbanization." Social Forces, 20:311-316.

U.S. Department of Commerce (1974). "Estimates of the population of metropolitan areas, 1972 and 1973, and components of change, since 1970." Current Population Reports, series P-25, No. 537.

WARNER, S.B., Jr. (1968). The private city. Philadelphia: University of Pennsylvania Press.

WATSON, J.W. (1970). "Image geography: The myth of America in the American scene." Advancement of Science, 27:1-9.

WEBER, A.F. (1899). The growth of cities in the nineteenth century. New York: Macmillan.

WELLS, H.G. (1962). Anticipations: The reaction of mechanical and scientific progress on human life and thought. London: Harper and Row.

2

Urban Environments in the United States and the Soviet Union: Some Contrasts and Comparisons

ROBERT G. JENSEN

□ IN THE CONTEMPORARY WORLD, no two countries are so often compared as the Soviet Union and the United States. In part this is because they have enormous and often competing influence in international affairs. In part it is simply because they are large countries, with vast resources and similar-sized populations, undergoing rapid modernization and technological change within quite different social-economic systems. By most definitions, both the Soviet Union and the United States are urban societies, and they become increasingly so with each passing year. It seems appropriate, therefore, to seek comparisons between the process of urbanization and the resulting urban environments in these countries even though they have evolved in quite different contexts. The author is well aware that such a broad and complex subject cannot be given adequate treatment in a short paper. In both countries the relevant literature is immense, far more than any one person could read or comprehend. It is doubtful, moreover, that any person would claim to fully understand the process and results of urbanization in his own country, much less in another. Yet, though a cross-national comparison is thus doubly difficult, it often has the benefit of providing a new perspective on that which is familiar or taken for granted.

Cross-national comparisons can be undertaken at many levels and with many methodologies. They can range from the casual observations of a traveler to the rigorous analysis of carefully prepared data by the scientist. This essay fits somewhere between those extremes. It is admittedly impressionistic, being based on a limited reading of the Soviet and American urban literature as well as on personal observation of urban environments in both countries. The impressions thus gained are highly generalized and are used here to suggest a framework for discussion and interpretation rather than to demonstrate specific data based conclusions.

URBAN ENVIRONMENTS AS A REFLECTION OF SOCIETY

It seems reasonable to assume that many characteristics and values of a society will be reflected in the spatial organization of the human landscape. This should be especially true in urban areas since they are the center of both the social and the spatial order. If this line of reasoning is followed in considering Soviet and American urban environments, one would expect to find spatial contrasts stemming from differences between socialism and capitalism as they are practices in those countries. The central business district with its massive corporate structures, to take one example, might be viewed as a symbol of the commercial and market forces which dominate in America. The less intensive commercial development in the center of Soviet cities, where emphasis is placed on large public squares and other public facilities, might be interpreted as representing the socialist character of Soviet society and the absence of market forces in the competition for land use. The central business district is literally the vertical peak of the American urban landscape, whereas the central part of many Soviet cities appears to be concave with respect to the surrounding built up area. Certainly there are many precedents for such analysis in both the Soviet and American literature. What is common among such interpretations, as Harvey (1971:1) has noted in another context, is the attempt to "show how the geographic organization of space is the result of human activity as it unfolds in a particular cultural, social, economic, political and technological setting."

In theory, the notion of some degree of correspondence between the socioeconomic and spatial order is rather simple. But in reality, of course, there is inevitably a substantial discrepancy between the existing spatial organization of society, which was developed incremently in the past, and the spatial organization which might seem to be required of present and future societies. Physical-geographic inertia (the staying power of already established physical patterns) is thus a factor confronting planners in all societies, at all levels. Equally important, the planned goals of a society, even if clearly identified, are likely to have internal contradictions and imbalances, at least in the short run. The latter are likely to have their greatest impact on urban areas. Even if there were no contradictions among the expressed goals of a society, there is likely to be disagreement about the spatial organization best suited to the achievement of those goals. Examples of the above, it seems to this writer, are abundant in both the Soviet Union and the United States.

Urbanization, of course, is closely related to industrialization and the more general process of modernization. Whether capitalism and socialism will move toward convergence as a result of these processes is a question which has been widely discussed in Western literature. A noted American economist has asserted that the industrial systems of the United States and the Soviet Union are converging at all fundamental points (Galbraith, 1967:391). Numerous writers in the Soviet Union and elsewhere have strongly disagreed with that view. The notion of convergence among societies is an old concept, however, which has typically stressed the cultural and organizational impacts of industrialization at the expense of other factors (a variety of cultural, social, political, and historical filters) that may lead to quite different results in both spatial organization and social outlook. Certainly this writer holds no special brief for convergence, but instead agrees with those who suggest that comparative studies should be alerted to convergence as well as divergence without a priori dismissal of either possibility (Weinberg, 1969:15).

Urbanization in the Soviet Union and the United States has produced a variety of similar trends and common problems. And it is clearly a force in both societies which transcends the number, size, and structural features of urban places. Still, comparative interpretations are difficult. The two countries are at different stages of development, are the products of different historical experiences, and operate under very different geographic and institutional settings. The urban system in each country is so complex, moreover, that any comparative generalization is bound to ignore significant elements of reality. Perhaps the best that one can hope to do is identify dominant themes.

ATTITUDES TOWARD THE CITY AND URBAN PLANNING

In a number of instances, urban problems in America and the Soviet Union are, in a relative sense, reverse images of each other. In America, large cities (in this case the political city) have serious economic and social problems and are declining in population, whereas the surrounding suburbs, which offer more attractive living environments, continue to attract a wide spectrum of the population. In the Soviet Union, the largest cities appear to be the most successful, whereas the surrounding suburban areas and small towns lag behind in the provision of services and various urban amenities. Some of the differences between suburban areas in America and the Soviet Union will be discussed later in this essay.

Such differences, no doubt, have their roots in what are quite different attitudes toward the city proper. In the Soviet Union a positive attitude toward the city and city life has prevailed. The city is in fact seen as an expression of the ideals of socialist society, in sharp contrast to rural living, which is to be gradually transformed by the infusion of urban life-styles. In America, on the other hand, the city and city life have been viewed with considerable uneasiness. Instead, rural and small town life have been idealized as the proper setting for maintaining the desired social values of family, community, and a personal sense of place. The city, although an economic necessity, is seen as promoting indifference and as being generally destructive of traditional social norms. The American "flight" to the suburbs can, among other things, be interpreted as an attempt to recapture the small town environment.

Although attitudes toward the city are important, it would be impossible to understand the Soviet and American urban experience without reference to fundamental differences in the planning context. The central articulation of planning objectives, combined with the central allocation of resources, clearly has a powerful influence on all aspects of urban growth in the Soviet Union. Comprehensive national planning means a commitment to specified goals at all geographic scales and requires that planning be integrated at the various levels of the territorial system. Thus in the Soviet Union it has been possible to think in terms of a national urban policy with potential for shaping the entire urban environment. The elements of such a policy, in fact, exist in normative documents relating to urban development and in national economic plans.

By planning the urban environment, the Soviet Union has hoped to avoid many problems which characterize cities in other parts of the world. In 1920 there were only a handful of large cities scattered across the country, so most urban development has occurred within the framework of a planned economy. However, the pressures of rapid industrialization, wartime destruction, consequent housing shortages, and other problems have made the ideal of a planned socialist city difficult to achieve.

In contrast to the above, American society has been suspicious of planning in general and comprehensive planning in particular. Urban growth has occurred with relatively little interference from national and local governments. It has been more or less presumed that demand for particular kinds of urban environments will be met by an array of often unrelated public and private institutions (builders, local governments, zoning boards, etc.). Only recently has there been much interest expressed in more comprehensive planning and goal formulation.

URBANIZATION AND CITY SIZE

The Soviet leadership, as noted above, as well as the majority of individual writers, has maintained a positive view of cities and urban life in general. It must be added, however, that such a view does not include big cities and sprawling conurbations. Since the 1930s, planners have underscored the desirability of limiting the growth of large urban places and instead have supported the development of medium or even small cities. Since 1958, at least, it has been officially recommended that new industry be located in small (under 50,000) and medium-sized cities (50,000-100,000), but only to an extent that such cities do not come to exceed a population of 250,000-300,000 (Pravila i normy, 1959). Later, the State Planning Commission recommended more than 500 medium-sized and small cities suitable for the location of new industrial enterprises (Khorev, 1972:196-199). The 1971-1975 plan called for the location of approximately 75% of the large new industrial enterprises in cities with a population under 250,000 (Gosudarstvennyi, 1972:249). In major cities (over 500,000) and very large cities (250,000-500,000), the establishment of new industrial establishments or the expansion of existing ones is not allowed unless they are directly related to municipal organizations or consumer services, and then only if the city labor force would not be increased (Cornell University, 1974:2-3).

The possibility of exercising this kind of control has been seen by Soviet planners as a means of avoiding many perceived negative consequences of uncontrolled urban expansion which have occurred in other parts of the world. Yet, as Harris (1970:1) notes, the Soviet Union has become a land of large cities, roughly equivalent to the United States in terms of the number of cities with over 100,000 population. By 1974 there were 35 cities of 500,000 or more population with nearly 41 million inhabitants, and there were 203 cities with a population of 100,000-500,000 with nearly 44 million inhabitants (Narodnoe, 1974:32). Thus, out of an urban population of 150 million (60% of the total population), more than 70% live in what are officially termed large cities. Currently, 13 cities have populations over one million, and another 10 or so are within easy reach of that mark (SSSR, 1975:13-14). Although no data are available for regions equivalent to the Standard Metropolitan Statistical Area, one might assume that the U.S.S.R. has at least 20 metropolitan areas in the million class.

During the 1960s, considerable work was done by geographers and others to determine an optimal range for city size. Depending on a city's location and functions, most studies suggested that the optimal size was in the 50,000-250,000 range, although in some cases the range might be 20,000-400,000 (Khorev, 1968a:108-122). The interest in optimal city size has been based on the notion that growth beyond certain limits would result in increasing per capita costs for providing essential services and maintaining a clean environment. At some lower limit the costs of essential services would become economical, but beyond a certain size the costs might exceed acceptable limits. Studies of per capita expenditures on housing, transportation, and other services have indicated that such expenditures increase rather slowly up to a city size of 200,000, but, beyond that, some costs, especially for transportation, rise more sharply (Davidovich, 1960, 1964). Limiting the growth of large cities, combined with the development of smaller urban places, has also been viewed as supporting the general aim of reducing the differences between town and countryside.

The problem of controlling urban growth in the face of rapid urbanization has produced a lively discussion among Soviet urban scholars. Perevedentsev (1970:38), an economist, maintains that restrictions on urban growth have been relatively unsuccessful because major cities offer a greater range of economic and social advantages. The economically optimal city, he suggests, is not one requiring the smallest per capita investment, but rather one which has the greatest difference between per capita expenditures and output. Using that definition, he questions the negative description of big cities and concludes that the concentration of population and production in the largest industrial centers is the most advantageous. To do otherwise, he concludes, would conflict with the aim of increasing the overall efficiency of the national economy. Urban plans, based on projected population growth, often have to be corrected, Perevedentsev points out (1975:10), precisely because they underestimate the powerful economic and social attractions of the largest cities.

During the last decade, however, discussion of the regulation of urban growth seems to have shifted from a focus on individual cities and their size to a consideration of urban agglomerations and systems of cities in the broader context of urbanization (Pivovarov, 1972:30-32). The formation of urban agglomerations is viewed as a relatively new trend in the Soviet Union, around which new concepts need to be elaborated to insure that the settlement process is directed toward a rational spatial organization of society. It has been suggested, for example, that the organization of data along the lines of the Standard Metropolitan Statistical Area would be useful for city and regional planning. Increasing stress, moreover, is being placed on the social and cultural role of large urban centers which, with modern technology, are seen as incorporating ever larger territories into their sphere of influence, thereby further reducing the difference between town and country (Strongina, 1974:135). While there is disagreement about the specific path future urban growth should take, most urban scholars seem to be directing attention to the possibilities for regulating the evolution of urban agglomerations.

Although attitudes about the formation of urban agglomerations have been varied, they are generally viewed as having a positive impact on social and economic development in the Soviet Union. Certainly there is no reason to assume, Lappo (1974:27-37) suggests, that in a planned economy the agglomeration process will produce the kinds of patterns which have characterized their growth in the United States (i.e., the merging of urban areas into vast conurbations). In contrast to the United States, absolute population growth has been concentrated mainly within the core area of Soviet urban agglomerations, and there has been relatively little merging of built-up areas. Indeed, one of the striking visual features of Soviet cities is the sharp edge which divides the built-up area from the surrounding or intervening countryside. Urban agglomerations in the Soviet Union, Lappo argues, are thus far from being overdeveloped or encountering ecological barriers. Instead it is suggested that agglomerations offer a "hidden" potential for economic development which can be used not only to regulate the growth of large cities but also to provide a "legimate" setting for their very existence (Lappo, 1974:540-541).

A better way to conceive of future urban patterns, Khorev (1968b:83-103) suggests, is to think in terms of a "unified system of settlement." The latter is defined as a "functionally delimited and structurally interrelated network of places over a large territory, forming several subordinated hierarchic levels, developing in a regular way, regulated in a planned manner for the benefit of society, and encompassed by a unified system of regional planning" (Khodzhayev and Khorev, 1972:95). According to Khorev, one of the advantages of a unified system, compared to a large urban agglomeration, is that it avoids the "inevitable" breakdown of a region into a privileged core and a backward periphery. Both the urban agglomeration and the concept of a unified settlement system depend on intensive interaction among places, but an agglomeration would be only one element in a unified system. The latter, moreover, would presuppose a higher level of interurban relations and the availability of more or less equal levels of public service to all residents of the regional system (Khodzhayev and Khorev, 1973:96-97). It is indicated, however, that the formation of a unified settlement system would require considerable improvements in transportation and the network of service centers so that such a system could not be a reality in the near future even in the European part of the country.

The urban system of the United States, in sharp contrast to that of the Soviet Union, has taken form with relatively little government intervention. To the extent that there has been concern for a national urban policy, that concern has been characterized by fragmentation with a focus usually on a variety of single issues such as housing, urban renewal, and, more recently, environmental quality. There has been little serious discussion of optimal city size or of limiting city growth, probably because the controls needed for such policies would be impossible to implement on a broad scale even if they were thought to be desirable. A large number of agencies exist to protect the public, but, with the exception of the location of government projects, decisions determining urban growth are made largely in the private sector.

By 1960 more than 90% of the American population lived within commuting range of central cities or within what Berry (1973:38) has termed the Daily Urban System. The implications of this new situation have been well described by two American planners who have interpreted the spatial structure of the country in terms of metropolitan areas and an intermetropolitan periphery (Friedmann and Miller, 1965). The latter, except for sparsely populated parts of the American interior, includes all areas that intervene among the metropolitan regions. In recent years, as a result of rapid outward expansion, the core-oriented character of metropolitan areas has declined, creating more uniform densities of population and economic activity across urban regions (Berry, 1973:45, fn. 24). What is under way, it has been suggested, is an extension of urban living space beyond the traditional boundaries of metropolitan areas, which will alter life in the intermetropolitan periphery, making it difficult to distinguish what is "properly urban and what is properly rural." Urban infrastructure and services will be easily accessible throughout this "urban field," providing residents with a wide choice of living environments all coexisting within a common network of communication (Friedmann and Miller, 1965, fn. 25).

The extension of urban life into the countryside and the consequent blurring of the traditional distinction between urban and rural living, whether by design or not, has accompanied urbanization in both the Soviet Union and the United States. The process and results of suburban growth, however, are quite different in the two countries. Suburbanization in America has been associated with population decline in the central cities as urban residents have moved into the suburbs. The dominance of individual detached housing along with widespread use of the private automobile have produced "urban sprawl" and the merging of once separate urban areas into continuous built-up regions served by scattered commercial centers. Suburban development in the Soviet Union, although still in a relatively early stage, is not likely to follow a similar path. To date, suburban growth has not resulted in a decline of the central cities. Emphasis on public transportation and apartment complexes, moreover, should mean that a more concentrated population pattern will be maintained within urban agglomerations. State control over the location of new economic activity combined with the designation of protected greenbelt areas, therefore, ought to minimize "urban sprawl." It will be interesting, at any rate, to observe the degree to which open space can be maintained between component parts of developing urban agglomerations. Finally, suburban growth in the Soviet Union is not likely to be associated with an image of a distinctive life-style such as has been characterized in the American literature on "suburbia."

COMMUNITY IN THE CITY

The future society of the Soviet Union will be one that is essentially urbanized. The government has expressed faith not only in the virtues of a properly organized urban living environment but also in the possibilities for designing urban environments that would be supportive of broad social objectives. As much as possible, such environments should provide equal opportunity for all members of society and reduce feelings of separateness between different social groups. Residential areas should therefore be relatively heterogeneous in terms of occupation, income levels, and other social features, but relatively homogeneous in their physical organization so as not to attract or repel different segments of society. In this context, urban physical planning has been considered a means of bringing about social change. Policies focusing on the residential community as a framework for social interaction and shared services have been especially interesting.

Over the years, Soviet planners have prepared and experimented with a variety of alternatives for organizing the urban residential environment. In all cases there has been a strong preference for apartment buildings over detached single-family houses. Apartment construction, of course, has a number of technological and economic advantages for rapidly increasing the urban housing stock. Apartment complexes have also been viewed as a better setting for shared residential services and broadening social contacts among residents, especially among those who have advocated the communal type apartment house *(dom-kommuna).* For the most part, however, the strategy has been to develop a hierarchy of relatively self-contained residential areas, composed of conventional apartment buildings, with conveniently located community services. The areas in the hierarchy from smallest to largest are the apartment complex *(kvartal),* the microdistrict *(microraion),* and the residential district *(zhiloi raion).* Beyond the residential district, cities are divided by the urban district *(gorodskoi raion)* with 100,000-300,000 population and the urban zone *(gorodskaia zona)* with a population around one million.

Since the late 1950s the microdistrict has been considered the primary spatial unit of the urban residential environment. As a rule the microdistrict should have 6,000-12,000 inhabitants. If high-rise structures (more than five stories) are used, the number may reach 18,000; and in areas of complex topography with low-rise structures, the population may be as low as 4,000. The residential district, consisting of several microdistricts, should range from 24,000 to 36,000 inhabitants, although, if high-rise construction is used, the number may be increased to 60,000. In all cases the 1967 norms suggest that the number of inhabitants should be calculated on the basis of 9 square meters of floor space per person (Cornell University, 1974:49-94, fn. 8).

The standards for the layout of microdistricts and residential districts suggest a major effort to develop relatively self-contained, clearly bounded geographical communities within the urban environment throughout the country. Services and facilities which are supposed to be provided within the microdistrict include day-care centers, primary and secondary schools, a variety of commercial enterprises, public eating facilities, and recreational space. The residential district should contain a community and commercial center with higher order services including a movie theater, library, public health establishments, and a variety of open space for public use. Norms describe the kinds and level of services and facilities which should be provided per thousand persons, along with standards for arrangement and spacing of the built environment. Since many of the essential urban services are provided within walking distance at the microdistrict or residential district level, it may be assumed that one objective has been to reduce an urban resident's dependence on other parts of the city by making it convenient to conduct many activities in a more limited geographical community.

In terms of concept and spatial organization, microdistricts seem fairly conventional. The grouping of apartment buildings around community facilities and the planned provisions of services, recreational space, and other amenities are found in residential complexes in many parts of the world, including America. What may set the Soviet experience with microdistricts somewhat apart is the central determination of design standards having nationwide application, their special role in developing a unified system of urban services, and the large number which have already been constructed (Osborn, 1970:249-251).

The microdistrict approach has had a clearly visible impact on the organization of the urban environment, has been associated with a rapid improvement of housing conditions, and seems also to have resulted in a greater dispersal of urban services. However, considerable doubt has been expressed about the effectiveness of the microdistrict concept in promoting a feeling of local community within the city. Some Soviet scholars have suggested that attempts to organize urban life around neighborhood units may conflict with the very nature of urbanization. Their research has suggested that social contacts at the neighborhood level are minimal, that feelings of neighborhood *(sosedstvo)* decline in large cities and socially differentiated communities, and that urban dwellers simply do not orient their lives around the apartment complex or larger neighborhood unit (Frolic, 1970; Kogan, 1967; Pokshishevskii, 1975). With growing affluence and mobility, Soviet urbanites have apparently sought their community in the diversity of the city as a whole, rather than in a more limited geographical neighborhood. This would suggest that physical planning at the microdistrict scale has had less impact on social interaction in the urban environment than might have been expected.

The American experience with residential communities and neighborhoods has been the subject of a vast literature. Nothing in that experience, however, is really comparable with the Soviet attempt to develop a unified system of relatively standardized geographical communities over the entire urban residential environment. To be sure, one can point to planned communities, neighborhoods organized around a shared focus of activity, and various other features associated with the Soviet microdistrict. But within the American urban region a dominant characteristic is the great variety of neighborhood and community settings which are distinguished by the relative homogeneity of their populations in terms of socioeconomic status, race, ethnicity, and general life-style as well as by the physical appearance of the area. This mosaic of relatively homogeneous communities, as Berry (1973:66) points out, "maintains different life styles that are internally cohesive and exclusive. . . . Mobility within the mosaic leads to a high degree of expressed satisfaction by residents with their communities, and the option for those who are dissatisfied to move to an alternative that is more in keeping with their life-style requirements." The degree of residential choice, of course, depends on affluence as well as mobility. The urban poor may be dissatisfied with their neighborhood and lack the resources to move, but, for the majority of urban residents, choice among different kinds of communities is possible.

The American urban resident, as in the Soviet case, also participates intensively in the larger urban environment. He may indeed be tied more closely to a variety of widely scattered "interest communities" based on occupation or other activities than to a single local community or neighborhood (Webber, 1966:29). Increasing mobility and ease of communication, combined with large-scale organization, have produced similar results on a national scale. Those who participate in the national society move from one urban region to another seemingly without roots in any particular place. Such moves may cause little change in life-style, however, because they are likely to occur between similar kinds of community settings. As Berry (1973:57) has observed, "the attachment to a type of environment that sustains a particular life style is the key to the way in which contemporary Americans have adjusted the need to retain a locally based sense of security and stability to the emergence of a nationwide high-mobility society."

CONCLUSION

The existing and future urban systems of America and the Soviet Union are often described in terms which are strikingly similar. Increased mobility along with ease and variety of communications have resulted in the incorporation of ever larger territories into the sphere of urban influence. Urban scholars in both countries emphasize the enlarged scale of urban life and the consequent blurring of traditional distinctions between city and country. In both countries, from a socioeconomic point of view, the urban environment is (or is rapidly becoming) the dominant environment. Organization of the intraurban residential environment has stressed the local neighborhood or geographical community, but in each country urban residents have sought their "community" through interaction within the diversity of the larger urban region. Physical planning at the neighborhood level, while it obviously affects the quality of urban life, would seem to have limited impact on social interaction in a modern urban setting.

The process of urbanization, though it may produce similarities in the scale of urban life along with a common set of challenging physical-environmental problems, is not likely to substantially reduce the distinctive characteristics of Soviet and American urban landscapes. In the Soviet Union, state planning designed to create the "city of socialist man" should lead to greater uniformity in urban living environments. In the United States, institutionally supported "privatism" should continue to maintain a more varied urban setting. The often stated Soviet objective of overcoming the difference between city and country, when linked with a generally positive attitude toward the socialist city, suggests an attempt to make rural life more urban. In America, on the other hand, the ideal of rural and small-town environments has resulted in a broadly felt desire to make urban life more rural. This difference in attitude has a bearing on the distinctive features of urban agglomerations and suburban development in the two countries. Urban sprawl, however it may be viewed by planners, is probably a permanent feature of the American landscape. Improved communication, widespread use of the automobile, and the availability of private land have made it possible for even the moderately affluent to choose a suburban life style. More comprehensive planning in the Soviet Union, with its emphasis on controlling urban growth and limiting sprawl, should lead to a different kind of urban setting, perhaps with quite different consequences for the nature of urban life.

REFERENCES

BERRY, B.J.L. (1973). The human consequences of urbanization. New York: St. Martin's.

Cornell University, Center for Urban Development Research (1974). Guidelines and standards regarding the planning, layout, and facilities of settled areas in the Soviet Union. Ithaca, N.Y.: Author.

DAVIDOVICH, V.G. (1960). Rasselenie v promyshlennykh uzlakh. Moscow: Gosstroiizdat.

--- (1964). Planirovka gorodov i raionov. Moscow: Stroizdat.

FRIEDMANN, J., and MILLER, J. (1965). "The urban field." Journal of the American Institute of Planners, 31(4):312-319.

FROLIC, B.M. (1970). "The Soviet study of Soviet cities." Journal of Politics, 32(9):680-681.

GALBRAITH, J.K. (1967). The new industrial state. Boston: Houghton Mifflin.

Gosudarstvennyi piatiletnii plan razvitiia narodnogo khoziaistva SSR no 1971-1975 gody (1972). Moscow: Politizdat.

HARRIS, C.D. (1970). Cities of the Soviet Union. Chicago: Rand McNally.

HARVEY, D. (1972). Society, the city, and the space-economy of urbanism (Commission on College Geography, Resource paper no. 18). Washington, D.C.: Association of American Geographers.

KHODZHAYEV, D.G., and KHOREV, B.S. (1972). "The conception of a unified system of settlement and the planned regulation of city growth in the USSR." Soviet Geography: Review and Translation, 13(2):95.

KHOREV, B.S. (1968a). Gorodskie poseleniia SSRL problemy rosta i ikh izuchenie. Moscow: "Mysl."

--- (1968b). Problemy gorodov. Moscow: "Mysl."

--- (ed., 1972). Malyi gorod: Sotsial'no demograficheskoi issledovanie nebol'shogo goroda. Moscow: Izdatel'stvo Moskovskogo Universiteta.

KOGAN, L. (1967). "Urbanizatsiia-obschchenie-mikroraion." Arkhitektura SSSR, (4):39-44.

LAPPO, G.M. (1974). "Problems in the evolution of urban agglomerations." Soviet Geography: Review and Translation, 15(9):539.

Narodnoe khoziaistvo SSR v 1973 (1974). Moscow: Statistika.

OSBORN, R.J. (1970). Soviet social policies: Welfare, equality, and community. Homewood, Ill.: Dorsey.

PEREVEDENTSEV, V. (1970). "Migratsiia Naseleniia i ispol'zovanie trudovykh resursov." Voprosy Ekonomiki, (9):38.

--- (1975). "Goroda-millionery." Literaturnaya Gazeta, (18):10.

PIVOVAROV, Iu.L. (ed., 1972). Problemy sovremennoi urbanizatsii. Moscow: Statistika.

POKSHISHEVSKII, V.V. (1975). "Social geographic problems in the regulation of settlement systems in a developed socialist society." Soviet Geography: Review and Translation, 16(1):34.

Pravila i normy planirovki i zastroiki gorodov (1959). Moscow: Gosstroiizdat.

SSSR v tsifrakh v 1974 (1975). Moscow: Statistika.

STRONGINA, M. (1974). "Sotsial'no-ekonomicheskie problemy urbanizatsii." (Obzor literatury), Voprosy Ekonomiki, (1):135.

WEBBER, M.M. (1966). "Order in diversity: Community without propinquity." In L. Wingo, Jr. (ed.), Cities and space: The future use of urban land. Baltimore: Johns Hopkins University Press.

WEINBERG, I. (1969). "The problem of the convergence of industrial societies: A critical look at the state of a theory." Comparative Studies in Society and History, 11:15.

3

Urban Britain: Beyond Containment

ROY DREWETT
JOHN GODDARD
NIGEL SPENCE

□ THE STARTING POINT FOR THIS STUDY came from the Political and Economic Planning (PEP) study of *The Containment of Urban England* (Hall et al., 1973). Using a national set of functional urban areas, the PEP study, inter alia, focused both on the broad internal dynamics of urban areas as well as on urban England as a whole.

From a research and policy viewpoint, there is increasing interest in treating urban Britain as a unified urban system—that is, from national perspectives. Traditionally, urban research and policy in Britain has concentrated on the physically defined urban areas, while regional research and policy has concentrated on broadly defined tracts of country not necessarily having an urban focus. While confirming the importance of a regional redistribution of population, the PEP study suggested that the dominant trend was one of local decentralization of population from built-up areas to their peripheries, a process which was steering the pattern of regional movement. A subsequent study by Johnson et al. (1974), which treated migration between standard metropolitan areas, confirmed the importance of population movement between urban areas in different regions. But until the PEP study no areal units had been available at the appropriate scale to identify this process of urban change.

AUTHORS' NOTE: *There is no senior author. The authors would like to acknowledge the cooperation and support of their research colleagues on the Urban Change Project at the London School of Economics and Political Science: Steve Kennett, Robert Pinkham, and Caroline Connock.*

Internationally, opinion has similarly been converging on the need to view the cities of a country as a unified system where growth or decline in one urban region has repercussions in a number of other towns. In view of the importance of urban areas in channeling regional development processes, the need to connect urban policy and regional policy at the national level has been widely emphasized (Cameron and Wingo, 1973; EFTA, 1973; Friedly, 1974; Goddard, 1974).

While in Britain the current focus of public policy is on the problems of the inner city, it is clear that a national context for such a focus is required, particularly one that emphasizes the changing position of individual urban areas. And insofar as the problems of decline in the inner city are counterposed by related but different problems of peripheral growth, it is necessary to see this process of change as a whole. It is in this context that the present study was commissioned as the Urban Change Project, Department of Geography, London School of Economics and Political Science, financed under government contract from the Urban Affairs Directorate, the Department of the Environment, London. The aim of this paper is to present some interim findings.

OBJECTIVES OF THE STUDY

The object of the present study has been to update the PEP findings to 1971 and to examine urban change in Britain in the 1960s in greater depth than was possible in that study. This has involved redefining the functional urban areas on the basis of more up-to-date information and collecting additional information on social and economic characteristics of these areas and their constituent parts. The emphasis of the study is very firmly upon changes in the characteristics of urban areas rather than upon their structure at any particular point in time.

The present paper contains four major sections. This introductory section concludes with a brief description of data and definitions. The next section, "National Perspective on Urban Change," identifies the major characteristics of the city system as a whole and the constituent zones of metropolitan areas. The two subsequent sections, "Metropolitan Growth Performance" and "Intra-metropolitan Decentralization," deal with these themes in more detail. Wherever possible, the concept of changes in population and area is used to examine the relationship between inter- and intrametropolitan change at the national, regional, and individual city scales. Only aggregate population and employment trends are examined here; in subsequent research, fuller consideration will be given to detailed population and employment characteristics.

DATA AND DEFINITIONS

Data for 1,732 local authority areas, adjusted for boundary changes, were allocated to 126 functionally defined urban areas for 1951, 1961, and 1971.

These data, therefore, provide a picture of urban population change in Britain that is, first, independent of modifications in the definitions of local authority areas and, second, related to statistical areas which have very real meaning in terms of the daily lives of individuals. These areas are basically employment cores and their commuting hinterlands. The cores are composed of local authority areas with an employment density of over five workers per acre or a total employment of over 20,000. The commuting hinterland is divided into two areas; first, those from which more than 15% of the economically active population commute to the core and, second, all of those areas sending more workers to a particular employment core than to any other core. The employment core plus the 15% commuting hinterland are defined as a *Standard Metropolitan Labour Area* (SMLA) which should normally have a total population of 70,000 or more. The whole urban unit is defined as a *Metropolitan Economic Labour Area* (MELA). The full list of standardized definitional criteria can be found elsewhere (Hall et al., 1973; London School of Economics and Political Science, 1974).

The structure of metropolitan Britain has already been well described (Hall et al., 1973) and will not be duplicated here. The main thrust of this paper is urban change, and it is on this theme that attention will be focused. However, it is worth emphasizing the importance of the structure of metropolitan Britain. The size of SMLAs and their regional location do have an important bearing on the interpretation of subsequent analyses. The SMLAs are very unequally distributed (Figure 1), and they vary considerably in size, area, and total population. There are seven SMLAs with a population of over a million, with the London SMLA by far the largest with a population of 8.8 million in 1971; and 32 SMLAs have a population over one-quarter of a million. The majority of the total of 126 SMLAs are, therefore, medium or small in size. The median SMLA size is 142,000.

NATIONAL PERSPECTIVES ON URBAN CHANGE

NEW AND EXPANDING URBAN AREAS IN BRITAIN

When the definitional criteria are applied to Great Britain, some 126 SMLAs are identified in 1971. This compares with the 111 SMLAs previously identified for 1961 (Hall et al., 1973; Westaway, 1974). Comparing the two definitions gives some idea of the nature and location of urban change during the 1960s. New SMLA cores have basically emerged in two situations:

(1) In areas previously unclassified as outside "urban Britain"—for example, Yeovil, Hereford, Canterbury, and Corby. This suggests that many free-standing towns are becoming increasingly urbanized.

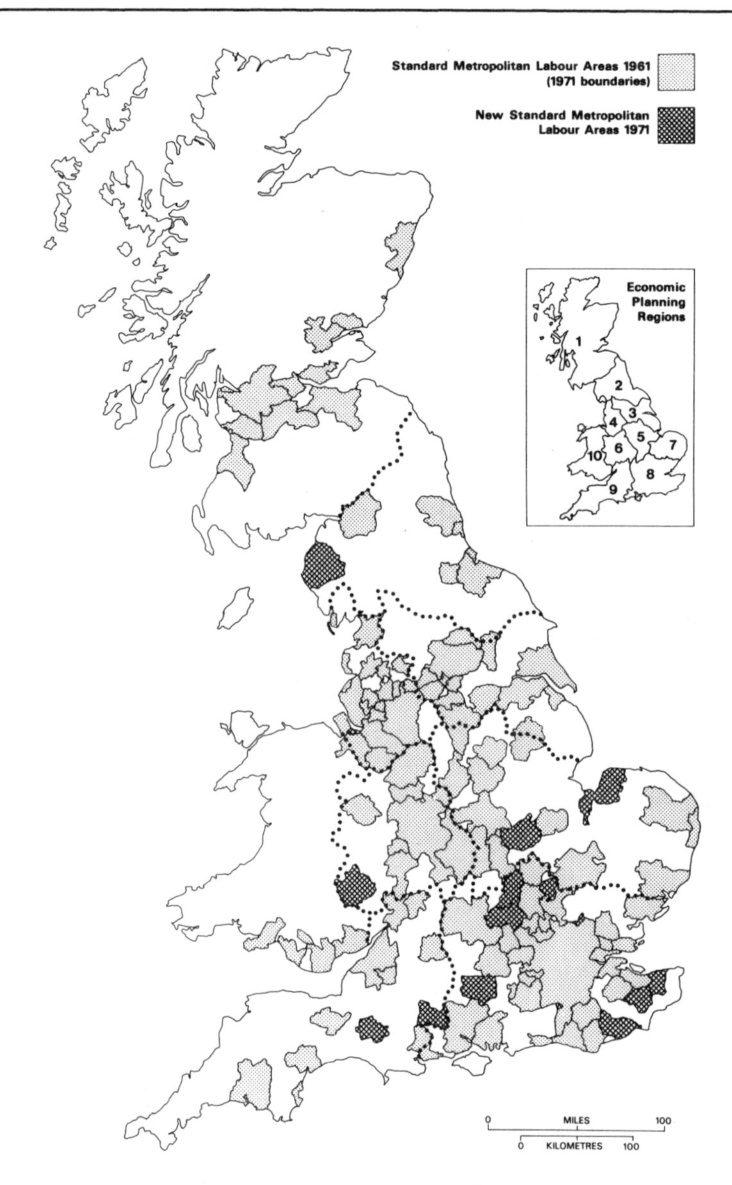

Economic Planning Regions: 1, Scotland; 2, North; 3, Yorkshire and Humberside; 4, Northwest; 5, East Midlands; 6, West Midlands; 7, East Anglia; 8, Southeast; 9, Southwest; 10, Wales.

(Note that the Stockton and Kirkcaldy SMLAs cease to exist in 1971, becoming part of the Teesside and Dunfermline SMLAs respectively.)

SOURCES: Hall et al., 1973; Westaway, 1974.

Figure 1: STANDARD METROPOLITAN LABOUR AREAS IN BRITAIN: 1961 AND 1971 DEFINITIONS

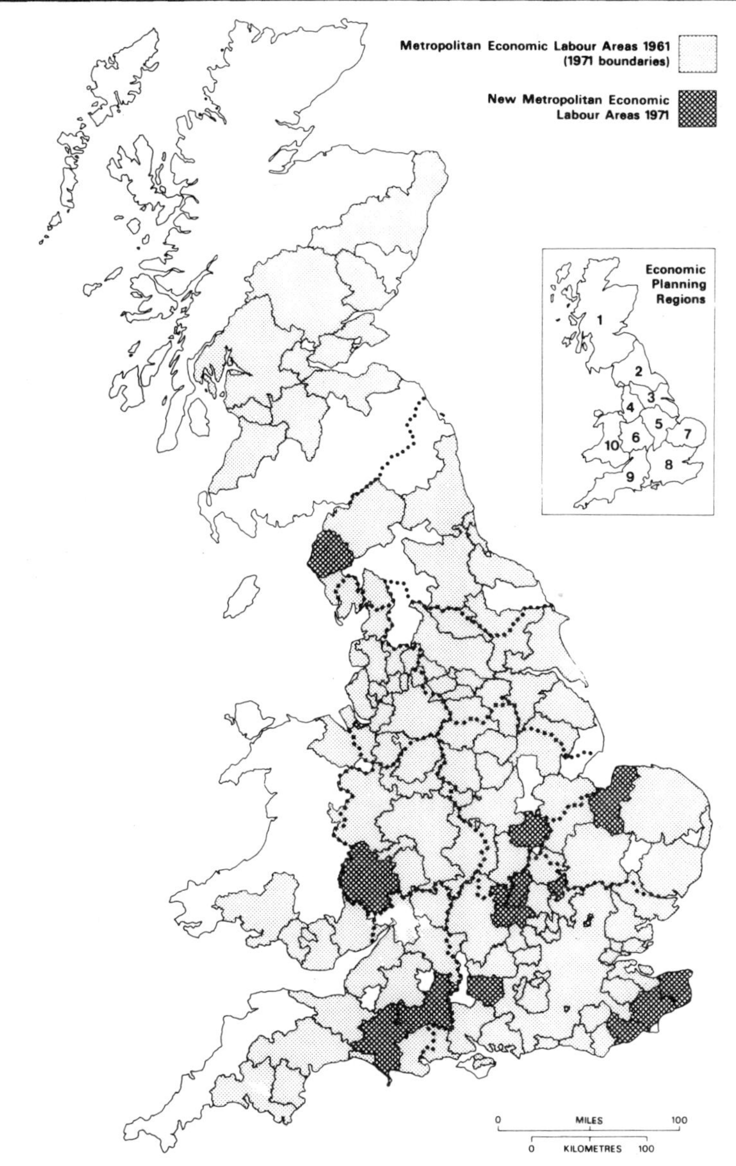

Economic Planning Regions: 1, Scotland; 2, North; 3, Yorkshire and Humberside; 4, Northwest; 5, East Midlands; 6, West Midlands; 7, East Anglia; 8, Southeast; 9, Southwest; 10, Wales.

(Note that the Stockton and Kirkcaldy MELAs cease to exist in 1971, becoming part of the Teesside and Dunfermline MELAs respectively.)

SOURCES: Hall et al., 1973; Westaway, 1974.

Figure 2: METROPOLITAN ECONOMIC LABOUR AREAS IN BRITAIN: 1961 AND 1971 DEFINITIONS

(2) From the outer metropolitan rings of other areas—for example, Aylesbury, Basingstoke, Harlow, Hastings, and Crawley. This suggests a process of satellite growth resulting from the decentralization of employment.

In both situations the commuting rings of the new employment cores may have encroached upon the commuting rings of established metropolitan areas.

In addition to the emergence of new urban areas, existing areas or their constituent parts may have *extended their boundaries.* Significantly, very few local authorities have been added to urban cores, indicating that there is little central expansion of employment. The most dominant pattern of change has been one in which the 15% commuting hinterland has been extended into the outer metropolitan ring. This process of extension apparently exists throughout the country. In addition, outer rings have been extending into previously unclassified areas.

The location of the new SMLAs (Figure 1) and MELAs (Figure 2) shows the majority to be in southeast England or on the periphery of the main axial belt of urban England. Noticeably, none are in Scotland; indeed, Kirkaldy no longer qualifies as a core on employment criteria and is now absorbed into the more dominant labor market of Dunfermline.

In addition to identifying newly emerging cores and metropolitan areas, it is also feasible, for the first time in Britain, to compare the proportion of the total population living in metropolitan areas using both a 1961 definition (PEP) and a 1971 definition (LSE)—i.e., a floating definition. Despite the existing high level of population concentration in urban areas in Britain, the proportion of the population in the MELAs has increased from 93.7% to 96.8%. Both fixed definitions show the same trend, though less markedly.

URBAN GROWTH AND DECLINE: THE CITY SYSTEM

When the overall structure of the city system is examined in each of the three years 1951, 1961, and 1971, in the form of a rank size graph (Figure 3), the system-wide structure gives the appearance of stability. There is some flattening out of the distribution over time as population growth has concentrated in medium and small-sized cities, and as new urban areas and New Towns, with high rates of growth, are added to the bottom of the urban hierarchy. These new entrants into the metropolitan system are responsible for the variability in the distributions at the lower end.

However, this graph conceals changes in the fortunes of individual cities. When the changing position of all SMLAs ranked according to total population in 1951, 1961, and 1971 is compared (Figure 4), the real dynamism and instability of the system is revealed. The diagram dramatically shows the relative stability in rank order of the top 20 or so SMLAs, but the remaining 106 rarely sustain their rank position over time. In nearly all cases, cities that were climbing

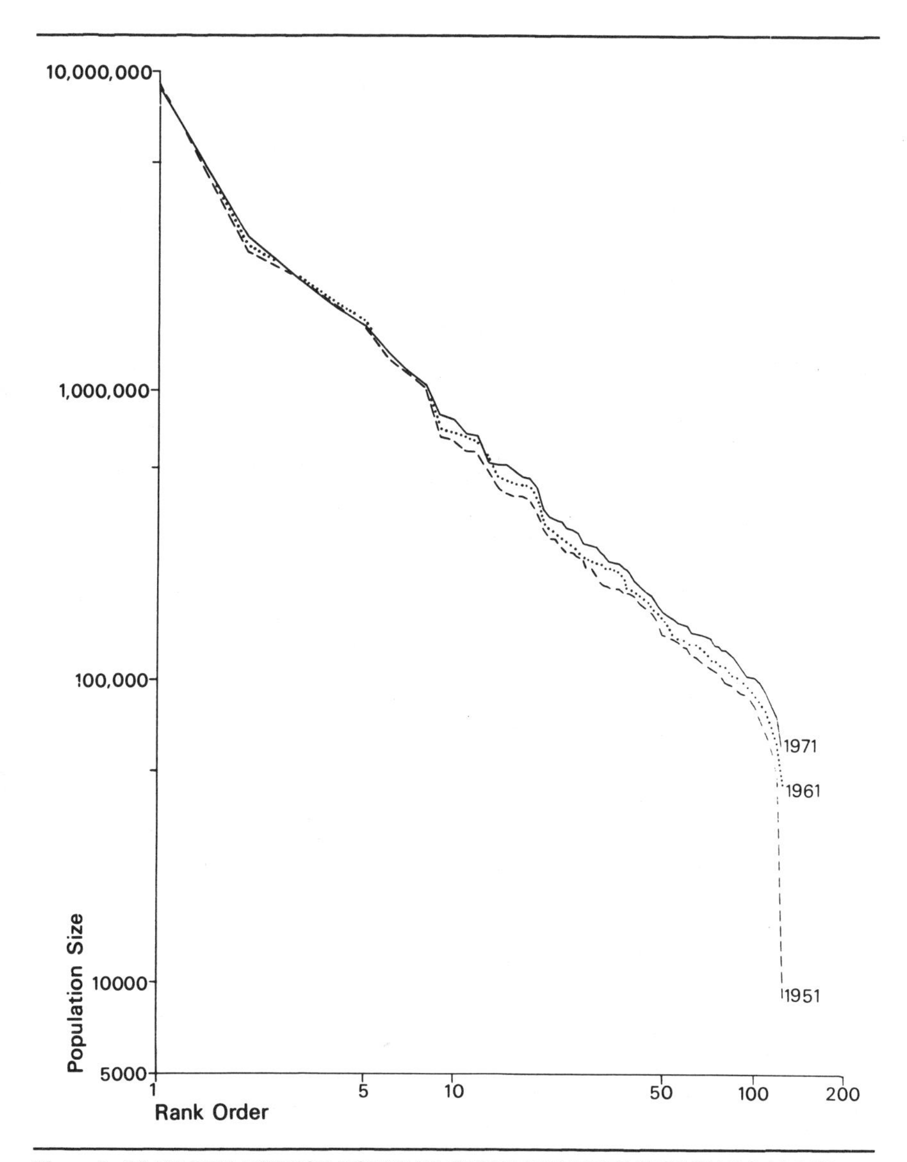

Figure 3: RANK-SIZE DISTRIBUTIONS OF STANDARD METROPOLITAN LABOUR AREA POPULATIONS, 1951, 1961, AND 1971

up or dropping down the hierarchy in the 1950s continued to follow the same trend in the 1960s. A most striking feature is this continuity of change between the two decades. Obviously there are differences in rate of ascent or descent in the two decades: as a city reaches higher levels, more population increase in absolute terms is required for each additional gain in place, while the reverse is true in the case of cities descending the hierarchy; hence the accelerating rate of decline in position recorded by many cities in the second decade.

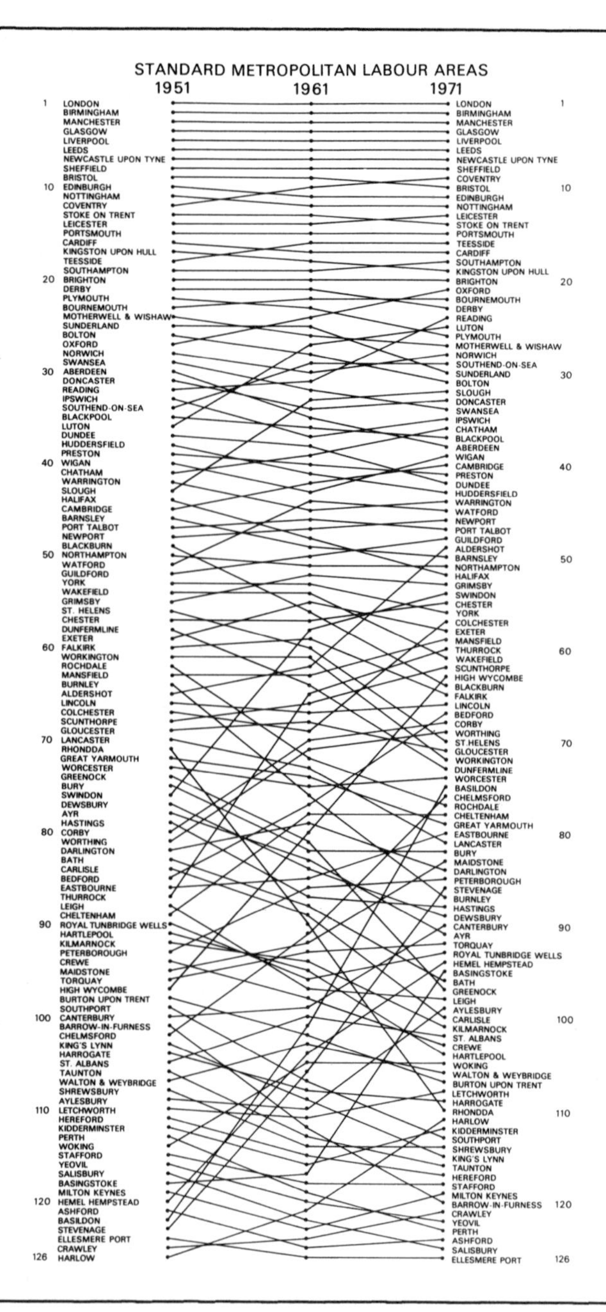

Figure 4: CHANGE IN POPULATION RANK ORDER OF STANDARD METROPOLITAN LABOUR AREAS, 1951, 1961, AND 1971

THE CONSTITUENT ZONES OF METROPOLITAN AREAS

An early insight into the changes in urban Britain can be obtained from the aggregate population and employment figures for 1951, 1961, and 1971, when they are disaggregated into the main components of metropolitan Britain, namely the cores, rings and outer rings (Table 1 and Table 2).

Taking population first, *the dominant trend is accelerating decentralization,* initially from urban cores to their 15% commuting hinterlands but subsequently spilling over in the 1960s into the outer metropolitan rings areas which are only weakly connected to the urban cores. The implication is that, while in the 1950s the SMLA rings were gaining population from urban cores *and* from the nonurbanized parts of Britain, in the 1960s some of this movement was directed into the outer metropolitan rings. So during this period the frontier of most active population change has moved progressively from the urban cores.

In detail, the proportion of an increasing national population living in urban cores has fallen from 53% to 47% since 1951, while the proportion living in suburban rings has increased from 26% to 32%. During the 1950s the population living in urban cores increased, but at a slower rate than the population as a whole. *By the 1960s this relative decline had been turned into an absolute loss of population.* In response, the rate of increase of population in the suburban ring jumped from 13.3% in the 1950s to 17.2% in the 1960s. During the earlier period, growth was concentrated in this ring, with outer rings showing a rate of population increase lower than the national average. However, in the sixties, population increase in the outer ring had also jumped dramatically from 3% to 10%, a rate nearly 5% above the national average.

For employment, consider first the earlier period 1951-1961. The principal finding is that employment became increasingly concentrated in the urban core areas. The proportion of employment increased in the cores from 60.5% to 61.4% in the decade. The metropolitan rings also increased their share, but very marginally. The only losses occurred in the outer rings and unclassified areas; but these losses totaled only some 71,000 jobs, so there was little apparent transfer of jobs between urban zones in the 1950s. The increase in jobs stemmed mainly from growth in the national economy. In great contrast, the second decade was a period of low national employment growth (1.69% compared to 5.06% in the 1950s), due mainly to the economic downturn at the end of the decade. In absolute terms, the effect of employment decline was felt greatest in the areas containing the majority of jobs—the urban cores. In aggregate, the cores lost 438,000 jobs, a reduction of 3.06%. Equally significant during such a period of slack growth was the marked increase in the growth of the metropolitan rings. Metropolitan rings increased by over 15% or 706,000 jobs to reach a national share of 22.87%. The outer rings also increased their share; although modest in total, it did substantially reverse the losses of the 1950s. The unclassified areas declined only marginally. *Overall a process of metropolitan decentralization of jobs in the 1960s can be clearly detected.* It is informative to compare the trends

TABLE 1
GREAT BRITAIN POPULATION AND EMPLOYMENT BY URBAN ZONE, 1951, 1961, 1971 (1,000s)

	Population						Employment					
	1951		1961		1971		1951		1961		1971	
	Absolute	%	Absolute	%	Absolute	%	Absolute	%	Absolute	%	Absolute	%
Urban cores	25,767	52.74	26,253	51.19	25,524	47.42	13,434	60.48	14,337	61.43	13,898	58.56
Metropolitan rings	12,914	26.44	14,635	28.54	17,147	31.86	4,425	19.92	4,719	20.22	5,429	22.87
SMLAs	38,681	79.18	40,887	79.73	42,671	79.28	17,859	80.40	19,055	81.65	19,327	81.44
Outer rings	7,808	15.98	8,053	15.70	8,838	16.42	3,327	14.98	3,313	14.19	3,443	14.51
MELAs	46,489	95.16	48,940	95.43	51,509	95.70	21,187	95.38	22,368	95.84	22,769	95.94
Unclassified areas	2,366	4.84	2,344	4.57	2,312	4.30	1,027	4.62	970	4.16	963	4.06
Great Britain Total	48,854	100.00	51,284	100.00	53,821	100.00	22,214	100.00	23,338	100.00	23,733	100.00

TABLE 2

GREAT BRITAIN POPULATION AND EMPLOYMENT CHANGE BY URBAN ZONE, 1951-1961 AND 1961-1971

	Population Change				Employment Change			
	1951-1961		1961-1971		1951-1961		1961-1971	
	Absolute	%	Absolute	%	Absolute	%	Absolute	%
Urban cores	486,061	1.88	–728,567	–2.77	902,387	6.72	–438,542	–3.06
Metropolitan rings	1,720,786	13.33	2,512,201	17.17	293,353	6.63	706,918	15.04
SMLAs	2,206,847	5.71	1,783,634	4.36	1,195,740	6.70	271,376	1.42
Outer rings	244,581	3.13	785,679	9.76	–14,235	–0.43	129,678	3.91
MELAs	2,451,423	5.27	2,569,313	5.25	1,181,505	5.57	401,054	1.79
Unclassified areas	–21,839	–0.92	–31,841	–1.36	–56,807	–5.53	–6,744	–0.70
Great Britain Total	2,429,589	4.97	2,537,472	4.95	1,124,698	5.06	394,310	1.69

of both population and employment, though careful note should be made of the declining participation rates from 45.5% in 1951 to 44.1% in 1971.

A comparison of population and employment shares across component zones of metropolitan areas reveals that employment is much more heavily concentrated in the cores than population (in 1971, the proportions were 58.56% and 47.42% respectively). The reverse is true of metropolitan rings, but not so marked in outer rings. These simple figures are not very surprising, particularly when one recalls that the SMLA concept is an employment-centered classification with commuting levels as the basis of ring definition. Of greater interest is the possible lead-lag relationships in the trends that both variables are undergoing in time and space.

For the national urban system as a whole (Table 3), the decentralization of population to the metropolitan rings and outer rings preceded any similar trends in employment. Clearly, the decentralization of population was leading the decentralization of jobs. By the 1960s, employment was rapidly following suit.

These lead-lag relationships are illustrated by the following ratio of population to employment for each urban zone.

$$D(I) = \frac{P(I)/P(T)}{E(I)/E(T)}$$

The ratio D, for zone I, is equal to the proportion of the total population, p(I)/P(T), in zone I divided by the proportion of the total employment, E(I)/E(T), in zone I. A zone which is "net importer" of workers will, as a result, have a value of less than one, and a zone which is a "net exporter" of workers will have a value greater than one. From Table 4 it is clear that the decentralization of population has led that of employment as the ratios for core areas decreased significantly over time. Over time this decrease in the ratios, which are all less than one, shows that the urban cores are becoming increasingly dependent for labor on the rest of the urban system–increasingly net importers of workers–despite the recent decentralization of jobs. Conversely, the metropolitan rings and the outer metropolitan rings have increasing ratios, all above one, reflecting a rate of population increase in excess of that for employment.

TABLE 3

GROWTH RATES OF POPULATION AND EMPLOYMENT BY URBAN ZONES, GREAT BRITAIN 1951-1961 AND 1961-1971

	1951-1961		1961-1971	
Urban Zones	**Population**	**Employment**	**Population**	**Employment**
Urban cores	1.88%	6.72%	–2.77%	–3.06%
Metropolitan rings	13.33	6.63	17.17	15.04
Outer rings	3.13	–0.43	9.76	3.91

TABLE 4
RATIOS OF POPULATION AND EMPLOYMENT BY URBAN ZONES, GREAT BRITAIN 1951, 1961, AND 1971

Urban Zones	1951	1961	1971
Urban cores	0.88	0.85	0.81
Metropolitan rings	1.31	1.35	1.40
Outer rings	1.07	1.11	1.14

NATIONAL PERSPECTIVE ON URBAN CHANGE

There are two major conclusions concerning national urban change in Britain over the two decades 1951-1971. First, metropolitan growth is still a characteristic phenomenon of the system with the emergence of new cores and the physical expansion of existing metropolitan areas. The rank size distribution shows relative stability of the major urban areas, but the remaining cities rarely hold their rank; they either decline or improve throughout both time periods. The growth of medium and small cities is clearly in evidence, particularly New Towns or those influenced by planning legislation. Conversely, many cities declined in the 1950s and continued to do so into the latter period. Second, the intrametropolitan system is equally dynamic. Population decentralization from metropolitan cores to rings has gathered apace and has now reached a stage of spilling over into the outer metropolitan rings which had traditionally weak allegiances to central core areas. Employment was meanwhile centralizing in the urban cores in the 1950s; but in jobs as well, the outward flow started to follow population trends by the 1960s.

After this national perspective, these various phenomenon will now be examined in their regional and metropolitan settings.

METROPOLITAN GROWTH PERFORMANCE

METROPOLITAN CHANGE RELATIVE TO THE NATIONAL AVERAGE

One of the main characteristics of population change in metropolitan Britain has been the dramatic change of fortune of some metropolitan systems in comparison with others. This inability of certain cities to sustain their ranking in the urban hierarchy while others improve their position seems worthy of closer inspection.

To help identify these variations, the following classification of SMLAs was adopted:

Class I: Those which have grown at a slower than average rate 1951-1961 *and* 1961-1971.

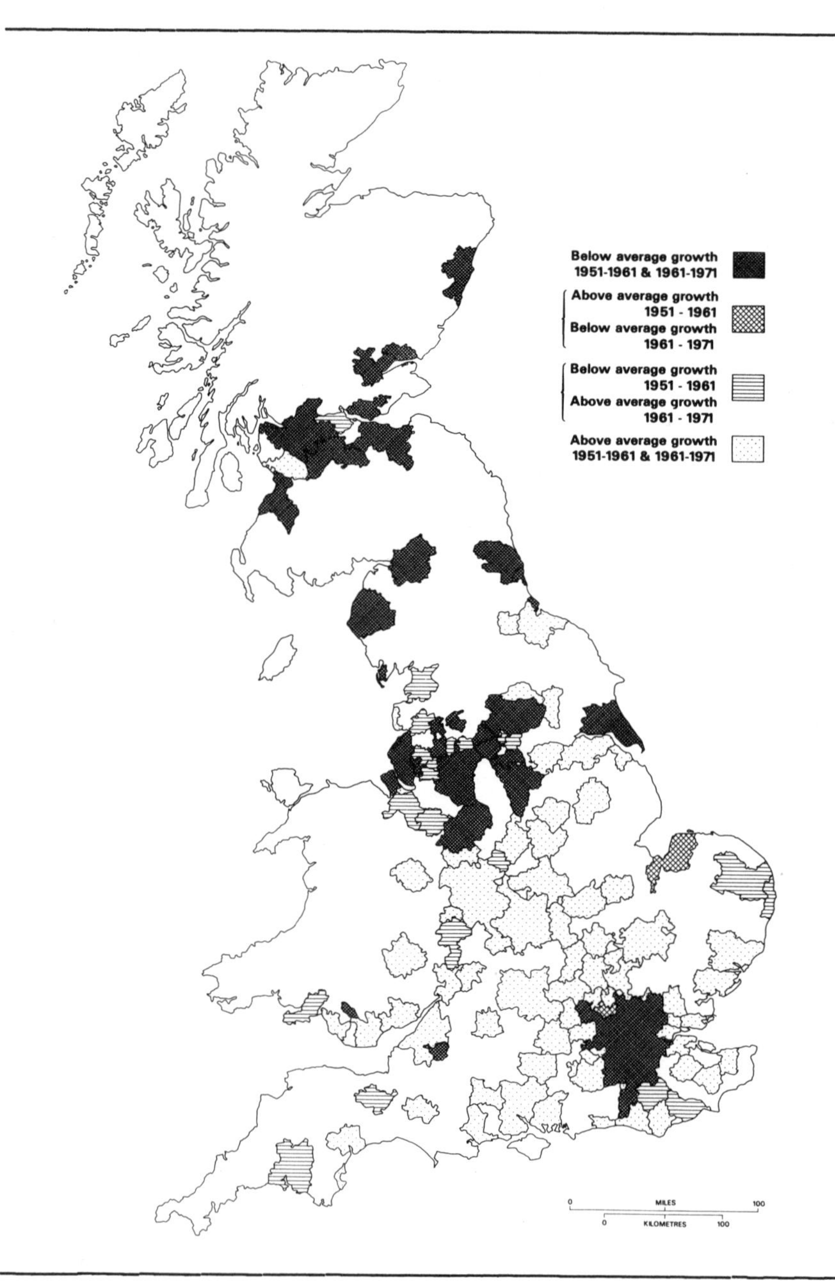

Figure 5: POPULATION CHANGE PERFORMANCE OF STANDARD METROPOLITAN LABOUR AREAS, 1951-1961 AND 1961-1971

Class II: Those which have grown at a slower than average rate between 1951 and 1961 and a faster than average rate between 1961 and 1971.

Class III: Those which have grown at a faster than average rate 1951-1961 *and* 1961-1971.

Class IV: Those which have grown at a faster than average rate between 1951 and 1961 and a slower than average rate between 1961 and 1971.

However, this classification is critically affected by how the *average rate* is calculated; different average rates produce a wide variation in the pattern of metropolitan growth performance. Two average rates were adopted. The *national SMLA rate* refers to the percentage change in the aggregate of SMLA population and employment values over the periods 1951 to 1961 and 1961 to 1971. For population the rates were 5.71% (1951-1961) and 4.36% (1961-1971) and for employment 6.70% and 1.42% respectively. The *individual SMLA rate* is an average and refers to the mean of the percentage changes in population and employment for the range of SMLAs in the system over the periods 1951 to 1961 and 1961 to 1971. For population the rates were 18.04% (1951-1961) and 11.58% (1961-1971), and for employment 20.84% and 10.48% respectively.

These two rules enable interesting comparisons to be made between different classifications and between population and employment growth performance. The conclusion is very clear: the difference between population and employment growth performance within each classification is minimal. The difference in trends for both population and employment is most marked when different growth rates are used, and the results serve to highlight different phenomena of change in the urban system. The major variation resulting from the use of different rates occurs in Classes I and III.

The similarity of trends for population and employment changes does not warrant their individual presentation. Therefore, the effect of using the national SMLA growth rate will be illustrated with respect to population change, and the effect of the individual average growth rate will be illustrated with respect to employment change.

When population change for each SMLA is compared to the national growth rate (Figure 5), those SMLAs with a persistently low growth rate (Class I), tend to be concentrated mostly in the North, the Northwest, Yorkshire, and Scotland. Notable exceptions to this generalization are London and the Rhondda. The broad mass of central and southern English SMLAs fall into Class III, enjoying a persistently high rate of growth throughout the period. The "unstable" Classes II and IV have a more widespread distribution, but a special note should be made of the south Lancashire area, where a reversal of the downward trend in the 1950s has become apparent during the second decade. This can be directly related to inter- rather than intrametropolitan decentralization, in this case from the Manchester and Liverpool SMLAs. This classification concentrates on percentage growth rates, so not surprisingly most of the large SMLAs–including Leeds, Liverpool, London, Manchester, Newcastle,

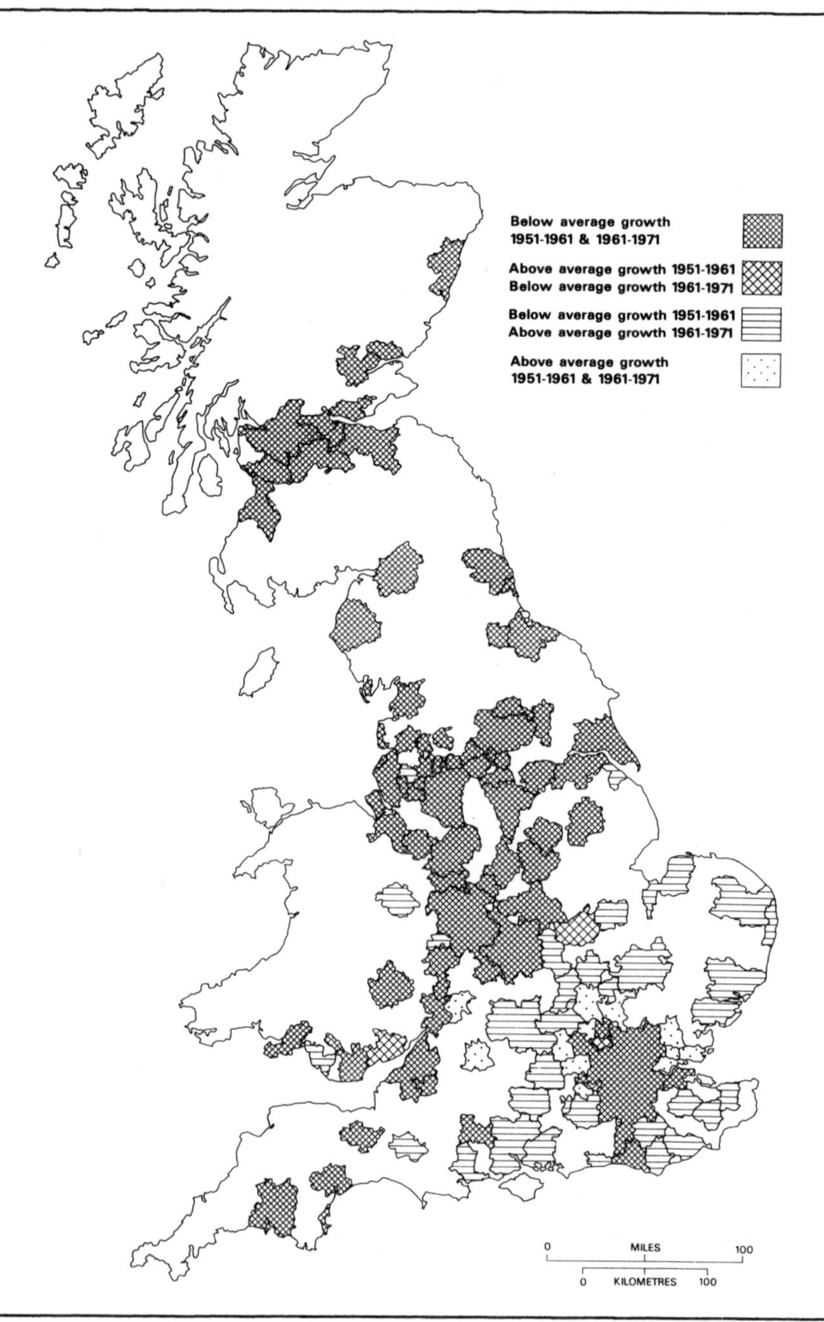

Figure 6: EMPLOYMENT CHANGE PERFORMANCE OF STANDARD METROPOLITAN LABOUR AREAS, 1951-1961 AND 1961-1971

Edinburgh, and Glasgow–fall into Class I, the low growth group. For the same reasons most new and expanded towns fall into the high growth Class III.

By comparison, the growth performance using the SMLA growth rate shows marked differences. This is illustrated with respect to employment change (Figure 6). The London SMLA, itself declining over both decades, is surrounded by a series of concentric rings of cities which fall into different classes. First, an inner ring of small SMLAs lying close to the edge of London, which reversed their fortunes in the 1960s (Class II), are themselves surrounded by a wide band of SMLAs further from London which have grown for both decades (Class III). Much of the rest of the country declined in jobs relative to the average for all SMLAs.

The difference between the two distributions does pose important methodological questions concerning the interpretation of such empirical analyses. The impact of varying the average has produced a different picture of urban change performance. Clearly the extremes of high and low growth SMLAs are unaffected, but the affect on the intermediate range of SMLAs is considerable. The use of each average does illustrate different phenomena. The national SMLA rate highlights the regional variations in the country but also detects the recovery of some small SMLAs in the Northwest, for example. This is not detected using an individual SMLA growth rate, but the concentric rings of varying performance around London is brought out, together with the inclusion of many large cities in the West and East Midlands in Class I. Clearly the various aspects of the urban change process can be detected only by the use of a variety of measures, although it must be stressed that the relationship between the detailed process and the individual measure is as yet unexplored.

This comparison of SMLA growth performance does indicate significant regional variation, but the main emphasis lay at the individual SMLA level, and the criteria used in the classification were national average growth rates. If we sum all SMLAs for their respective region, we can see more clearly the regional variability itself in SMLA population and employment growth.

REGIONAL VARIATION IN METROPOLITAN CHANGE PERFORMANCE

For the purpose of this and some subsequent sections, SMLAs were allocated to the Economic Planning Regions on the basis of the region in which their cores lie, irrespective of whether parts of their metropolitan rings or outer metropolitan rings fall in other regions. This decision was made because the SMLA is the basic unit of analysis for the study, and it was felt that to divide up these units for a regional analysis would be contravening the value of the SMLA as a functional base. Figure 1 shows the relationship of SMLAs to the boundaries of the Economic Planning Regions of the country.

The regional picture that results from such a grouping procedure (Table 5) shows that during the 1950s all regions enjoyed absolute growth in aggregate for both population and employment. However, there are regional variations in

TABLE 5
REGIONAL VARIATION IN SMLA PERCENTAGE POPULATION AND EMPLOYMENT CHANGE, 1951-1961 AND 1961-1971

Economic Planning	1951-1961		1961-1971	
Regions	Population	Employment	Population	Employment
Southeast	7.29%	9.77%	4.43%	2.94%
West Midlands	7.93	8.91	6.58	1.69
East Midlands	8.66	8.56	9.65	3.87
East Anglia	8.66	10.42	11.76	14.51
Southwest	8.19	9.31	9.48	8.36
Yorkshire & Humberside	3.39	4.77	3.58	–2.04
Northwest	1.89	0.73	1.88	–3.59
North	5.84	4.85	1.66	2.37
Wales	5.31	7.75	4.83	4.66
Scotland	2.48	1.58	0.40	–2.71
Great Britain	5.71	6.70	4.36	1.42

relative growth performance. In general, the pattern is of southern regions performing better than northern regions; this is even more striking in terms of employment, with Yorkshire and Humberside, the Northwest, the North, and Scotland relatively declining (less than the national growth rate).

An even greater regional variation is apparent in the 1960s. All regions grew in population, with four regions increasing their share compared to the 1950s (East Midlands, East Anglia, the Southwest, and Yorkshire and Humberside); the *range* of growth rate was higher–11.76 to 0.40 (compared with 8.66 to 1.89 in the 1950s). The differences in employment are the most striking. Some four regions actually declined in employment: the Southeast, Yorkshire and Humberside, the Northwest, and Scotland. The Southeast figure conceals the heavy decline of London. East Anglia (14.5) and the Southwest (8.4) clearly lead in employment growth. Although not presented here, very similar trends are to be found in the SMLAs.

SIZE VARIATIONS IN METROPOLITAN CHANGE

Various attempts have been made to explore the relationship between urban growth performance and urban size. Core and SMLA population and employment have been correlated with decennial change (Table 6). Some interesting similarities and differences between the core and SMLA analyses should be noted. First, the relationship between size and growth performance for both population and employment is generally stronger at the scale of urban cores than SMLAs. This is because cores are characterized by a much greater homogeneity of change performance than exists for SMLAs; at the same time there is a greater variation in size among SMLAs than cores. Second, size is more strongly correlated in most instances with absolute population and employment change than with relative change. Although this is by no means unexpected, it is

TABLE 6
RELATIONSHIP BETWEEN CORE AND SMLA SIZE AND DECENNIAL GROWTH

Variables	Core (r)	SMLA (r)
Absolute Population Change, 1951-1961 & Log Population Size, 1951	–.44	+.43
Absolute Population Change, 1961-1971 & Log Population Size, 1961	–.61	–.27
Relative Population Change, 1951-1961 & Log Population Size, 1951	–.41	–.19
Relative Population Change, 1961-1971 & Log Population Size, 1961	–.47	–.39
Absolute Employment Change, 1951-1961 & Log Employment Size, 1951	+.45	+.48
Absolute Employment Change, 1961-1971 & Log Employment Size, 1961	–.64	–.53
Relative Employment Change, 1951-1961 & Log Employment Size, 1951	–.45	–.40
Relative Employment Change, 1961-1971 & Log Employment Size, 1961	–.47	–.39

reassuring to note that, when relative change is used, most of the relationships with size are similar in direction to those involving absolute data.

Table 6 indicates that the relationship between urban size and change performance is, in the main, inverse. In other words the larger the urban area, in terms of either population or employment, the smaller the growth or the larger the decline. For the urban core areas, each of the four temporal pairs of analyses shows that the negative relationship between size and change is becoming stronger over time. This is also true of three out of the four parallel analyses for SMLAs, while the fourth differs in the opposite direction by only 0.01. In addition to the strength of the negative relationship increasing in the sixties, in all eight analyses relating to the sixties, a negative relationship holds. In the fifties, although this is generally the case, it does not apply to absolute population change. While cores exhibit the usual negative relationship, the relationship between total population and absolute population change for SMLAs is markedly positive. The interpretation of this contrast involves the notion of centralization-decentralization. In the fifties, the core areas of large cities were indeed experiencing decentralizing population, but this process was contained within the SMLA boundary. The result was a contrasting relationship for cores and SMLAs which by the sixties had disappeared in direction although not in degree as decentralization began to spill over the SMLA boundaries of the larger cities into smaller SMLAs or outer metropolitan rings. For total employment the size-change relationships are positive for both cores and SMLAs. It seems, then, that the process of decentralization that was apparent for population has not begun for employment even for the core areas, let alone SMLAs. It does, however, make more remarkable the dramatic switch in the size-change relationship for employment in the sixties. Finally, all the relative change analyses result in a negative relationship, indicating that the interpretation so far enunciated of declining population and employment fortunes associated with the larger urban areas is substantially correct.

METROPOLITAN GROWTH PERFORMANCE

It has been noted that SMLA growth performance varies significantly according to the size of the metropolitan system and its regional setting. In addition, location relative to other cities is important.

This is illustrated by comparing the results of shift in share analyses of population trends for the two decades. If the 1951-1961 percentage share of the shift is subtracted from the equivalent figure for 1961-1971 (i.e., if the two proportions of the national shift are subtracted for the two time periods), the difference in the percentage shares can be either positive or negative (relative improvement or not) while the SMLA may be either growing or still declining.

The SMLAs in the negative shift group (relatively worse off) fall into one of three distinct groups. They are among either the very largest SMLAs in the country (London, Liverpool, Glasgow, Newcastle, Birmingham, and Coventry), London's New Towns (Crawley, Harlow, Hemel Hempstead, Stevenage, and Basildon) or the small SMLAs located very close to the London SMLA itself (Slough, Thurrock, Luton, St. Albans, etc.).

The SMLAs in the positive shift group (relatively better off) also form distinct groups: they are either medium-sized, free-standing SMLAs (Bristol, Leicester, Oxford), outer London SMLAs (Colchester, Reading, Chatham, Maidstone), outer Manchester-Liverpool SMLAs (Wigan, Warrington, Bury, Rochdale, etc.), or the second-phase New Towns and Expanded Towns (Northampton, Peterborough).

The implication of these findings is clear. The size and stage in the metropolitan life cycle determine very clearly the rates of SMLA growth and intermetropolitan growth. This metropolitan decentalization from large metropolitan centers to nearby smaller satellite SMLAs, whether they be New Towns or existing cities, is the most dominant process which currently forges the changing shape of metropolitan Britain.

The principal conclusions in the analysis so far bear out those of the PEP report (1951-1966), namely, that two main influences are important for SMLA change: first, a regional effect and, second, a local decentralization effect from dominant SMLAs centered on the conurbations to nearby satellite or subdominant SMLAs, usually smaller in size. Size is, therefore, not the only controlling factor but is strongly coupled to an SMLA's functional relationship to nearby metropolitan areas.

INTRAMETROPOLITAN DECENTRALIZATION

The national perspective on urban change showed the intrametropolitan system of urban zones to be exhibiting a clear dynamism: that of aggregate decentralization from urban cores. This phenomenon is now considered in more detail.

REGIONAL RATES OF CHANGE FOR CONSTITUENT URBAN ZONES

To discover whether or not decentralization of population and employment is all-pervasive, the relative change of each variable has been plotted over two time periods for the three component parts of the metropolitan system (Figure 7). For population, the aggregate trend for cores in all regions shows a decline in relative growth. In only one region (the Northwest) were cores declining absolutely in the 1951-1961 period; but by 1961-1971, cores in all regions were declining absolutely with the exception of East Anglia, the Southwest and the East Midlands, and their population growth rates were dropping rapidly. In contrast, the metropolitan rings in all regions exhibited positive growth rates for both time periods, and only two regions showed relative decline, namely, the Southeast and the North. This relative decline in the Southeast was compensated for in the outer ring, where the Southeast had the fastest growth rate. All regions had increasing growth rates in this zone, although two (the North and Wales) were growing but still negative.

For population, there was a very clear uniformity of decentralization at the regional level. For employment, the overall trends are very similar, the only difference being the greater range of growth rate values and the fact that the degree of decentralization was less intensive. All the regions exhibited positive growth in the cores in 1951-1961, but by the second decade, all but three (East Anglia, the Southwest, and Wales) were declining absolutely. The angle of decline was greater than population due to the higher centralization of employment in the 1950s. In the metropolitan rings, the outward push of jobs follows that of population. The majority are positive in the 1950s, and they are all positive by the 1960s. The decentralization trend of jobs into the outer rings is less clear. With the exception of the North and Wales, all regions are increasing their growth in the outer rings, but this is only in relative terms in four regions. The one clear exception is the Southeast, where a rapid decentralization of jobs is occurring in the outer ring, similar to the trends in population. The main feature of these trends is that although the rates of change vary, the decentralization trend remains steady and persistent throughout all regions, despite the varying size of regions and the number and size of SMLAs contained in each.

Having established the decentralization trends at the regional level, we may now classify all SMLAs according to their decentralization characteristics.

MEASURES OF DECENTRALIZATION

There are a number of ways of measuring the decentralization of population and employment from urban cores to SMLA rings and the outer rings of MELAs. The simplest approach is to compare relative population change rates for the constituent areas; some six patterns of relative change can be identified for SMLAs:

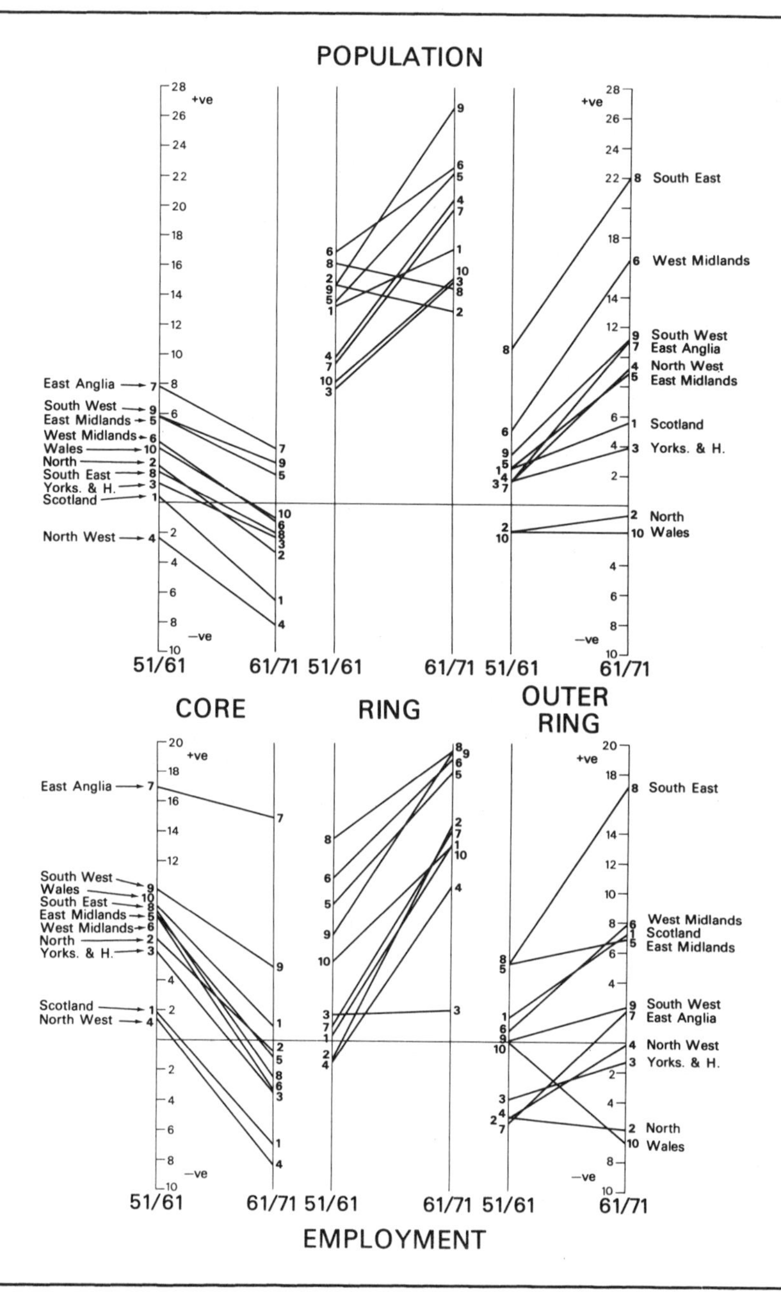

Figure 7: RELATIVE POPULATION AND RELATIVE EMPLOYMENT CHANGE FOR THE CONSTITUENT ZONES OF STANDARD METROPOLITAN LABOUR AREAS AVERAGED FOR ECONOMIC PLANNING REGIONS, 1951-1961 AND 1961-1971

(1) *Absolute centralization:* core growing but metropolitan ring declining.

(2) *Relative centralization:* core growing faster than metropolitan ring.

(3) *Absolute decentralization:* core declining but metropolitan ring growing.

(4) *Relative decentralization:* metropolitan ring growing faster than core.

(5) *Centralization during decline:* metroplitan ring declining faster than core.

(6) *Decentralization during decline:* core declining faster than metropolitan ring.

When population and employment are classified in this way and compared over time, it is apparent that SMLA population tended to decentralize in the 1950s (67 SMLAs), while the trend in employment was one of centralization (73 SMLAs). By the 1960s the employment trends were reversed, and decentralization was more dominant (67 SMLAs). During the same period, population decentralization was very much stronger than before with 101 SMLAs decentralizing in one form or another.

The question now arises: which SMLAs are decentralizing faster than others? Does decentralization reflect the size of the system, the density of the core, or the stage in the "life cycle" of an evolving metropolitan system? All three possible factors are now considered briefly.

When centralization and decentralization are tabulated against SMLA size groups (Table 7), it is very clear that the larger groups tend to decentralize while

TABLE 7

SMLA CENTRALIZATION-DECENTRALIZATION CLASSIFICATION, 1961-1971 FOR POPULATION AND EMPLOYMENT RELATED TO SMLA POPULATION SIZE, 1961

Class of Centralization and Decentralization		Population Size Groups						Total SMLAs*
		1	2	3	4	5	6	
Absolute and relative centralization	Population	–	1	1	3	3	10	18
	Employment	–	2	5	11	12	16	46
Absolute and relative decentralization	Population	1	24	24	22	18	12	101
	Employment	1	23	20	14	9	6	73

Size group: 1 9,232,700 and over
2 257,900 to 9,232,699
3 151,100 to 257,899
4 113,500 to 151,099
5 88,500 to 113,499
6 less than 88,500

*All SMLAs are not included. Some 7 SMLAs are unclassified because they have no metropolitan rings.

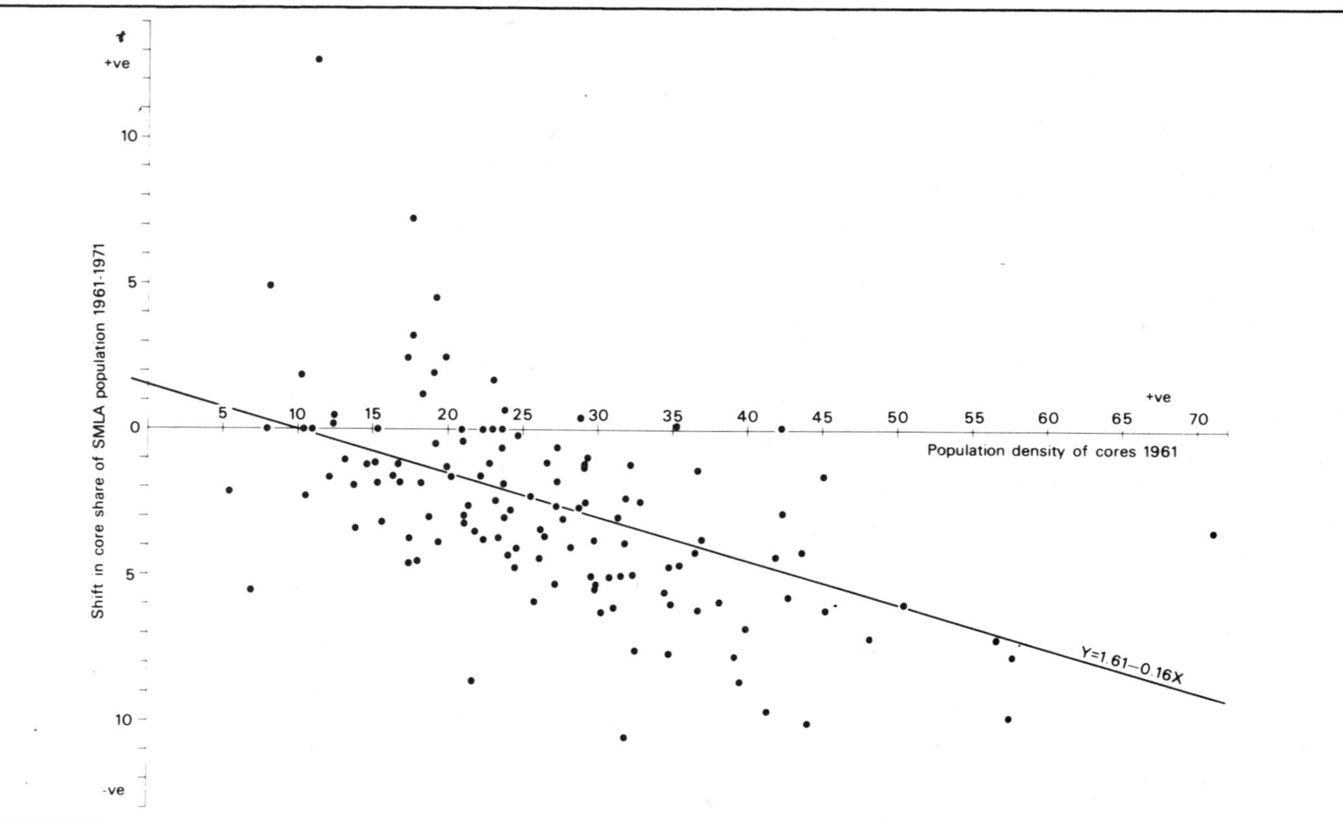

Figure 8: POPULATION DENSITY OF URBAN CORES (1961) AND THE SHIFT IN CORE SHARE OF STANDARD METROPOLITAN LABOUR AREA POPULATION (1961-1971)

smaller SMLAs have a greater tendency to centralize. This trend is more accentuated for population than for employment.

The density of population at the urban core, though positively related to city size with bigger cities having higher core densities, does predict well the variations in rates of decentralization. If another measure of decentralization, the shift in core share of SMLA population (1961-1971), is related to population density of the core (1961), the effect of urban core density on decentralization can be measured. The nature of the relationship (Figure 8) is that SMLAs with the highest density cores have relatively high rates of decentralization and vice versa, and a reasonable correlation is apparent. Simple linear regression confirms this statement; with the dependent variable, Y, indicating a shift in core share of SMLA population 1961-1971, and the independent variable, X, indicating a population density of core 1971, the regression equation is $Y = 1.48 - 0.16\,X$, with a coefficient correlation of -0.54.

Finally, it can be postulated that centralization and decentralization in both the SMLA and MELA systems are dependent upon the stage in the "life cycle" of individual cities, which in turn may be related to their location relative to other cities in the urban system. At this point no connection with actual time is assumed, for at any one point there may be leaders and laggards amongst the set of cities. The sequential stages are three.

Stage 1. Initial growth of an area to urban status, whether as a result of planned expansion or natural growth, will tend to be located in the center or core area. This stage will therefore be characterized by centralization of both the SMLA and the MELA. Core growth may come from the metropolitan ring or the outer metropolitan ring or from outside the MELA altogether. The conditions of this stage are:

(a) The percentage population growth of the core exceeds the percentage population growth of the metropolitan ring.

(b) The percentage population growth of the SMLA exceeds the percentage population growth of the outer metropolitan ring.

There exists a special case of this stage in which, typically, a rather diffuse old basic industrial area is collapsing in upon its center. In this way, although the whole system is declining, the outer areas may be shrinking faster. This is apparent in mining areas such as in the Northeast or South Wales.

Stage 2. Some time after the initial stage, decentralization tendencies set in for a variety of reasons. It seems reasonable to expect that decentralization will begin with areas nearest to the core receiving population. As a result, the conditions for Stage 2 are as follows:

(a) The percentage population growth of the metropolitan ring exceeds the percentage population growth of the core.

(b) The percentage population growth of the SMLA exceeds the percentage population growth for the outer metropolitan ring.

The outcome is a decentralizing SMLA existing within a MELA which is continuing to centralize.

Stage 3. This final stage sees the decentralization movement spreading to the outer metropolitan ring and both the SMLA and the MELA area decentralizing. To date, this can be considered the final stage for Britain, although there is some North American evidence to suggest that further stages are possible. The conditions for Stage 3 are as follows:

(a) The percentage population growth of the metropolitan ring exceeds the percentage population growth of the core.

(b) The percentage population growth of the outer metropolitan ring exceeds the percentage population growth of the SMLA.

A special case of Stage 3 exists where large metropolitan areas generate such massive decentralization that they overflow into neighboring MELAs. It is quite possible, therefore, that a decentralizing metropolitan area could be losing population in total.

Testing this model against the actual data revealed that of the 91 MELAs having the three spatial components, we found that only 10 in the fifties and four in the sixties proved unclassifiable into one of the three life cycle stages. The exceptions were peripheral MELAs like Taunton, Kings Lynn, Dunfermline, and Kilmarnock, which still have largely rural outer rings, and MELAs containing large and growing free standing towns in their outer rings such as Hastings with Bexhill in its outer ring. Given the sequential nature of the classification, it is expected that MELAs progress from Stage 1 through Stages 2 and 3 over time. In fact, 35 MELAs "progressed" into a further stage in their life cycle during the two decades, and the remainder stayed in the same stage. New Towns, free standing market towns on the periphery of megalopolis, and large city satellites characterize the early periods, while the larger cities that form the backbone of urban Britain are generally in the middle or later phases of development.

INTRAMETROPOLITAN DECENTRALIZATION

Decentralization of population and latterly of jobs is a *process* that characterizes all British cities, irrespective of their size, region, or location relative to other cities. What does vary is the *rate* at which this process is operating. Cities in the prosperous regions of the country are decentralizing more rapidly than those in the peripheral and less prosperous regions. Size of city is also an important factor in explaining rates of decentralization, principally because size is related to the density of the urban core, which is in turn linked to the stage that the city has reached in its life cycle. Large cities record the highest core densities and have progressed most toward rapid decentralization; smaller cities, especially New Towns, have been able to grow at the core before growth has spilled over into the urban ring. Nevertheless, if growth is to continue in small and medium-sized towns without wholesale central urban renewal, it is inevitable that this growth must occur in the metropolitan ring, thereby resulting in net decentralization of population.

NATURAL CHANGE AND NET MIGRATION COMPONENTS OF POPULATION CHANGE

Urban change in Britain over the two decades under review here between 1951 and 1971 has several distinctive features: the interregional variation in SMLA growth performance; the regional variation of rapidly expanding small and medium-sized SMLAs and the decline of associated large SMLAs usually centered on the old conurbations, and the now very familiar intra-SMLA decentralization of both population and jobs from cores to rings. These aggregate shifts in population and their lagged effects on employment within and between urban areas can be attributed only to variations in natural change and net migration. Their relative importance in explaining variations in population change at the regional, SMLA, and intra-SMLA levels are now considered.

NET MIGRATION AND POPULATION CHANGE

At the national level, the study has examined the relationship between natural change, net migration, and population change for all SMLAs. If natural increase is taken as a constant (6.18%) on a graph of net migration change (Y) against total population change (X), then deviations from this line will indicate urban areas with above- or below-average rates of natural change, all other population changes being attributable to variations in net migration rates. The graph shows a striking conformity (Figure 9). The great majority of areas vary only slightly on either side of the mean rate. The exception to this general relationship can be easily identified. Those to the far left of the mean natural change line are almost in every case major retirement centers; conversely, those to the right of the mean line are almost all associated with New Towns or Expanded Towns. So, with the exception of these predictable deviations, net migration rates can be readily equated with population change rates and vice versa.

It follows, therefore, that the net-migration fortunes of all remaining individual SMLAs would closely reflect their population change—usually a function of regional location, stage in the metropolitan life cycle, and functional relationship to other SMLAs.

Most of these characteristics can be summarized by an SMLA classification of the constituent components of population change. The various combinations

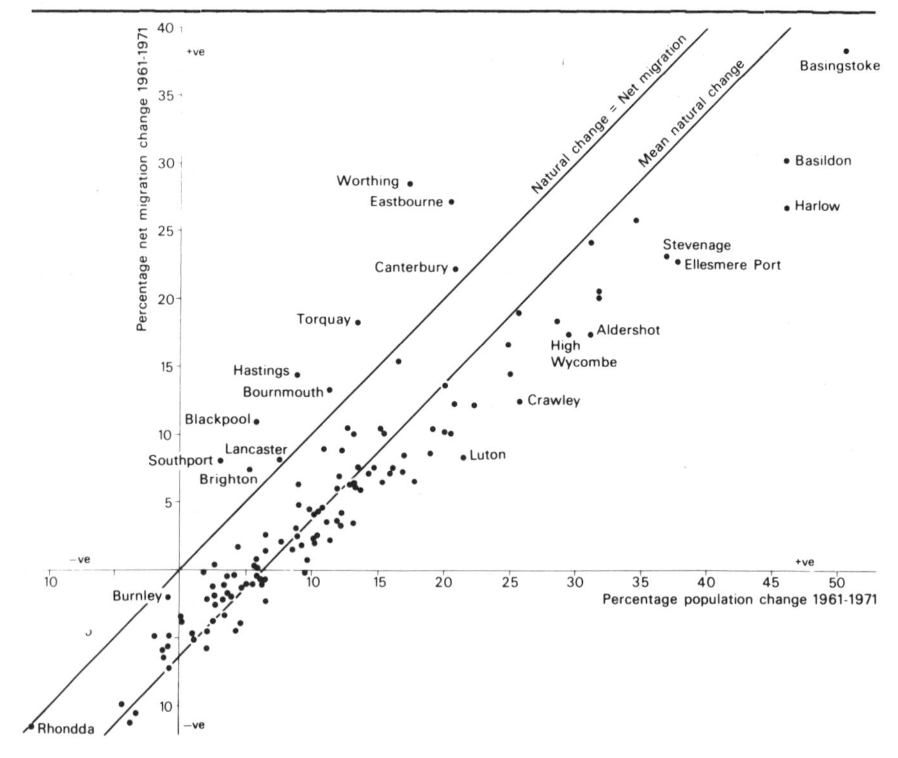

Figure 9: RELATIVE NET MIGRATION AND RELATIVE POPULATION CHANGE FOR STANDARD METROPOLITAN LABOUR AREAS, 1961-1971

produce an eight-way classification (after Webb, 1963, and first applied to SMLAs by Kennett, 1975). The classes are as shown in Table 8.

When the classification is applied to SMLAs, clear-cut regional and sub-regional variations emerge in the relationship between the components. Urban areas with in-migration greater than natural increase (Class I) are concentrated in the Southeast, while the more important role of natural increase in the growth of cities in Wales, the Midlands, the North, and Scotland is readily apparent. Again, the resorts stand out with positive net migration and a natural decrease. The main exception to this pattern is, of course, London; natural increase fails to offset the rates of net out-migration.

REGIONAL VARIATION IN NATURAL CHANGE AND NET MIGRATION

To assess their relative importance in determining urban change at the regional level, the mean rates of both natural change and net migration are

TABLE 8

Class	Description	Result
1	Net migration increase exceeds natural increase	Population gain
2	Natural increase exceeds net migration increase	Population gain
3	Natural increase exceeds net migration decrease	Population gain
4	Net migration decrease exceeds natural increase	Population loss
5	Net migration decrease exceeds natural decrease	Population loss
6	Natural decrease exceeds net migration decrease	Population loss
7	Natural decrease exceeds net migration increase	Population loss
8	Net migration increase exceeds natural decrease	Population gain

compared (Table 9). Two significant characteristics emerge: the first is the far greater variation in mean migration rates than mean rates of natural change. The range for natural change is from 4.1% to 7.7%, while the range for migration is from 14.1% down to –3.9%. There are no negative values for natural change, while four regions had a net loss through out-migration. Second, the regional pattern is clear: Scotland, Wales, the North, and Yorkshire and Humberside all recorded an average net loss of population due to migration, while SMLAs elsewhere gained population by both migration and natural change. The Southeast has by far the highest mean value for in-migration, followed by East Anglia, the Southwest, and the West Midlands. The Southeast had the highest rate of natural change, followed more closely by the East Midlands and the West Midlands.

At the aggregate level it would appear that urban change seems to be a clear response to variations in migration. However, these net figures conceal a particularly interesting characteristic in interregional migration variations. It is not simply a case of gainers and losers or "drifts" between regions. The relationship between in-migration and out-migration by SMLA (Figure 10) illustrates three main features. When interpreting this graph, one should note

TABLE 9

REGIONAL VARIATION IN MEAN SMLA POPULATION CHANGE DUE TO NATURAL INCREASE AND MIGRATION, 1961-1971

Economic Planning Region	Mean % Natural Increase	Rank	Mean % Migration	Rank
Southeast	7.2	3	14.1	1
West Midlands	7.7	1	4.0	4
East Midlands	7.6	2	3.7	6
East Anglia	5.4	7	5.6	2
Southwest	4.6	8	5.4	3
Yorkshire & Humberside	5.9	5	–0.7	7
Northwest	4.1	10	3.8	5
North	5.6	6	–3.7	9
Wales	4.5	9	–1.2	8
Scotland	6.0	4	–3.9	10

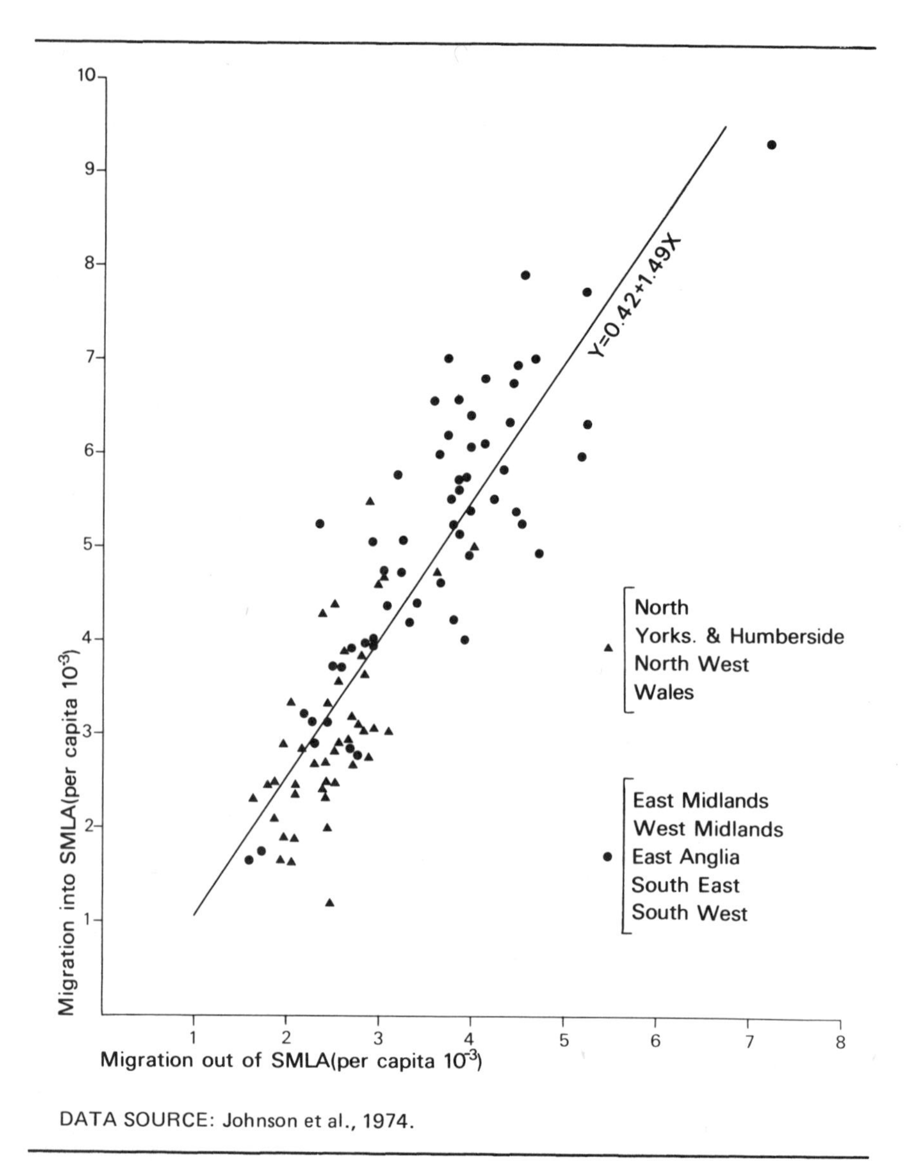

DATA SOURCE: Johnson et al., 1974.

Figure 10: PER CAPITA IN-MIGRATION AND PER CAPITA OUT-MIGRATION FOR STANDARD METROPOLITAN LABOUR AREAS (1961 DEFINITIONS) IN ENGLAND AND WALES, 1965-1966

that each axis has been standardized in the same way, i.e., by dividing through the local population to derive the per capita values.

(1) A strong relationship exists between in-migration and out-migration of individual SMLAs; i.e., the highest gainers are also the highest losers, and, conversely, those receiving few migrants also export the least.

(2) The SMLAs that gained or lost through migration (see Table 9) also cluster into two main groups: the SMLAs in the regions that gained through net migration (shown by circles) also had the highest population turnover due to both in-migration *and* out-migration; in the declining regions (triangles), the SMLAs experienced a low turnover in both respects.

(3) One further feature of interest emerges. It is readily apparent from the full scatter of points that out-migration and in-migration are seldom in balance. To illustrate this more clearly, a regression line is drawn through the points. It was considered necessary only to regress out-migration (X) on in-migration (Y), since the correlation coefficient is relatively high (0.87), the standard error relatively low (0.83), and the scatter of points relatively linear. No causal relationship is implied. This line helps to identify two phenomena. First, any given proportional increase in out-migration is accompanied by an even greater proportional increase in in-migration. This confirms the conclusion made earlier in this paper that migration accounts for the major variations in population change. Second, the regression line ($Y = 0.42 + 1.49 X$) *diverges* from a line where in-migration is balanced by out-migration (the 45° line). In terms of migration, this means that SMLAs with the highest migration turnover have the highest net balance of in-migration.

Because the data base for the graph is in-migration and out-migration from each SMLA to every other SMLA and to outer rings and nonmetropolitan areas, it is interesting to consider the implications of these findings in terms of earlier conclusions on urban change. One would anticipate the out-migration from the larger SMLAs to be mainly directed into their own MELA ring or to newly smaller SMLAs. For smaller SMLAs, the in-migration would be greater than out-migration, since they act as receiving areas for intermetropolitan decentralization. These conjectures cannot be substantiated at this stage, but further research should clarify the position when the full origin-destination migration matrix is available for all urban zones.

At this time, however, it is possible to draw upon previous research to illustrate the basic patterns of inter-SMLA migration flows. To do this, a matrix of gross migration flows between the 100 SMLAs that formed the basis of the PEP study of England and Wales for 1965-1966 was subjected to a principal-components analysis in which each migration destination was compared with every other destination in terms of the origins of its migrants (European Free Trade Association, 1973). Metropolitan areas were then classified according to similarities in their pattern of immigration. The destinations in each group can then be linked with their principal common origin (Figure 11).

Figure 11 clearly demonstrates that, from the point of view of receiving areas, the dominant pattern of interurban migration flows is represented by relatively short distance movements outwards from the largest SMLAs. Such flows are much more significant than a regional redistribution between widely separated

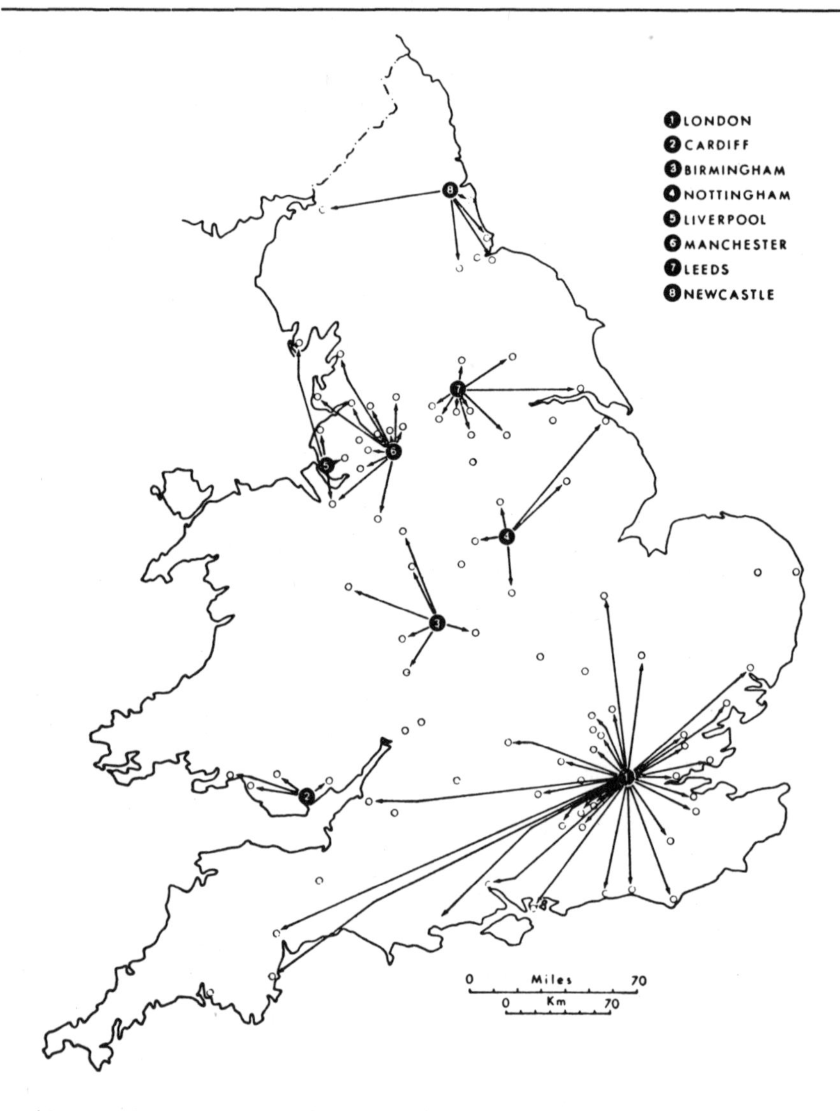

(Functional subsystems are defined by linking migration destinations with their principal common origin.)

DATA SOURCE: Johnson et al., 1974.

ANALYTICAL SOURCE: European Free Trade Association, 1973.

Figure 11: DOMINANT INTERURBAN MIGRATION PATTERNS (IDENTIFIED THROUGH FACTOR ANALYSIS), ENGLAND AND WALES, 1965-1966

SMLAs. Movements outwards from the London labor market area represent the leading pattern of decentralization. Amongst these movements, the most significant are those to a ring of SMLAs up to a 100 kilometers away from the capital (including all the New Towns), together with some long-distance movements to the Southwest region. The same pattern of metropolitan decentralization from each major SMLA to surrounding smaller SMLA systems can be seen for Birmingham, Manchester, Leeds, and so on.

When the data are examined from the point of view of migration origins rather than destinations, almost the converse of the above situation applies. If each migration origin is linked to the destination SMLA to which it directs the most emigrants, it is apparent that major provincial centers like Manchester and Birmingham attract immigrants from surrounding labor market areas, while on an interregional scale these centers in turn also direct migrants toward London. London itself is also an intraregional focus for its surrounding labor market areas.

It therefore appears that the metropolitan areas of England and Wales are linked together by a complex pattern of hierarchically ordered but reciprocated migration flows. But the important point that is highlighted by the components analysis is that the pattern of outward migration from London and the subnational centers is much stronger than the return flows.

COMPONENTS OF INTRAMETROPOLITAN POPULATION CHANGE

So far it can be concluded that aggregate shifts in population between urban areas can be attributed mainly to variations in net migration. Does the same process operate at the intrametropolitan level? The rapid decentralization of population over the two decades 1951-1971 and of jobs in the latter decade was a major conclusion of the core ring analysis earlier in this paper. What was the relative importance of the two components of population change in explaining population decentralization?

At the scale of constituent metropolitan zones there is greater variation in average rates of net migration than in rates of natural change. The average percentage population change in urban cores due to net migration is only 0.60%; about this mean there is a considerable spread, as is indicated by a coefficient of variation of 21.01. In contrast, metropolitan rings experienced a massive 14.73% population increase due to net inward migration, which, unlike the cores, was uniformly characterized by high rates of in-migration (coefficient of variation 0.78). Outer rings, while having an average 5.05% population increase due to net inward migration, exhibit a slightly higher degree of variation in this characteristic than rings (coefficient of variation 1.78).

The differences between cores, rings, and outer rings in natural increase, while less than for net migration, are still significant. Obviously, differential age-specific net migration of the order described will have an impact on rates of natural increase or decrease. With their high rates of net outward migration, it is,

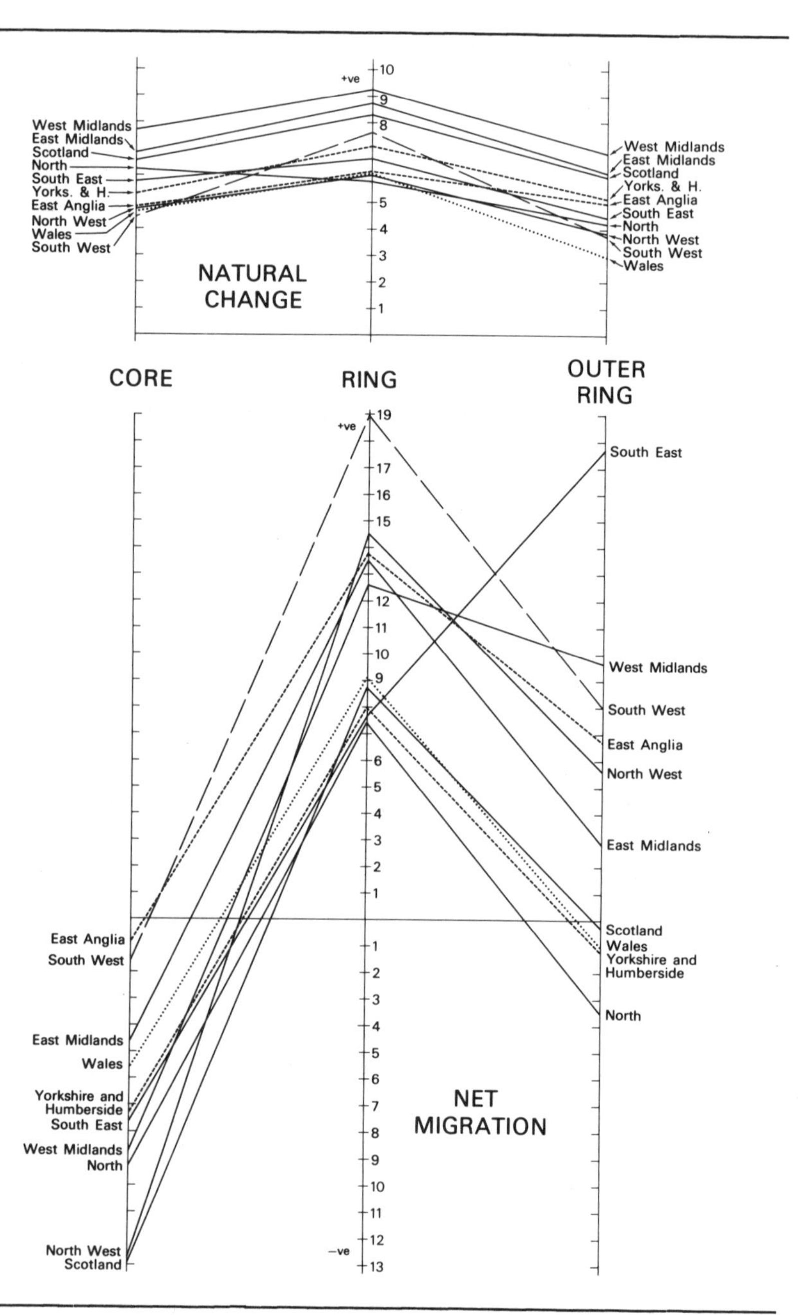

Figure 12: RELATIVE NATURAL CHANGE AND RELATIVE NET MIGRATION FOR THE CONSTITUENT ZONES OF STANDARD METROPOLITAN LABOUR AREAS AVERAGED FOR ECONOMIC PLANNING REGIONS, 1961-1971

therefore, not surprising that urban cores exhibit lower average rates of natural increase than metropolitan rings (5.86% compared with 6.35%). Outer rings, many of them reaching out to still largely rural areas, also exhibit low average rates of natural increase (4.38%), but, as with the urban cores, there is a relatively high degree of variation about this average.

In view of the figures quoted earlier concerning national average rates of natural change and net migration for cores, rings, and outer rings and the regional variations in SMLA and MELA performance that have just been described, it would be surprising if there were not significant differences between these two components of change for constituent urban zones in different parts of the country. Figure 12 indicates, however, a remarkably consistent pattern. With the exception of the North, natural increase is highest in the metropolitan rings with the most pronounced difference relative to cores and outer rings occurring in the Southwest region. The higher rates of natural increase in the West and East Midlands prevail across all three zones.

Although exhibiting a much greater range, net migration change follows a broadly similar pattern. Urban cores in all regions are exhibiting a net loss of population, although the figures do vary between –1% in East Anglia to nearly –13% in Scotland. Highest rates of net inward migration occur in the metropolitan rings, with the exception of the Southeast, where the very high rate of growth in outer rings can be attributed to the 21% net migration increase of London's outer zone. While the rankings of rates of change for constituent zones do differ between MELAs in various parts of the country, the fact that regional trends can be restated in terms of the internal dynamics of urban areas is clearly evidenced. In general, then, a higher rate of core decrease, due to migration, is associated with a generally lower rate of migration gain in the metropolitan ring and possibly population loss due to migration in the outer ring.

CONCLUSIONS AND FUTURE RESEARCH

The object of this paper has been to present some results of ongoing research into urban change in Britain. The main conclusions were given at the end of each section. A brief summary of the conclusions is now presented together with some indications as to the direction of further research.

Two major themes of urban change have emerged. They relate to changes in the inter-SMLA system, viewed both nationally and as a series of subregional groups, and also to changes at the intra-SMLA level.

When the urban system is considered nationally, the most important conclusion is the marked differential growth and decline of individual SMLAs. This can be explained by a number of factors. The regional location has been of overriding importance, with SMLAs reflecting strongly the fortunes of the region in which they are located; the size of an SMLA also influences the propensity to grow or decline. The observed decline in both population and employment of nearly all major cities has been offset by the growth of small and medium-sized SMLAs. The spatial relationship between the large, declining SMLAs and the smaller, expanding ones is by no means independent. The declining large cities are frequently centered on the old conurbations which are surrounded by smaller satellite, metropolitan areas. It is in these satellites and in other medium-sized, free-standing SMLAs that growth is now so disproportionately concentrated, particularly in the Southeast and in central southern England.

These trends raise important questions about the underlying causes. Several factors would seem to be important, but, of particular importance would be the stage that an SMLA has reached in its natural life cycle and the impact on the system of cities of consciously planned new urban growth. In the former case, a descriptive movel of metropolitan growth showed how the sequence of growth, decentralization, and decline can explain the evolution and current state of the majority of SMLAs in Britain. This natural change in the system has, however, been dramatically affected by the New Town and Expanded Town legislation. The highest growth rates of all SMLAs occurred in these planned communities, although they, too, have started to exhibit, not surprisingly, the characteristics of older cities.

The impact of regional policy on urban change should also be given consideration. There is evidence that regional policy, particularly in the North and the Northwest, is reinforcing the intermetropolitan decentralization from larger SMLAs to nearby smaller ones. These latter SMLAs, particularly in central Lancashire, are showing clear signs of being resuscitated, albeit the growth is only relative in many cases. This serves to reemphasize strongly the direct relationship that is now recognized to exist between the traditionally unrelated fields of regional economic planning and urban policy.

At the intrametropolitan scale, the dominant process is one of decentralization in both population and employment, particularly in the 1960s. The only exception occurred in New Towns after their inception, but they now follow the general trend. This now means that virtually the whole metropolitan system can be characterized by declining cores and expanding rings, with growth now spilling over into the outer rings, areas having little allegiance to the central cores. It is of considerable significance that the process is operating *irrespective* of metropolitan size or regional or subregional location. The only variation in decentralization occurs in the *rates* of change, with the largest SMLAs having the highest rates of decentralization.

These changes in population distribution can be attributed only to natural change and net migration. The analysis shows conclusively that net migration is mainly responsible for regional and intrametropolitan growth or decline. Natural change was seen to vary only slightly throughout the whole system and its component urban zones.

The research described in this paper has so far concentrated on an overview of urban change using aggregate population and employment data. Future research will be directed toward achieving a better understanding of the economic and social implications of the overall changes described. A detailed investigation is currently under way into the impact of urban change on the social and economic structure of metropolitan areas and the implications that this has for migration and journey-to-work. The effect of these changes on the urban system is expected to show differential benefits to various social groups, and these variations will provide a meaningful empirical input to the continuing debate on the efficacy of national urban policy in Britain.

REFERENCES

CAMERON, G., and WINGO, L. (eds., 1973). Cities, regions and public policy. London: Oliver & Boyd.

European Free Trade Association (1973). National settlement strategies: A framework for regional development. Geneva: Author.

FRIEDLY, P. (1974). National policy responses to urban growth. London: Saxon House.

GODDARD, J. (1974). "The national system of cities as a framework for urban and regional policy." In M. Sant (ed.), Regional policy and planning for Europe. London: Saxon House.

HALL, P. GRACEY, H., DREWETT, R., and THOMAS, R. (1973). The containment of urban England (Volume 1, Urban and metropolitan growth processes). London: Allen and Unwin.

JOHNSON, J., SALT, J., and WOOD, P. (1974). Housing and the migration of labour in England and Wales. London: Saxon House.

KENNETT, S. (1975). "A classification of the components of intra-urban and inter-urban population change in England and Wales between 1961-1971." Graduate School of Geography Discussion Paper No. 54, London School of Economics and Political Science.

London School of Economics and Political Science (1974). Standard Metropolitan Labour Areas and Metropolitan Economic Labour Areas: Part 1, Definitional notes and commentary (Working report no. 1). Urban Change Project, London School of Economics.

WEBB, J. (1973). "The natural and migrational components of population changes in England and Wales 1921-1931." Economic Geography, 39.

WESTAWAY, J. (1974). "Contact potential and the occupation structure of the British urban system, 1961-1966: An empirical study." Graduate School of Geography Discussion Paper No. 45, London School of Economics and Political Science.

4

The Changing Nature of European Urbanization

ELISABETH LICHTENBERGER

☐ PUBLICATIONS ON THE EUROPEAN CITY are quite diverse. On the one hand, there is an abundance of descriptive monographs. But on the other hand, there is a paucity of sound explanations of urban structure and processes. Several years ago, I attempted to sketch the general elements of this structure and process, to serve as a backdrop against which the future development of European cities might be gauged; I focused particularly upon historical city types, governmental roles in urban planning, contrasts between single-family and apartment-house construction styles, and the social-ecological patterns of the major cities (Lictenberger, 1970, 1971, 1972). It is the purpose of this paper to explore urban structure and processes from another complementary point of view, a comparison based upon the political division of Europe into two spheres of influence. Therefore, the political variable underlying the process of urbanization and its impact on the physical structure of cities is taken as the point of departure.

GENERAL DEVELOPMENT PRINCIPLES

URBANIZATION AND CITY SIZE: THE COMEBACK OF THE MEDIUM-SIZED CITY

In contrast to the United States, where concentration of growth in major metropolitan areas continued throughout the 1960s (Adams, 1972), the growth pattern of European cities reveals some major differences. It should not be surprising at all, of course, that the capitals of the socialist countries gained only 2.5 million inhabitants between 1946 and 1970, while the big cities below 1 million grew by 8.5 million people (Blazek, 1972), for in the East there is a complicated system of restrictions curtailing undesired migration. Since 1960, even the classical primate city, Budapest, increased by only 7.5%, and its rate of growth was surpassed by the large and medium-sized Hungarian cities, which grew by more than 20% (Lettrich, 1975; Toth, 1972). See Figure 1. It is the similarly decreasing attractiveness of the primate city of classically the most-centralized country in Europe, France, as revealed by the 1968 census, that came as the great surprise, together with a similarly unexpected shifting of immigration toward cities with 250,000 inhabitants in the German-speaking countries. The Paris agglomeration increased by only 8% in the 1964-1968 period, whereas the large and medium-sized cities with more than 50,000 inhabitants reached a growth rate of 12 to 14% (Beaujeu-Garnier, 1974; Klasen, 1972; Pailhe, 1973). Meanwhile, the so-called *metropoles d'équilibres,* such as Marseille, Toulouse, Lille, and Strasbourg, which the planning authorities had expected to paralyze the growth rate of Paris, did not differ at all in their growth from other medium-sized cities.

Yet should there have been surprise? After all, it is the cities of intermediate size that have been able not only to retain an attractive townscape, in most cases, but also to offer a multiple choice of jobs, a more or less perfect technical infrastructure, and extensive facilities for shopping, education, and recreation. Moreover, they are not burdened with environmental and transportation problems as serious as those in the bigger cities. In this context, the results of a survey in the German Federal Republic in 1972 should be mentioned, showing that 50% of the urban population wanted to live in medium-sized cities (Puls, 1973). In fact, it is becoming more and more apparent that it is this city size which gives planning the greatest chance to succeed in mastering the organization of urban life, insofar as physical structures, conflicts of land use, the transportation network, and the provision of utility services are concerned. Furthermore, new guiding principles for urban migration are emerging, partially influenced by the fact that both in the German-speaking countries and in France there is a very slow growth in the overall population, which is caused almost exclusively by the immigration of foreign workers. During a period of overemployment and simultaneously increasing leisure for the individual, the evaluation of residential location is based no longer upon job structures and wage levels but, increasingly, upon the attractiveness of the city and primarily upon its potential in terms of recreation and leisure facilities. A first outcome of these changed conditions can be observed. Since the end of the sixties in Germany, a polarization of migration from the then-predominant northwestern to a southeastern direction has been the result. Munich has become the main center of attraction.

As part of these shifts, traditional locational factors, such as resources, energy, and transportation facilities, have lost their importance, and regional

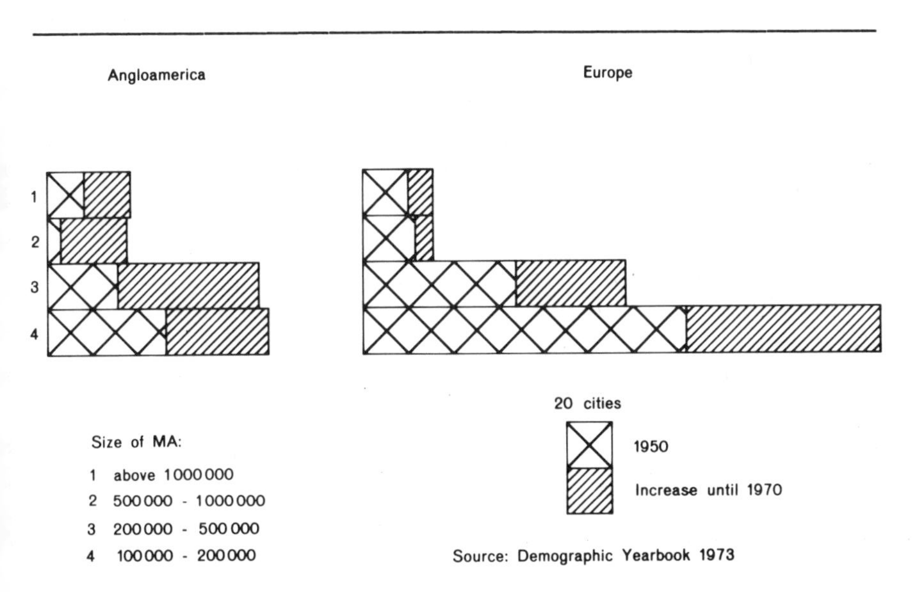

Figure 1: THE INCREASING NUMBER OF METROPOLITAN AREAS 1950-1970(72)

growth is increasingly influenced by the development of tertiary activities. Thus, by 1980 in the German Federal Republic the number of jobs in services is expected to increase by 3 million, but jobs in industry are expected to increase by only 1 million (Ganser, 1970). This generalization does not, however, apply to internal migration in southern Europe, where the supremacy of the national metropolis remains unbroken. This is especially true of Greece, where the two cities of Athens and Saloniki grew by more than 40% from 1960 to 1970, while the number of inhabitants of regional centers increased by only 20% (Pechoux, 1972).

Despite the similarities in recent growth of the medium-sized cities on both sides of the Iron Curtain, there are, however, differences as to the small towns. A comparison of the German Federal Republic and the German Democratic Republic reveals this (Schöller, 1974). In the former it is one of the favorite ideas of planning to promote small towns with 8,000 to 12,000 inhabitants, because they are central places of local importance, serving a rural hinterland with four to five times as many inhabitants. Therefore, public institutions such as grammar schools and hospitals are administered from them (Bobek, 1975). But things are different in the German Democratic Republic; as in most socialist states, East German cities of less than 20,000 inhabitants have been decreasing in population (Grimm et al., 1974). There are several reasons for this. Many of the small towns lost their functions because of recent administrative reorganization. Moreover, nationalization of trade and services has resulted in a loss of economic potential, because the distribution of tertiary activities in any centrally planned economy always starts at the top of the hierarchy. For this reason, little is normally left for the lower ranks. Besides, in these states the whole consumer sector occupies only a secondary position relative to the production of industrial goods. As a consequence of the collectivization of agriculture, it is true, many villages have been established as new local centers with small industries, repair shops, machine stations, etc., but many neighboring small towns, however, have been condemned to atrophy. Quite recently, planners in Poland have recognized this fact with much regret (Dziewonski, 1973). It seems to have been forgotten that in most socialist states the political intent has been to weaken the administrative and economic structure of many small towns, where the petty bourgeoisie was deemed an unwanted relic of the former social system to be replaced by a new class-free society in new industrial and agricultural cities as quickly as possible.

DICHOTOMIES IN THE DEVELOPMENT PROCESS: METROPOLITAN GROWTH VERSUS THE DEVELOPMENT OF SMALL CENTRAL PLACES

These last remarks suggest a problem that has gained importance as urbanization has progressed—namely, the dichotomy between the development of metropolitan areas and large agglomerations on the one hand and the development of the countryside with small central places on the other hand. There are, however, considerable differences between the East and the West as far as planning is concerned. In the West the contrast mentioned is reflected in the existence of two agencies—one responsible for urban planning and one responsible for regional planning—reflecting the traditional division between city and country that has long been characteristic of continental Europe north of the Mediterranean. Apart from northern Europe and the Netherlands, where national planning authorities could gain all-encompassing powers, these two types of planning agencies are cooperating hardly at all, being based upon two different legislative levels and having separate budgets. Consequently, laws applying to urban design apply to large and medium-sized cities only, whereas there are specific regulations for the small towns. Regional planning is supposed to have as its main goal the development of an efficient network of central places that administer definite public services from certain towns. This task should not be underestimated, because the provision of free school places and free hospital beds and subsidized public transportation and technical infrastructures are part of the social overhead of most European citizens. Because of the predominant concept of regionalism in Europe, regional planning boards are at a loss to contend with phenomena originating in the large urbanized areas, such as commuting and the second-home movement, both increasing in significance rapidly during the last decade.

Furthermore, a problem is posed by the growing metropolitan areas, since no metropolitan planning agency (with the exception of the Netherlands and the Paris agglomerations) is endowed with legislative powers and a budget of its own.

The situation is quite different in the East. There, state planning covers both urban and rural communes. Investment politics from the very outset have confined urban planning agencies to a subordinate role and set close limits to their powers of decision. Initiatives by municipal councils as to the allocation of economic enterprises, especially industrial estates, are possible only on a very small scale, whereas it is precisely there were the chance lies for the cities in the West, not least the small towns, to attract enterprises by offering building sites and infrastructures. Moreover, their chances are bettered by the fact that in the West many small and medium-sized factories are being established, while in the East large industrial plants are favored.

URBANIZATION OF THE COUNTRYSIDE: COMMUTERS AND SECOND HOMES

One of the most remarkable changes of the postwar period is the spread of commuting over vast areas of Europe, in both the West and the East. Unlike the situation in North America, this is the result of a process of suburbanization to a much smaller extent; in part, commuting has replaced traditional forms of seasonal labor migration. Moreover, it has put a stop to the rural exodus and has brought about a transformation of agricultural villages into commuters' settlements on the Mediterranean. Today, the absence of commuting is frequently used as an important indicator for delimiting underdeveloped agricultural regions. The expansion of commuting still continues in rural areas. This is connected with the fact that, contrary to all the discussions and concepts of planners, the distance covered between home and place of work has gradually increased during the past decade. At present, planners in the German-speaking countries consider acceptable a commuting distance of half an hour for medium-sized cities and one hour for cities with more than half a million inhabitants. This corresponds to an interval of approximately 50 kilometers and is, according to American standards, rather moderate. In several remote areas, daily commuting is replaced by weekly commuting. Burgenland in Austria may serve as an example. In this province 40% of all resident employees commute to places outside its boundaries (Osterreichisches Institut für Raumplanung, 1968).

In studying the structure of the commuting areas of medium-sized and large cities, one discovers interesting similarities between the East and the West. As a general principle it must be born in mind that there is social differentiation between the city and its commuting area in all cases. Drawing upon rough data fro the German urbanized regions, one can document that, within the city, the proportion of self-employed tradesmen, professional people, and white-collar employees—namely, those employed in tertiary activities—is considerably higher than outside the city boundaries. On the other hand, the outer commuting areas supply the majority of the skilled and unskilled workers needed by industry.

These differences are even more marked east of the Iron Curtain, especially in states with a considerable surplus of agricultural population, such as Rumania, Bulgaria, and Yugoslavia (Crkvencic, 1974).

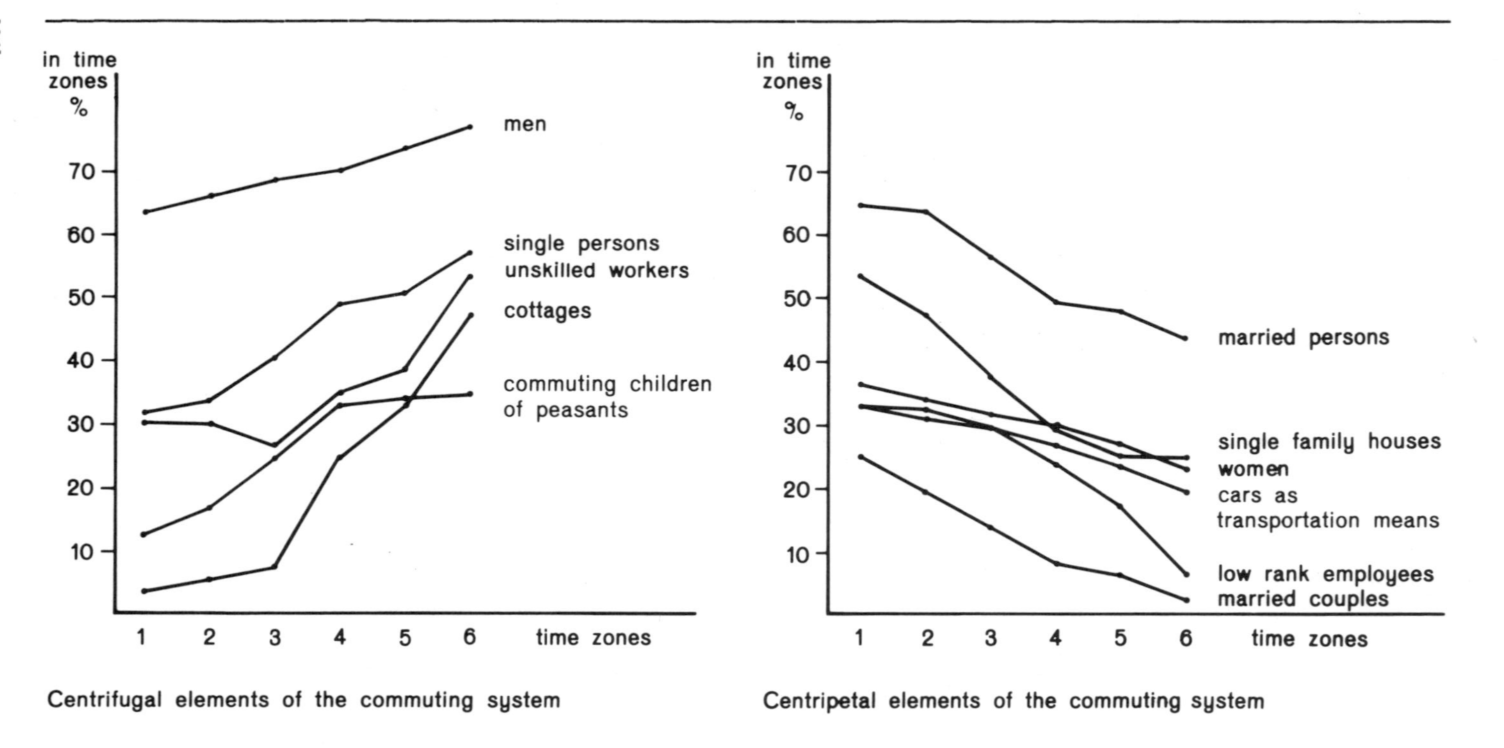

Source: cf. bibliographical note 17)

Figure 2: SCHEMATICAL DIAGRAM OF THE COMMUTING AREA OF A MEDIUM-SIZED CITY

Figure 2 represents the important centripetal and centrifugal elements of the commuting system in a schematic form:

(1) Male workers, particularly the unmarried and unskilled, assume increasing importance with increasing distance from city boundaries. At the periphery of the commuting areas, the residential structure is dominated by farm houses, showing the expansion of commuting as a process that reduces the surplus agricultural population, who, though converted into commuters, continue to live in their cottages.

(2) Approaching the city boundaries, the proportion of family houses increases, as does the proportion of low-rank employees and female commuters. The commuting of both partners in marriage also is gaining in importance (Satzinger, 1974).

Whereas research into commuting belongs to the standard inventory of regional planning, the second-homes movement did not attract attention until a few years ago, though the future significance of this process initiated by the metropolitan leisure society cannot be overestimated. Undoubtedly it will constitute the most important process of urbanization in the years to come. Its scope, however, is largely dependent upon the physical attractiveness of an area. The ensuing transfer of land will bridge the traditional gap in property and capital between city and country north of the Mediterranean. Serious problems arise where the sale of real estate goes beyond national boundaries and results in a selling-off of rural property to foreigners, however. In Austria, the western provinces of the Tyrol and Salzburg have put a stop to this development by passing real estate transfer laws. This measure was taken fairly late; in some of the important tourist centers, such as Seefeld in the Tyrol, no less than 50% of the area is already owned by foreigners (Haimayer, 1973).

The communes face other serious problems in the second-homes movement:

(1) No efficient means exists for controlling and regulating this phenomenon.

(2) So far, no attempt has been made to make use of it as a new source of communal revenue.

(3) As land values rise, it is more difficult for the communes to acquire lots for public use. Real estate prices are already $15 a square meter in small Austrian villages with second homes.

(4) Scattered settlement takes place wherever there is no master plan, and the pattern of agricultural land use is being fossilized. In this way the chaotic layout and inadequate infrastructure of the urban fringe of the early and interwar periods sprawls into the rural areas. It is not only commuting but the second-homes movement that cause deplorable chaos in many a rural settlement.

(5) Communal budgets are burdened by temporary populations that do not pay any taxes, but claim the use of the infrastructure provided by the communes, such as the road network and the water supply.

There is a highly diversified range of housing types connected with this phenomenon. It comprises remodeled farmhouses, huts on mountain pastures, small bungalows, rented flats, and freehold apartments. For the last type the well-known vacation centers of the German Federal Republic—e.g., in the Harz and on the shores of the Baltic Sea—may serve as examples (Uthoff, 1975). Here especially it is hardly possible to distinguish clearly between weekend traveling and seasonal mass tourism.

This movement does not stop at the Iron Curtain. It has spread widely in the industrialized and urbanized countries of long standing, such as the German Democratic Republic and Czechoslovakia. Regulations as to the form of buildings remind one, however, of the allotment gardens of the urban fringes in the West. The intensity of this movement, comprising more than 10% of the households of the German large cities and even up to 30% of some in Scandinavia, can be understood by studying residential structure. It is mainly the people living in apartment houses that are attracted by this new way of life, ready to put up with a new form of weekly commuting between their "working apartments" in the cities and their "leisure homes" in the country, causing more congestion problems on the main roads leading out of the cities in fine weather than intraurban traffic does during rush hours.

THE PROGRESS OF THE APARTMENT HOUSE AT THE EXPENSE OF THE SINGLE FAMILY HOUSE

European residential design is characterized by a traditional dichotomy between single family houses and apartment houses. Their core regions are found in northwestern Europe (Great Britain) and the Mediterranean (Italy), respectively. Originating from the Italian Renaissance cities, apartment houses have succeeded in occupying Europe gradually in a very complicated historical process, starting in the 16th and 17th centuries in France and Austria, accelerating during the founders' period. This progress was interrupted during the interwar period only. After World War II the apartment house quickly acquired a leading position and is advancing now toward the center of the single family home region in northwestern Europe, as well as expanding into small towns and more recently even into rural settlements.

Several events have furthered this success:

(1) Generally, land values have risen very quickly. The German Federal Republic may serve as an example. In the 1962-1971 period the average price for developed land rose from 14 DM to 33 DM, that is, by 2.5 times (Polensky, 1974), while it increased from 48 DM to 363 DM in Munich, that is, by almost 8 times. These figures predominantly mirror the situation at the periphery, where most of the construction activity is located.

(2) There has been substantial technological progress, involving prefabrication and assembly on site. In this context the socialist countries,

among them the German Democratic Republic, have been well ahead of the West.

(3) Powerful organizations, such as public housing agencies, huge cooperatives, and anonymous holding companies that construct freehold apartments, have applied themselves to the erecting of apartment structures, the construction of which has been and still is subsidized through public loans.

(4) The founding of new towns and satellite towns, starting in northern Europe, has paved the way for the apartment house.

(5) At an early date, planning authorities stressed that only the construction of apartment houses secures a population density high enough to maintain public transport. Therefore, integrated projects of large housing estates and subway construction can be found in both the West and the East.

As for the dimensions of buildings, the *grand ensembles* in France are paralleled by housing estates in the socialist countries–e.g., in East Berlin, Prague, Budapest, Belgrade, Zagreb, Bucharest, and Sofia. Comparing the sizes of the apartments, however, one observes a gradient from West to East. Whereas the majority of apartments in the surroundings of Paris (Sarcelles) each comprise four rooms, those in Frankfurt (Nordweststadt) and Berlin (Märkisches Viertel) consist of three rooms only; in the socialist countries apartments with two rooms are predominant. It seems worth mentioning that the differences in West-East housing are much smaller than the differences in living standards and gross national products. It is an enormous achievement of the socialist states that they have included housing in the social overhead for many years. Today, greater stress is being laid on cooperative and freehold apartments. A convergence with western trends seems obvious. City planning at last has realized that public housing *à fonds perdu* is too heavy a burden for any national budget.

There still are prejudices voiced against these gigantic residential blocks, which often comprise more than one thousand apartments, having a very high density of 250 or more inhabitants per hectare. Many sociologists have characterized them as "tomorrow's slums." This statement seems to have been precipitated by several things. Sarcelles (Paris), for instance, was regarded only a few years ago as an object lesson in the difficulties faced by a population having to adapt to new housing conditions, with a high crime rate and the growth of juvenile gangs. But Sarcelles has since developed quite favorably. The progress toward modern conveniences, when compared with former living conditions, must not be underrated either. This is especially true for the socialist countries, where the people moving into the cities have come mainly from rural areas with very primitive housing.

In the meantime, there has been a shift in the evaluation of the apartment house even in countries with some tradition of single family homes. Thus the survey undertaken in the German Federal Republic in 1972 showed that three-quarters of those interviewed preferred to live in apartment houses. As is to be expected, people were able to adapt themselves least of all in those countries where the big apartment houses still constitute a rare phenomenon, such as in the Netherlands. In 1968, a survey in Rotterday indicated that single family houses were still considered the ideal form of housing by two-thirds of those interviewed, a preference that can actually be realized by only one-third of the population anymore.

As far as economic aspects are concerned, their influence is dictated by a rapid rise in two price levels, namely, the price of construction and the price of building materials, which have not been offset by an advance of wages. West of the Iron Curtain, the metropolitan population is affected by the first cause. Therefore a large proportion of the middle classes can no longer afford to buy or build a single family home, though they may wish to do so. Little by little it is becoming a privilege of the well-to-do. Only in small towns and local centers as well as in commuting zones outside the large metropolitan areas can we still see the working classes building their houses on the principle of "do it yourself," with the help of relatives and neighbors. Their efforts are affected exclusively by the prices of building materials, which are increasing much faster than wages. This divergent movement has also come into play east of the Iron Curtain in the past two years, where countries like Czechoslovakia, Rumania, and Bulgaria are now having to face the same factors.

Here is an interesting phenomenon to consider: namely, the toleration of private property in the form of single family houses in most of the socialist countries so long as they do not exceed a certain size (normally five rooms). This can be interpreted as a means of tying the population to the small settlements and stopping the migration into the large cities, which are unable to meet housing demands. In spite of many restrictions against metropolitan growth there is a tremendous housing shortage. Extensive subtenancy ensues, involving up to 20% of the population, as in some parts of Budapest.

In comparing single family homes in continental Europe and North America, one may note the following differences:

(1) Continental Europe lacks organizations that specialize in financing and marketing single family homes.

(2) Except for a few construction innovations in northern Europe (especially the development of terrace houses, which began in Great Britain and are now occasionally found on the Continent), there has been little progress in technology and standardization. The traditional techniques of craftsmanship are still prevalent.

(3) Noneconomic variables, such as property rights, generational ties, and social prestige, are responsible for investment behavior in construction, renovation, adaptation, etc. Americans, who carefully take stock of possible profits, cannot understand why Europeans fail to question whether or not additional investment will increase the selling price of a house. This lack of a commercial spirit is especially obvious among

Yugoslavians, who work years in foreign countries, yet erect spacious houses on the periphery of such large cities as Zagreb, or even in remote parts of Yugoslavia, without knowing whether they will ever reside or get a job in such places. For all these reasons the construction of single family houses in Continental Europe is tying people to certain locations and cutting down mobility.

(4) Because labor is substituted for capital in large areas of Europe, especially outside the metropolitan areas, the size of houses in any particular country seems by no means determined by the economic wealth of that country. On the contrary, traveling in southeastern Europe, one is surprised by the considerable dimensions of the houses, especially in Yugoslavia, where the state sets no limit on size and where, moreover, the substantial savings accumulated abroad are invested in building. The preference for two-storeyed buildings in southeast Europe is an outcome not only of still-existing patriarchal family structures but also of oriental urban traditions. Accordingly, the upper floor is reserved for guests and is scarcely used; the family lives on the ground floor. Bulgaria offers many examples of this.

(5) In some parts of Europe the single family house has acquired additional functions. For instance, in the Austrian Alps, renting upper floors to tourists has gained great importance. In this way, tourist traffic not only stimulates construction activities but also gives large population groups a chance to participate in its yields (Osterreichisches Statistisches Zentralamt, 1974).

In spite of the spread of apartment houses, small towns and commuters' villages are still dominated by single family homes. Therefore, they differ markedly from urban agglomerations, where the further expansion of single family houses has been stopped at the periphery. Their construction, however, is still going on, where former weekend houses are being replaced.

In the German-speaking countries and in eastern Europe, there have been shantytowns on the edges of cities, comparable to the squatter settlements going back to the interwar period in France *(pavillons);* these have had counterparts in southern Europe, especially after World War II. These *bidonvilles* (literally, "tin-can towns") on the outskirts of large Spanish and Italian cities now belong largely to the past due to the extensive public housing programs of these countries (a notable exception is the primitive Andalusian quarter south of Madrid). Only Athens and some other Greek cities continue to develop single family houses at the periphery. Though their infrastructure is still insufficient, they cannot be classified as slum settlements, however, because they are well kept and nicely designed, owing to the earnings of their owners abroad.

Hardly any attention has been paid to, and no studies have been made of, another phenomenon of rural commuting areas and the urban fringe—namely, the houses for two families. They can be observed mainly in the German Federal Republic, where they comprise nearly one-sixth of the additional dwellings built since 1970 (Statistiches Bundesamt, 1974). Where does this form come from? When rent controls were abolished late in the sixties, farmers, tradesmen, and others discovered that the rental of apartments could be a new source of income, and they invested their capital in construction. This can be considered an interesting sociological phenomenon, since the era of apartment houses constructed and owned by the bourgeoisie, a situation very typical of the founders' period, is finally over everywhere else in Europe. Of course there is no possibility of similar profitable rentals in the socialist countries. For this very reason, the spacious houses around Zagreb, owned by workers employed abroad, are not inhabited, whereas all the apartment houses are overcrowded, and, on an average, two people have to live in one room.

CHANGING STRUCTURES OF BIG AND MEDIUM-SIZED CITIES

IMPACT OF URBAN PLANNING UPON CITY GROWTH AND DEVELOPMENT

The growing influence of city councils since the liberal era and, simultaneously, restrictions on free decisions in the private economy are characteristic of urban affairs even west of the Iron Curtain. But as for the goals of, and the ways and means applied by, the planning authorities, several stages can be distinguished:

(1) In a few cities, such as Vienna, control of the infrastructure began before World War I. Waterworks, gasworks, and so on became the property of the municipality, which succeeded in acquiring the greater part of public transport as well.

(2) During the crisis of the interwar period, social housing attained first priority in many cities. After World War II it was combined with the foundation of satellite towns, and it still counts among the most important activities of many city councils all over Europe. West Berlin may serve as an example, where at present social housing contributes about 60% a year to all residential construction activities.

(3) Long-range traffic programs, especially subways, were started at the end of the fifties. In this respect the capitals of the socialist countries, above all Budapest, have succeeded not only in catching up with the West but even in surpassing it as to the quantity and quality of public transport provided.

(4) For a long time, planning concentrated on housing and public transportation, but during the past decade concepts of an intraurban central system have become an indispensible item in all of the master plans, both in the West and the East. City planning has turned to the problems of the urban core, for a long time neglected in favor of the periphery. Programs for the renovation of historical cores are being enforced by preservation laws.

(5) Only recently has interest focused on structures built during the founders' period. A remodeling of the urban fringe is on the way, while the clearance of deteriorated quarters of flats is still under discussion in most cases.

(6) A fundamental change in basic city models seems to be much more important than widening the range of detailed programs. Now an entirely different approach to important physical variables such as building density and land use is being taken. Whereas during the sixties city planning ideas were best characterized by such slogans as "breaking up the row-house structure," "filling in the green areas," "cutting down population," things today have changed. It has been realized that any further waste of space is incompatible with a continuously increasing demand for room for all kinds of urban functions and that both the extension of the cities and their remodeling have to result in greater density. This means that the plot ratio has to be increased. The intrusion of skyscrapers into the urban skyline (discussed below) is among the most important phenomena of change in urban design. Simultaneously, the idea of increased separation of functions (living, working, recreational, educational, etc.) that was considered valid for centuries has been dropped. Discussion of a new kind of land-use model with specific types of mixtures are still going on.

(7) City planning and regional planning, having been set apart from one another for a long time, are now being combined in metropolitan planning. Some metropolitan areas have approved comprehensive planning concepts. This seems to be a first step only, however, because legislative and financial powers are weak everywhere except in the Paris agglomeration with its special legal position and in the Netherlands.

As for planning instruments, there are similarities among the Continental European cities in the West. During recent years city councils have attached more weight to land policy to ensure greater influence in the real estate market. Several goals are quite obvious:

(1) The right of compulsory purchase is being expanded wherever public projects for social and cultural infrastructure are concerned. For housing, this right can currently be executed in the Netherlands only.

(2) The right to acquire houses and other objects not properly taken care of in connection with renewal programs is also being expanded. These measures remind us of the right of the medieval urban community to expropriate abandoned houses and vacant land and to offer them to applicants. It cannot be denied that there is strong opposition to these trends. For this very reason, the majority of German citizens, including many in the working class, voted conservative at the last general election, since property rights in private homes seemed to be endangered by the left wing of the Social Democratic Party.

(3) The British model of a betterment levy, cashing in on the profits of private landowners who benefit from master plans, has been imitated so far only partly in the Netherlands. However, it has long been a general practice to levy a tax on profits resulting from development. The ill effects of this model, however, should not be overlooked: betterment levies can result in immobilization of the real estate market.

(4) In this context it seems necessary to point out that property taxes in Continental Europe have become minimal as a result of the two World Wars. This has important effects on the suburbs, since they cannot rely on a considerable tax budget even if they possess a highly exclusive residential structure. Since communal budgets are dependent mainly upon the local business tax, the communes try hard to acquire industrial enterprises. Especially in small towns, this brings about the surprising phenomenon that those with prospering industry, in spite of air and water pollution, possess a highly developed infrastructure that the neighboring residential settlements cannot afford.

Some sectors of the economy have not been integrated into city planning as yet—for instance, the tertiary sector (as far as it is not directly consumer-oriented) comprising administration and the marketing of industrial products, as well as the administration of hotels and the catering business in general. Accordingly, plans for remodeling the city centers focus on making them attractive for residential uses and shopping. There are no planning concepts yet for an optimal structure of the central business district as such. With regard to residential locations, surprising parallels between the East and the West can be stated, despite the fact that in the West there are no social models in urban planning. With increasing prosperity, increased segregation can be observed everywhere, although income and ethnic differences are the determining factors in the West, whereas demographic variables (age, structure of households, etc.), are the determining factors in the East.

A NEW GUIDELINE FOR THE MASTER PLANS: THE PLOT RATIO

Since the Middle Ages, building regulations have determined the physical appearance of European cities. But the interwar tradition of height zoning combined with the development of row-house structures has now been replaced by the concept of a plot ratio, in which the vertical dimension is defined in proportion to the ground floor area only. The former row-house structure has been broken up, and a new urban design has emerged. Originating from the British neighborhood idea, housing estates have been propagated as urban cells on a human scale. This break in urban design has wide-reaching consequences:

(1) Because of the diversified housing built in estates by various organizations with different rent levels, the population is even more

segregated than before, when certain strata were confined to specific urban quarters.

(2) The locational pattern of economic activities, especially retail trade, is strongly influenced (cf. below).

(3) The flexibility of the plot ratio as to the vertical dimension has paved the way for the appearance of high-rise structures.

THE INVASION OF THE SKYSCRAPERS

Late in the sixties, statements describing the "quiet horizontal skyline of Continental European cities" were still justified. High-rise structures, vehemently opposed by the daily press, appeared only now and then in million-plus cities. But the scenery now has changed. Skyscrapers are integral elements of the image of many cities. The question then arises whether Europe will follow the example of American urban design, with a slight lag. When taking a closer look, however, one discovers fundamental differences as to height and locational patterns. Skyscrapers do not invade the old towns. The newly dominant structures of banks and insurance companies, the large headquarters of industry, and the huge international hotels all keep well away from the old landmarks like churches, town halls, and palaces. They prefer former city boundaries marked by much open land and spacious premises. Railway terminals are builg built over, on the example of Montparnasse in Paris (similarly in Mühlheim-Ruhr and Franz-Josef-Terminal in Vienna), a fact pointing to the importance they still possess as cornerstones of central business district development.

In most cities in southern Europe (Seville, Barcelona, Madrid), and more recently in central Europe as well, a reversal of building heights from the center to the periphery is apparent in such a manner that the skyline of a city resembles a shallow bowl with a raised rim. The impressive contrast offered by single family houses being replaced gradually by high-rise apartments is characteristic of the periphery of cities in both the West and the East.

Nowhere, however, are there guiding principles regarding a desirable pattern of skyscrapers in the layout of a city. Detailed planning is still a long way from any overall concept of urban design.

Figure 3 demonstrates schematically the distribution of skyscrapers in cities of different sizes in the German-speaking countries (and in France). In the Million-plus city they are characteristic of the growing edge of the central business district and of slum clearance areas in the interior suburbs. They mark the entrances to exterior suburbs as well as the urban fringe. In large cities the skyscrapers are at the edges of the historical core areas and further out in the housing estates at the periphery. In medium-sized cities, where the historical nucleus often is protected by preservation laws, skyscrapers can be found only on the outskirts.

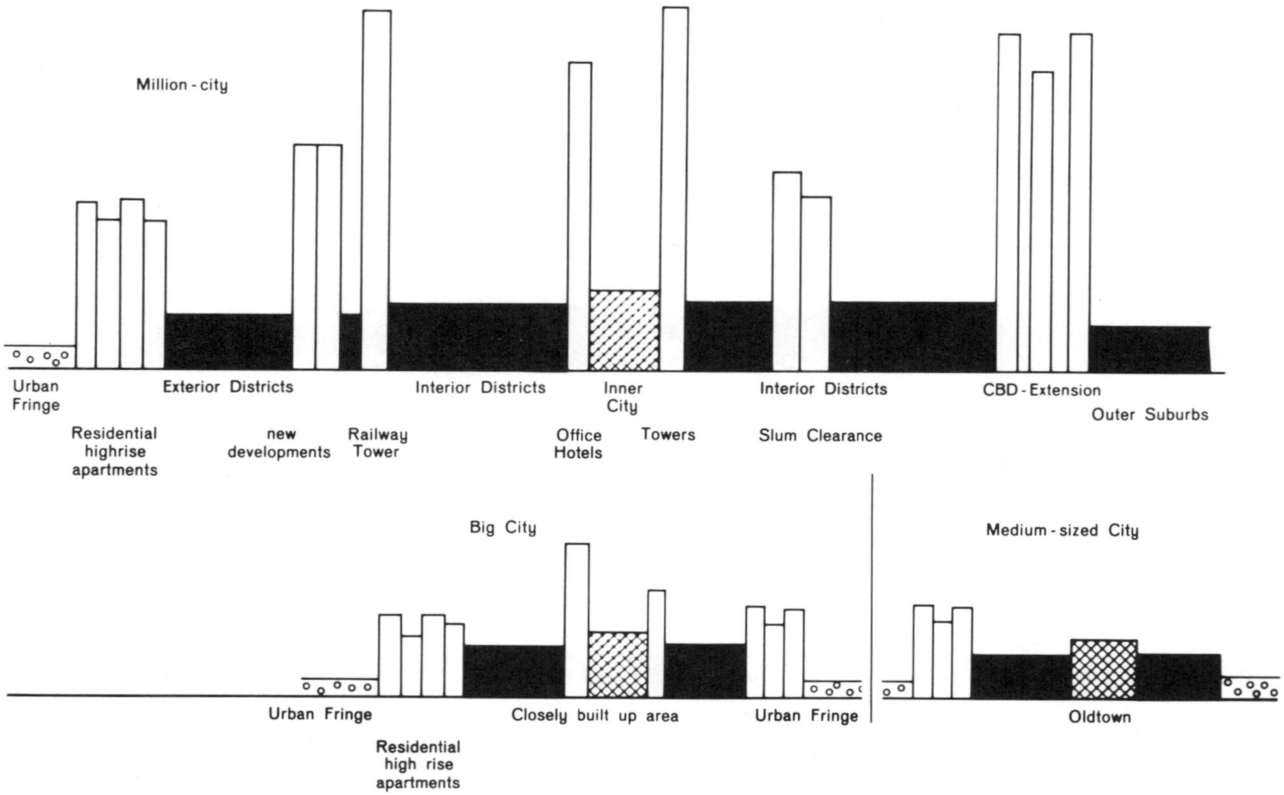

Million - city

Urban Fringe
Residential highrise apartments
Exterior Districts
new developments
Railway Tower
Interior Districts
Inner City
Office Hotels
Towers
Interior Districts
Slum Clearance
CBD - Extension
Outer Suburbs

Big City

Urban Fringe
Residential high rise apartments
Closely built up area
Urban Fringe

Medium - sized City

Oldtown

Figure 3: CITY SIZE AND LOCATION PATTERN OF SKYSCRAPERS

CITY REGIONS AND THE HETEROGENEITY OF MASTER PLANS

European city growth is still deeply influenced by the specific administrative structure of the rural areas. These are organized in so-called communes, tiny territorial divisions still reflecting the village pattern of the Middle Ages. Attempts to abolish and replace them by bigger administrative units has succeeded only in some socialist countries, for instance, in Yugoslavia. In the West these communes possess sovereign authority to plan within their boundaries and have to conform only to basic regulations that are issued by the provinces and the states. Wherever no master plan exists, decisions are still made by the mayor or the communal council in each individual case.

Building activities in an agglomeration, therefore, are subject to three different types of regulations:

(1) A modified-height zoning system still exists in the closely built-up areas of the founders' period, interspersed with only a few skyscrapers.

(2) On the outskirts of the city construction normally is regulated by a plot ratio.

(3) Outside the city boundaries the shaping of physical structures depends largely upon the historical layout of the communes and their socioeconomic pattern. For this reason, European suburbs can be described best as a very heterogeneous mosaic with haphazard land-use patterns and development. The establishment of new city outposts differs from that of the founders' period. Then, the erection of row houses progressed along the thoroughfares, whereas nowadays the interstices are filled up first. The Paris agglomeration is a model case for this. Only in the socialist countries and, partly, in southern Europe, where representative boulevards still form an integral element in urban design, domain roads leading out of the cities become the guidelines for the layout of large housing estates (Barcelona, Madrid, Bucharest, Budapest, Sofia, Warsaw).

CONTINENTAL TRANSFORMATION OF THE BRITISH NEW TOWN CONCEPT

No other topic has been more widely discussed in literature than the efforts of urban planning authorities to work with the New Town concept, devised as a means to cope with the uncontrolled, disorganized growth of cities. The British model as such was hardly ever imitated on the Continent. Rather, it was Stockholm that created the prototype for the countries west of the Iron Curtain: this was the satellite town, situated at the edge of a city region, equipped with supermarkets, schools, and recreation facilities, and connected with the city center by efficient public transportation systems. Numerous dormitory settlements of this type have been constructed in the German Federal

Republic. In the German Federal Republic and in France, and in the Benelux countries as well, there has been a continuous transition from satellite towns to planned city extensions. The sizes of both are especially large east of the Iron Curtain, where New Towns are situated much closer to the existing physical structures of the cities than they are in the West (Neo-Zagreb, Neo-Belgrade).

During the past two decades the concept of an optimal size of such extended towns has changed. Gradually the thresholds have been raised. There clearly is a connection with the increasing size of apartment houses mentioned above. A balance between residential population and services has been achieved only in northern Europe, however. In both eastern and southern Europe, where there are great housing shortages, the construction of apartment houses has priority.

In the socialist states, yet another type of New Town has been created, namely, the new industrial city. It had, however, forerunners in the Third Reich (Wolfsburg with its Volkswagen works, Salzgitter with its metalworks and steel mills). The cities in the German Democratic Republic may serve as examples (Schöller, 1074). Eisenhüttenstadt-Hoyerswerda-Schwedt and Halle-Neustadt characterize the phases based upon different priorities in national economies, from metal works to large-scale chemical plants. To found new towns in rural areas obviously posed fewer organizing problems and was more economical and ideologically more important than the reconstruction and renewal of cities. Their connection with old small towns facilitated the use of existing central functions. These industrial cities represent the achievements of the new society, and during the Stalinist era Nova Huta (Poland), Dunaújvaros (Hungary), Dimitrovgrad (Bulgaria), and Zenica (Yugoslavia) were founded on the model of Eisenhüttenstadt.

When considering Continental Europe as a whole, it is true in both the West and the East that in spite of all the efforts on behalf of the New Towns they constitute only an accessory element in the urban reality; and their importance, given the number of apartments provided, is far less than that of the planned city extensions. This is true above all in southern Europe, where building activities have always involved the densely built-up areas and where experimental satellite settlements such as the Ciudad Lineal of Madrid, dating back to the interwar period, have long since adopted the compact apartment house.

METROPOLITAN GROWTH VERSUS CITY REDEVELOPMENT

Because of concerted efforts to solve the growth problems of the cities, for a long time no attention was paid to the development of city centers. Therefore, all of a sudden, city councils were faced with problems of disastrous deterioration during the sixties, especially in France. The question was raised whether the American experience would be the fate of the European city too. Such eastern metropolises as Prague and Budapest that had not suffered extensive destruction during World War II were in a similar position. It seems to be worth noting that almost simultaneously both the West and the East advocated the renovation and regulation of old towns. In the West, France led the way by passing the Lex Malraux in 1962, which provided generous public funds for the modernizations of many old towns. Even earlier, a parallel movement started in the socialist countries, following Poland's example in reconstructing her demolished cities, like Warsaw, Poznan, and Gdansk, in every minute historical detail. Also Plovdiv, Bulgaria's capital during the Middle Ages, might be mentioned as one of the early examples of perfect renovation. Thus, preservation has become an integral part of urban city planning in the socialist countries, as can be seen clearly in the rehabilitation of the old towns of Budapest and Prague as well as numerous medium-sized cities. In the German-speaking countries, Austria was the first to pass a preservation law (Salzburg and Krems).

On the whole, there are differences between the West and the East as far as city sizes are concerned. In the West the small towns have been the most successful ones, especially in the German-speaking countries. Partly based upon the identity of house *and* shop property in the towns' centers, the private initiative of their citizens has been decisive. It is self-evident that the million-plus cities have to solve the biggest problems. There, the goal of a preservation of the old towns necessarily conflicts with central business district development founded upon economic power, which aims at a complete remodeling using high-rise structures.

The emerging social problems are considerable. Ardent discussions and intensive reactions on the part of the inhabitants affected by renovation have been triggered off mainly by left-wing intellectuals (Heinemeyer and Van Engelsdorf, 1972). It must be realized that the goals of social politics have changed since the times of Prefect Haussmann, who could construct his broad boulevards right across Paris, regardless of public opinion, as a radical means of clearing the slums, and give the bourgeoisie a chance to invest their capital in elegant apartments. Nowadays, a social conscience defends the economically underprivileged from being pushed out willy-nilly from the city center in the course of rehabilitation.

In general, though, renovation has brought the middle classes back into the city centers. The distinction between "working apartment" and "leisure apartment" may be given as one of the reasons for this development. The trend can also be considered a continuation of the processes current during the founders' period, when reconstruction was joined with the social upgrading of cities and with the removal of the lower classes to the urban peripheries. Paris, the largest metropolis of the European continent, is the best illustration of this trend.

PEDESTRIAN AREAS AS A REMEDY FOR THE RENOVATION OF DOWNTOWN AREAS

The creation of pedestrian precincts is one of those measures in city planning that have become a fashion all over Europe during the past decade. Whereas only a very few cities, first among them Stockholm, set an example during the sixties, this movement has spread both to medium-sized and large cities throughout Europe today. In the German Federal Republic alone, more than a hundred cities had a pedestrian mall in 1972 (Monheim, 1973). An optimistic prognosis seems to be that in the near future most of the cities with more than 50,000 inhabitants will have a central pedestrian area. It should be noted, incidentally, that large department stores encourage the trend because retail trade has improved substantially with the opening of pedestrian malls. Conversely, the opening of department stores strengthens the value of the urban core.

LOCATIONAL PRINCIPLES

No other aspect of urban development shows greater differences between the West and the East than retailing. The nationalization of trade in the socialist countries practically extinguished small shops. They sit empty and boarded up in the core areas of many cities and have been replaced (in part) by the big nationalized chain stores and department stores. Here and there, craftsmen have moved into the small shops, because one-man enterprises are generally exempt from nationalization—even in the German Democratic Republic, where nationalization has been strictest and where only 143,100 craftsmen, about 2% of the population, remain (Statistisches Jahrbuch, 1974). Nationalization hit the small towns hardest, and also, in countries with a low degree of urbanization, like Yugoslavia, Bulgaria, and Rumania, the medium-sized cities.

In the East, retail trade is highly concentrated in the inner city. There has been scarcely any decentralization of retailing from the city center to the periphery, as there has been in the West (Figure 4). Two facts are responsible for this. During the founders' period, no business worth mentioning developed on the outskirts of eastern cities because the working classes had scarcely any money; today urban planning authorities still concentrate more on housing than on anything else.

A comparison of Vienna and Budapest makes these differences obvious (Figure 5). Whereas Vienna possesses a well-defined intraurban central place, with smaller regional centers and community centers outside the central business district, which provide efficient supply for the closely built-up area, Budapest, on the other hand, though more populous, has poorly developed regional centers. The shopping streets in central Budapest have no more window display than those in Vienna, but they have to supply a far larger number of customers.

This centralization of retailing is even more pronounced in cities below one million inhabitants, where, apart from convenience stores, almost all retail trade is concentrated in city centers. Even such new city extensions as Neo-Belgrad

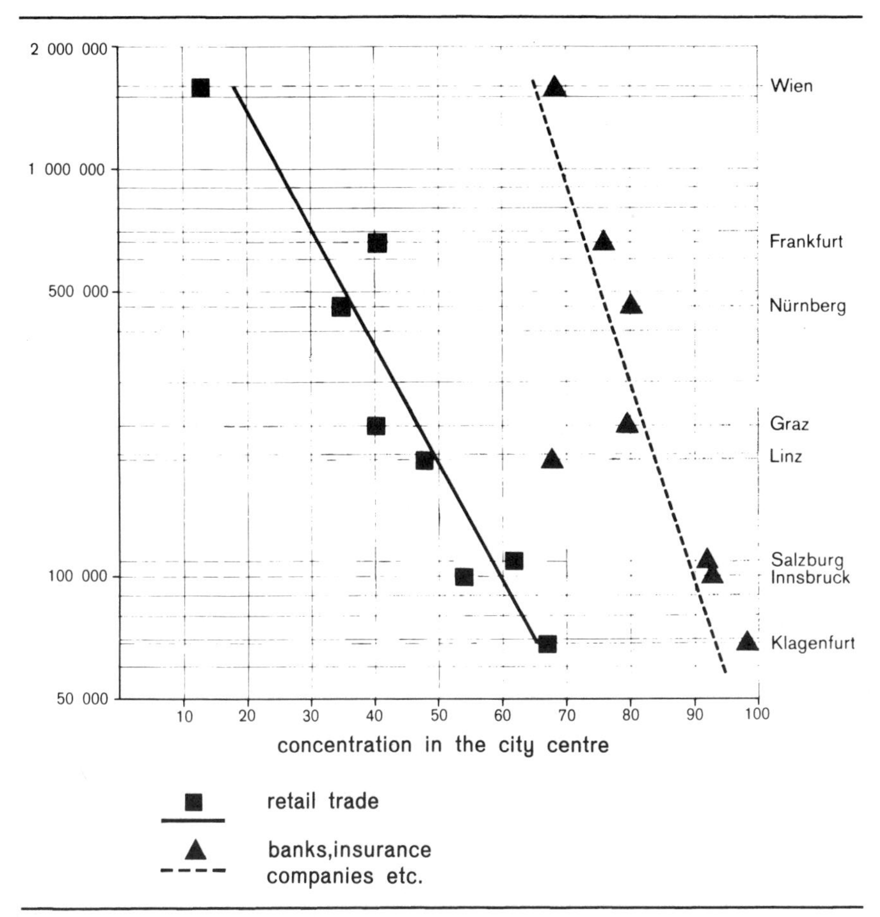

Figure 4: CITY SIZE AND DECENTRALIZATION OF TERTIARY ACTIVITIES

and Neo-Zagreb, each housing more than 100,000 people, have developed shopping facilities that would be considered sufficient for no more than 10,000 to 30,000 inhabitants in the West.

In western Europe the dualistic physical structure of the cities mentioned earlier is reflected in the pattern of retailing. In the densely built-up areas centripetal tendencies still predominate. This means that the radial shopping streets, often many kilometers long, form gravity centers close to the central business district and expand toward the periphery. The rank of the shopping streets is determined by the ratio between short-term goods (convenience goods) and goods in periodical demand (clothes, jewelry, etc.). Apart from the war-damaged West German cities, where large business organizations were able to acquire leading positions, small shops still predominate. They also contain a greater variety of branches than their counterparts in North America.

In connection with the outskirts of the cities some historical remarks seem necessary:

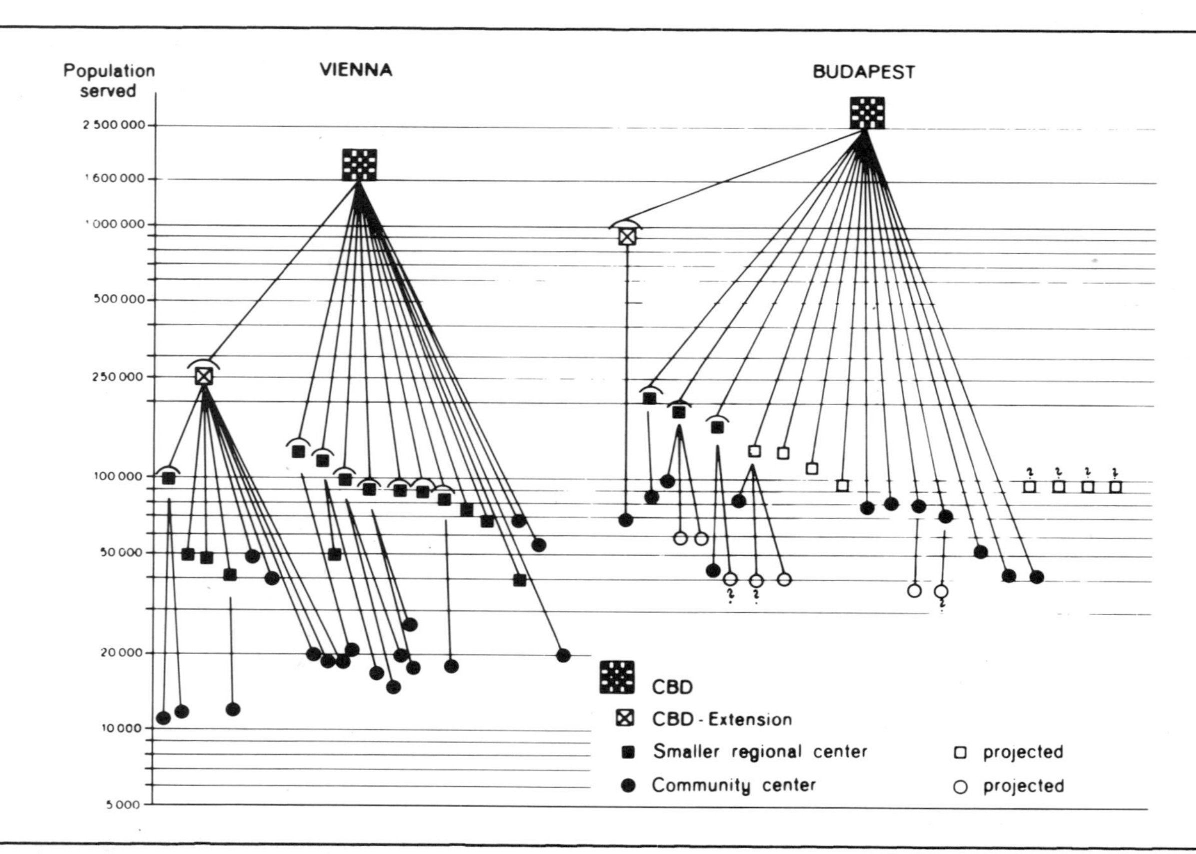

Figure 5: HIERARCHIC STRUCTURE OF RETAILING IN VIENNA AND BUDAPEST

(1) As early as the founders' period, some old market towns and villages were able to improve their positions as local shopping centers. In spite of the considerable expansion of single family houses during the interwar period not many new stores were opened.

(2) As late as the development of new housing estates in the postwar period, small groups of convenience stores appeared. Only in the planned northern European suburbs were department stores introduced at an early date. Everywhere else the population living at the periphery had to buy shopping goods in the city center.

(3) During the 1960s the first large shopping centers appeared, situated next to highways or other thoroughfares in the outermost fringes of the agglomerations. They were looked upon skeptically. In 1964, though, the French government gave permission for the development of regional centers for the Paris agglomeration, and, to date, four large shopping centers, each with more than 100 shops and at least two department stores have been completed, comprising a selling-floor space of 46,000 to 83,000 square meters. At present, they represent the largest of all shopping centers in Europe. Most of those situated at the periphery are much smaller, however.

In a historical context, European urbanization has to be understood as a process of innovation that started first in Great Britain and then spread to the west and the south of Europe during the 19th century. This gradual expansion involved many urban phenomena. Shopping centers on the peripheries of West European cities, for instance–termed "hypermarkets" in Great Britain–have gradually spread from north to south; today they are common around many large cities of the German-speaking countries and France, but appear near only a few of the million-plus cities of Spain and Italy. Due to the different incorporation policies of European states, these hypermarkets are situated either within the city limits, as in the German Federal Republic, or outside them, as in France. Consequently, they are partly integrated into city planning concepts; in Hamburg, for instance, where regional centers for 150,000 to 250,000 people are planned, they are governed by the decisions of regional planning boards; in Paris they are part of national planning. Despite the increasing numbers of hypermarkets, there has been no serious decline in the shopping potential of the downtown areas, and even the individual stores and shops on the outskirts of the agglomerations and in neighboring rural areas, have improved. It is to be expected that the balance of retailing between downtown and the suburbs aspired to by city planners can be maintained in the future. A division of functions will likely develop: the highly specialized shops of the inner city will continue to be patronized by the upper classes, whereas the vast residential areas of the fringes will develop shopping-streets to satisfy mass consumption.

QUO VADIS URBAN EUROPE?

Differences in the degree of urbanization, in historical city types, and in political systems prevent any general statement on the future development of European urbanism. It seems possible to concentrate only upon particular phenomena and to interpret them as innovation processes by making use of the concept of core and peripheral areas.

(1) About 1820, urbanization shifted from Great Britain to the Continent (Belgium and northern France), following industrialization, and then spread eastward (the Ruhr area, 1850-1870, and the Upper-Silesian manufacturing district, 1870-1880), transmitting the problems of older industrial areas. Before World War I offshoots had reached the present socialist countries, such as Hungary (Budapest) and Poland (Lodz, with its textile industry). In these eastern countries, as well as in the South, industrialization is taking place mainly in the present, after a tentative start in the interwar period. While urbanization based on industrialism thus is still occurring in the peripheral urban areas in the South and the East, the location principles of the tertiary sector have long since acquired predominance in the industrial states of long standing in the West, determining the future of city growth there.

(2) Currents of migration have always been dualistic. One component transferred people from smaller to larger cities, corresponding to the central place system, while, as early as the founders' period, intercontinental migration from east to west supported the development of industrial districts (such as the Ruhr area). Stopped during the crisis of the interwar period, masses of workers from the overpopulated agricultural regions of southern and southeastern Europe again started moving in the fifties, somehow skipping the small cities and migrating to the medium-sized and large urbanized areas of western and central Europe, where they occupy certain sectors in the job structure (building, catering, cleaning, and the like). Despite the recurrence of economic recessions, a return of all the foreign workers to their mother countries is most unlikely.

(3) The situation in the housing market is closely connected with the age and the character of urbanization. Four phases of demand can be distinguished.

Phase 1: Demand for more housing results from an increase in the number of inhabitants.

Phase 2: The number of households multiplies as the three-generation families of former times break up into smaller households. Again, the demand for housing increases.

Phase 3: The demand for quality housing grows with the greater demand for modern conveniences.

Phase 4: The demand develops for apartments much smaller in size than the average ones in North America. (Such northern European cities as Copenhagen are typical of this trend.)

France and the German-speaking countries are generally in phase 3 or, in some sectors, phase 4, whereas southern and eastern Europe are still progressing from phase 1 to phase 2. For this reason, quantitative housing will remain the primary demand for a long time in the South and East.

(4) The neighborhood principle as the basic model of the physical subdivision of cities has been widely adopted—largely as a consequence of the New Town concept—but not everywhere. In northern Europe, the German-speaking countries, and France it has long dominated planning ideologies. In the socialist countries the concept has found its way into planning offices, but it has been eclipsed by the large-scale technology now in use in such million-plus cities as Budapest, Zagreb, Prague, and East Berlin. In the European South, whose traditions favor density and intensity of urban life and to which such a molecular fragmentation is fundamentally foreign, the neighborhood principle has never caught on in urban design.

(5) The symbols of a consumption-oriented society, with its central pedestrian malls and peripheral shopping centers, doubtlessly will continue to spread in the West, moving from north to south.

(6) In central and western Europe urbanism has developed against a highly diversified rural background. The predominance of small farmers and fragmented holdings has tended to promote urban sprawl overall. Commuting and the second home movement have extended the sprawl even further. The agricultural tenancy system of the Mediterranean world, however, has tended to inhibit the transformation of peasants' villages into commuters' settlements or the transfer of land from rural to urban ownership; thus, urban sprawl seems to have little future in that region.

Countries east of the Iron Curtain have experienced some of these changes and are likely to experience even more in times to come. As the East produces more consumer goods, small cities get a chance to participate more in their distribution. Generally speaking, their development depends entirely on how much importance will be attached to regional development programs after a long period dominated by a centrally planned economy. Regionalization programs have already been inaugurated by the academies of sciences, and they may indicate a new approach in economic politics.

Because of undisputed primacy of mass transport, the city centers in the East remain the places of easiest accessibility, and accordingly they garner a very high concentration of business activities. Peripheral shopping centers are missing, and no planning policies exist to initiate decentralization; the pattern of development corresponds to that of the old western cities of the founders' period. Because segregation is limited, the former social gradient from the center to the periphery has been leveled everywhere. Quarters of crime and prostitution have vanished (e.g., in Budapest). There also are no problems with foreign workers as there are in the highly developed countries of the West.

Ironically, it has been the East that has been most successful to date in maintaining the traditional European concept of the city core as the center of urban life and economic activity.

REFERENCES

ADAMS, R.B. (1972). "Metropolitan area and central city population, 1960-1970-1980." Annales de Geographie, 81:171-205.

BEAUJEU-GARNIER, J. (1974). "Toward a new equilibrium in France?" Annals of the Association of American Geographers, 64(1):113-125.

BERRY, B.J.L. (1967). Geography of market centres and retail distribution (Foundations of Economic Geography Series). Englewood Cliffs, N.J.: Prentice-Hall.

--- (1973). The human consequences of urbanisation: Divergent paths in the urban experience of the twentieth century (The Making of the Twentieth Century Series). New York: St. Martin's.

BLAZEK, M. (1972). Les tendences de l'urbanisation dans l'Europe de l'est. International Geography (W.P. Adamas and F.M. Helleiner, eds.). 22nd International Geographical Congress, 2:876-880.

BOBEK, H. (1965). "The hierarchy of central places and their hinterlands in Austria and their role in economic regionalization." Pp. 139-144 in Proceedings of the fourth general meeting of the Commission on Methods of Economic Regionalization of the International Geographic Union, Brno, Czechoslovakia.

CRKVENCIC, I. (1974). Die Folgen der Urbanizierung in Jugoslawien am Beispiel der sozial-ökonomischen Struktur der Pendler und des Stadtrandes von Zagreb. Münchner Studien zur Sozial- und Wirtschaftsgeographie, 4:57-65.

Budapest Statisztikai Evkönyve (1973). Statisztikai Evkonyve. Központi Statisztikai Hivatal. Budapest: Városi Igazgatósága.

Deutscher Städtebau (1968). Die städtebauliche Entwicklung von 70 deutschen Städten (Hrsg.: Ak. f. Städtebau u. Landespl., Bearb.). Essen: W. Hollantz.

Die Mittelstadt 2 Teil (1973). Untersuchungen Ausgewählter Mittelstädte. Forschungs- u. Sitzungsber. 69, Stadtforschung 2. Veröffentlich. d. Adademie f. Raumf. u. Landesplanung, Hannover.

DZIEWONSKI, I. (1973). "The geographical differentiation of contemporary urbanisation." Geographia Polonica, 27:31-41.

ELKINS, T.H. (1973). The urban explosion. Studies in Contemporary Europe.

GANSER, K. (1970). "Image als entwicklungsbestimmendes Steuerungsinstrument." Stadtbauwelt, 26:104-108.

GRIMM, F., KRONERT, R., and LUDEMANN, H. (1974). "Aspects of Urbanisation in the German Democratic Republic." Paper presented to the conference on National Settlement Systems and Strategies, IASA, Laxenburg/Vienna, December 16-19.

HAIMAYER, P. (1973). Seefeld: Bevölkerung, Siedlung und Grundbesitzverhältnisse unter dem Einfluss des Fremdenverkehrs. Jber. österr. geogr. Ges., Zwigverein Innsbruck.

HEINEMEYER, W.F., and VAN ENGELSDORF, G.R. (1972). "Conflicts in Land Use in Amsterdam." Tijdschrift voor Econ. en Soc. Geografie, pp. 190-199.

HELLER, W. (1974). "Zum Studium der Urbanisierung in der Sozialistischen Republik Rumanien." Die Erde, 105(2):179-199.

KLASEN, J. (1972). "Urbanisierung in Frankreich: Bevölkerungs-, wirtschafts-, und sozialgeographische Aspekte." Münchner Studien zur Sozial- und Wirtschaftsgeographie, 8:109-116.

LETTRICH, E. (1975). "Urbanisierungsprozesse in Ungarn." Münchner Studien zur Sozial- und Wirtschaftsgeographie, 13.

LICHTENBERGER, E. (1963). "Die Geschäftsstrassen Wiens: Eine statistisch-physiognomische Analyse." Mitteilungen der Osterreichischen Georaphischen Gesellschaft, 105(3):405-446.

--- (1970). The nature of European urbanism. Geoforum, 4:45-62.

--- (1971). "Okonomische und nichtökonomische Variable kontinentaleuropäischer Citybildung." Die Erde, 102:216-262.

--- (1972). "Die europäische Stadt-Wesen, Modell Probleme." Berichte zur Raumforschung und Raumplanung, 16(1):3-25.

--- (1975). The eastern Alps (Problem Regions of Europe Series). London: Oxford University Press.

MONHEIM, R. (1973). "Fussgangerbereiche in deutschen Stadten-Neuersheinungen zu einem aktuellen Thema." Informationen, 23(2):27-41.

Osterreichisches Institut für Raumplanung (1968). Landesentwicklungsprogramm Burgenland (3 vols.). Vienna: Author.

Osterreichisches Statistisches Zentralamt (1974). Der Fremdenverkehr in Osterreich im Hahre 1973. Beiträge zur österr. Statistik 353: 285,500 beds in 48,300 private homes.

PAILHE, M.J. (1973). "L'urbanisation en France à travers les données statistiques." Bull. Assoc. Geogr. Franc., 406-407:461-472.

PARKER, A.J. (1975). "Hypermarkets: The changing pattern of retailing." Geography, 60(2):120-124.

PECHOUX, P., and ROUX, Y.M. (1972). "Stagnation demographique et mouvement d'urbanisation en Grece." Méditerranee, 10(2):1-10.

PENKOV, J. (1971). "Die neuen Städte in der Volksrepublik Bulgarien." Geogr. Ber., Gotha-Leipzig, 60(3):233-239.

PERENYI, I. (1973). Town centres: Planning and renewal. Budapest.

POLENSKY, TH. (1974). "Bodenpreise in Stadt und Region Munchen." Bauwelt, 65(12):37-40.

PULS, W.W. (1973). "Ballungsnahe Randzonen als Magnet: Aus dem Raumordnungsbericht 1972 der Bundesregierung." Gegenwartskunde, 22(2):205-212.

SATZINGER, F. (1974). Der Pendlereinzugsbereich von Klagenfurt. Wiss. Veroff. d. Landeshaupstadt Klagenfurt 1.

SCHOLLER, P. (1974). "Die neuen Städte der DDR im Zuzammenhang der Gesamtentwicklung des Städtewesens und der Zentralität." Veröffentlichungen der Akademie für Raumforschung und Landesplanung, 88:299-324.

SCHULTZ, J. (1971). "Les villes nouvelles hongroises." Bull. Soc. Languedocienne de Geographie, Montpellier, pp. 135-168.

Statistisches Bundesamt (ed., 1974). Statistisches Handbuch der Bundesrepublik Deutschland, 1973. Bonn.

Statistisches Jahrbuch der Deutschen Demokratischen Republic (1973). Staatsverlag der DDR, Berlin, 1974.

TIETZE, B. (1974). "Einzelhandelsdynamik und Siedlungsstruktur." Raumforschung und Raumordnung, 32(3-4):114-124.

TOTH, J. (1972). "Characteristic features of the population growth in the central settlements of Hungary in 1960-1970." Acta Geographica, 12(1-7):63-75.

UTHOFF, D. (1975). Ferienzentren in der Bundesrepublik Deutschland, wirtschafts- und sozialgeographische Analyse einer neuen Form des Angebots im Freizeitraum. Innsbruck: Verh. Dtsch. Geographentag.

VIELZEUF, B., and FERRAS, R. (1972). L'urbanisation de la Bulgarie: Tendences et methodes. Méditerranée, 2:11-34.

Part II

THE WESTERN MARGINS

5

Changing Urbanization Patterns at the Margin: The Examples of Australia and Canada

L.S. BOURNE
M.I. LOGAN

☐ AUSTRALIA AND CANADA have only recently awakened to the highly urbanized fact of their national character and well-being. It is now not uncommon to hear politicians argue that our cities, acting as a system, are the organizing force in articulating national economic growth and in distributing the benefits (and costs) of that growth. More critical perhaps, both governments now see potentially serious problems in the paths which urbanization seems to be taking or may take in the future. As one consequence, both have now joined the international debate on the need to consciously shape their urban systems through national planning and policy coordination. One *ex post* consequence of the belated public awareness of urban issues is a paucity of urban political and management experience and basic policy research. Our evidence is thin.

This paper undertakes a parallel, and sometimes comparative, analysis of urbanization patterns and policies in Canada and Australia. It attempts to document the current state of, and recent changes in, urban development in both countries and, more particularly, to speculate on what new trends and problems seem to be emerging. Emphasis is given, in order, to the following: (1) a brief overview of the historical underpinnings of the urbanization process, (2) current settlement patterns, city size distributions, and population concentration, (3) components of urban growth and change, with emerging trends, and (4) the urban policies and national political debates which have arisen in

AUTHORS' NOTE: *This paper owes a great deal to the parallel research by our colleagues C.A. Maher, J. McKay, and J.S. Humphreys at Monash University and J.W. Simmons at the University of Toronto. Simmons also provided valuable comments on an earlier draft.*

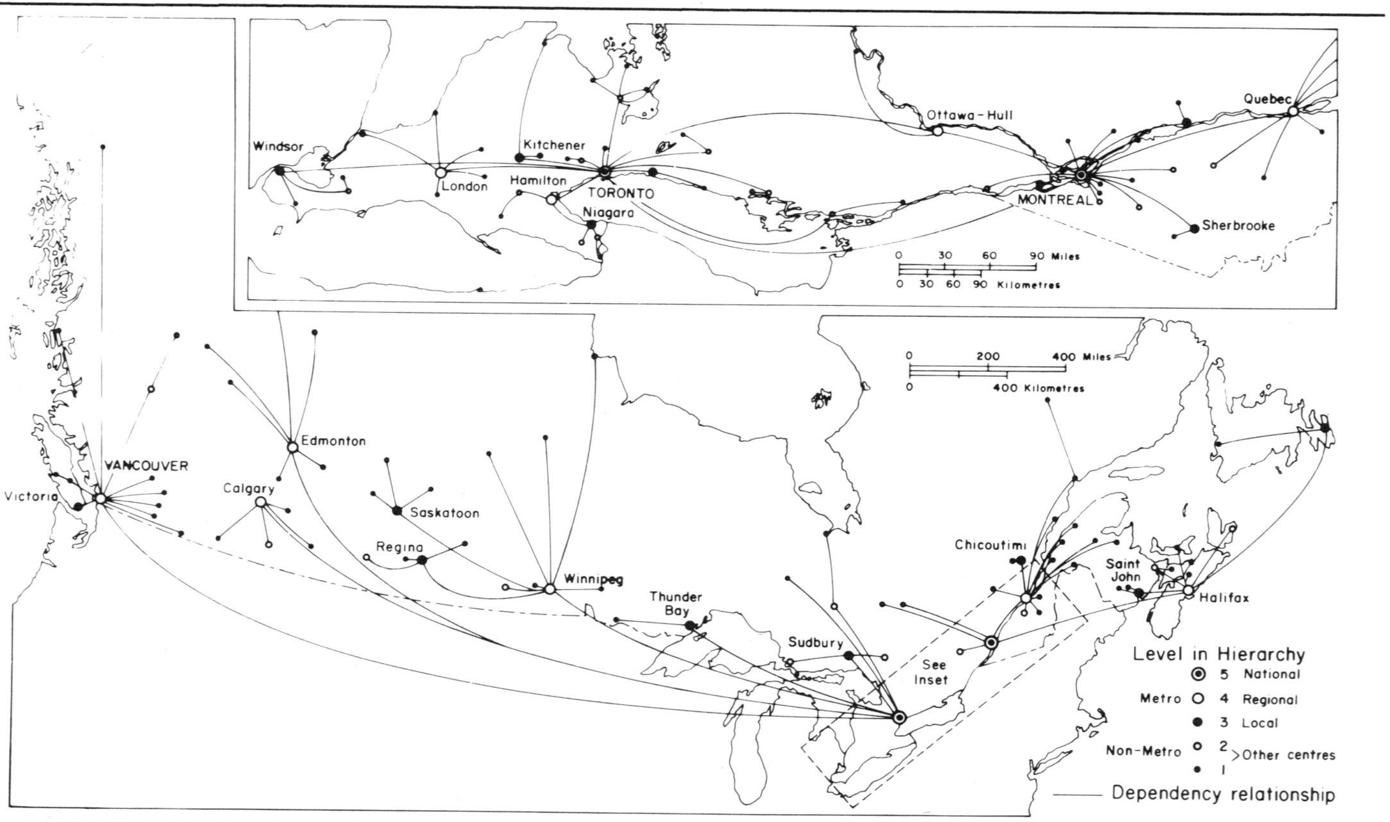

Figure 1: THE CANADIAN URBAN SYSTEM

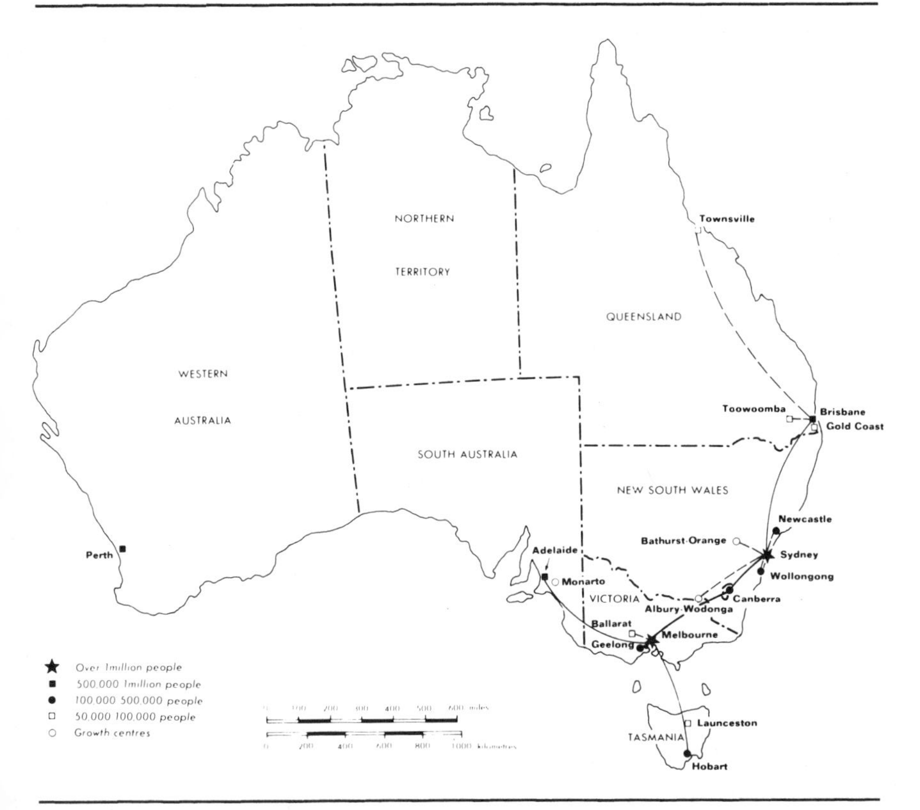

Figure 2: THE AUSTRALIAN URBAN SYSTEM

response to these patterns and trends. The paper concludes with an assessment of the possible success of these policies.

The guiding concept in our review is that of the urban system, which views a nation's settlement pattern in terms of an interacting set (or interacting subsets) of cities and urban economies (Pred, 1973; Goddard, 1974). Figures 1 and 2 display the structure of the urban systems of Canada and Australia in 1971. There are 23 urban areas with over 100,000 population in Canada, compared to 10 such urban areas in Australia. Because a number of recent descriptions of urban development for both countries are available elsewhere, their details will not be repeated here. (For Canada, see Lithwick, 1970; Ray, 1971, 1972; Bourne and MacKinnon, 1972; MacNeill, 1972; Science Council of Canada, 1974; Nader, 1975; Simmons and Simmons, 1974; Simmons, 1974, 1976; Yeates, 1975. For Australia, see Burnley, 1974; Stilwell, 1974; Logan et al., 1975.)

Why compare Canada and Australia? One obvious rationale is that as urban nations they hold in common many properties of historical background, economy, and environment. These common denominators reduce the number of difficulties inherent in any cross-national comparative study. Conversely, selective differences may assist in highlighting critical variables for research and policy. Their images abroad are certainly similar: both are relatively young, sparsely population, and federated political states; both are characterized by British institutional and cultural roots, capitalistic economies, and increasingly pluralistic social structures and by the presence of center-periphery extremes in development typical of the geography of colonial economies (Friedmann, 1973). Both also harbor essentially conservative, laissez-faire social philosophies and continue to be preoccupied with filling in national territory. Having noted these properties, however, it remains to be seen what effects they have on recent patterns of urbanization and on the direction of national urban policy debates.

Why the "frontier" connotation in the title, urbanization at the margin? We are not proposing to reintroduce the Turner approach to urban history, although that might be appropriate given Australian and Canadian attitudes toward their respective frontiers. Instead, to understand urbanization in either country, it is argued, one must be aware of the following preconditions: (1) the peripheral location of both countries in relation to world and continental markets, (2) their heavily staple-based economies and thus their extreme sensitivity to changes in world commodity and capital markets, (3) the increasing dominance of their economic plants by large-scale, foreign-based organizations, and (4) equally important, but internal to the country, the historical imprint of a continually unfolding settlement margin on the national character and on the current structure of the urban system. As Harold Innis (1957) once wrote, "the economic history of Canada has been dominated by the disparity between the centre and the margins of western civilization."

Among the many imprints of these conditions are the uncertainties of planning in a turbulent external environment, the limited and declining scope for devising national plans and strategies, and the preoccupation with events at the periphery rather than with the core of national development.

THE URBANIZATION PROCESS

The Australian and Canadian settlement patterns have been strongly influenced by the nature and timing of their initial European settlement and by their contrastingly harsh physical environments. The origins of the present urban systems are evident early in the colonial period. In Australia, the aridity of the central, northern, and western areas of the continent has tended to perpetuate the dominance of the initial coastal settlements. Urbanization continues to be a maritime phenomenon. Canada, in contrast, followed a more continental pattern. Settlement unfolded westward from the initial eastern coastal ports, between the U.S. border and the inhospitable northern frontier.

HISTORICAL SETTLEMENT SEQUENCE

Both countries were settled as appendages of Great Britain (and in Canada, of France before 1759), largely during periods of rapid industrialization and economic growth in the mother country (Buckley, 1975). Thus, the timing of settlement and colonial economic demands and policies shaped the structure of economic and urban life, as well as political institutions.

The Australian states, for example, emerged directly out of an autocratic system of government administered by the British colonial office. The six states did not attain self-government status until 1890, over 100 years after the initial colonization, by which date the settlement pattern had been firmly established. In particular, the six port cities, described by McCarty (1973) as "pure products of the nineteenth century expansion of capitalism," had emerged as the dominant features of the settlement system, each with a largely independent hinterland. The relative size and economic strength of the states ensured that their dominance would continue. Similarly, although for different reasons, the Canadian provinces to a certain extent partitioned the early settlement pattern into semi-independent regional systems and continue to do so today despite the integrating pressures of the national economy.

Geography had other effects. The argument developed by Geoffrey Blainey (1971) that the physical distance separating Australia from Britain significantly influenced its social history is even more pertinent to understanding the distribution of population and of economic activities. The port cities (Sydney, Melbourne, Brisbane, Adelaide, Hobart, and Perth), which would subsequently become the metropolitan centers, developed during the 19th century as the interface between the British market and the essentially open but state-dominated regional economies. Because of the great distances and costs involved in transport to the British market, the goods produced in the colonial economies had to be of high value. The staple products which evolved out of these circumstances—fine quality wool, gold, wheat, and, more recently, basic minerals—were of critical importance to the Industrial Revolution in Europe and provided the initial stimulus for the settlement of Australia (Logan, 1968). Sheep farming and mineral production also meant a dispersed rural population and encouraged the growth of only small local service centers. Expansion in staple production stimulated the growth of the initial port cities because they were export-oriented and because they experienced the cumulative effects of import substitution. These cities in turn quickly extended their dominance over their respective hinterlands through the construction of largely independent radial transport systems.

It may be argued that the emphasis on the commercial use of land in Australia, from the beginning of settlement, contrasts with the pattern in much of North America, particularly those areas settled at an earlier date. Such areas tended to develop a more community-oriented and self-contained agricultural economy (Andrews, 1964), which in turn encouraged higher population densities and a stronger focus on local centers for supplies and as market outlets.

But in most of Canada, except for Quebec and very early Ontario, the conditions of initial settlement (timing, environment, and economic stimulus) resemble those in Australia. Here, too, the settlement pattern is dispersed with a few centers dominating extensive rural hinterlands of low population densities. In the Maritimes, settlements served fishing and lumber needs. Grain exports stimulated the growth of Ontario, and wheat and furs mapped the service centers on the prairies.

One outcome of this historical sequence, in most of Canada and Australia, is the absence of a well-articulated hierarchy of centers. In neither case, however, is this due to the establishment of only a few towns early in the settlement process. For example, studies by Jeans (1965) and Williams (1968) of the history of settlement in New South Wales and South Australia, respectively, show that the colonial authorities established an abundance of villages and towns. In the Canadian West particularly, the railways established strings of towns at regular intervals across the prairies. But the density of rural settlement was insufficient to generate widespread and rapid town growth outside of the initial port cities in Australia and of major rail terminals or break-in-bulk points in Canada (Warkentin, 1968). In any case, as Robinson (1962) has pointed out, the growth of towns in inland Australia–unlike Canada–often postdated major developments in transportation, especially in railway construction. This late development may have prevented these towns from generating a strong community sense, such as arises in a long period of relative isolation, and from developing a set of linkages with their adjacent hinterlands. To a large extent the capital cities performed the role of regional centers. Preliminary studies by Williams (1965) support the thesis that the port cities had such strong functional ties with their hinterlands that the basis for intermediate-sized cities was not present (Bensusan-Butt, 1965). Small cities were simply bypassed en route to the growing metropolis. Interestingly, some studies of the urban system on an intrastate basis have identified a number of discrete levels in the urban hierarchy (Scott, 1964; Smailes, 1969; Bourne, 1974b).

STAPLES AND POLITICAL BOUNDARIES

The importance of staples in the development of the urban system in each country cannot be overemphasized, although the consequences differ between the two countries. Specialization in the production and export of a few basic primary products did not lead to dense agricultural settlement and the eventual generation of secondary and tertiary activities. It also, as noted earlier, renders the economy in question open to wide fluctuations in international commodity markets.

Canada, for example, passed through a wavelike sequence of basic staples, from furs and fish to pulp, grain, basic metals, and oil. Each product has its own market schedule and its own geography. Each also imposed its own urban system, varying in density, form, and connectivity: scattered linear settlements to serve the fishing trade; river networks for furs and timber; random nodes at mining sites; and more fully developed central place systems for agricultural regions.

Towns sprang up on the frontier to service each product, but only the largest centers responded to each and every wave. Smaller centers stagnated or died if their staple declined. But the major centers grew in any case. Growth stimuli then moved largely up the urban hierarchy, reinforcing the control of the initial settlements on trade and strengthening their capacity to withstand the boom and bust of staple markets. On the frontier, and in smaller towns, uncertainty and instability prevailed. One obvious consequence of uncertainty is greater locational concentration (Simmons, 1976).

In Australia, trade with Britain was particularly important from 1860 to 1890, a period of prolonged economic expansion based primarily on a boom in the export of wool, wheat, gold, and other minerals. Population growth was rapid, and although the primary staples remained the mainspring of the economy (Fitzpatrick, 1949) the unique ingredient was the relatively massive investment in urbanization. A leading economic historian (Butkin, 1964) describes the position thus:

> As early as 1891 this new and underdeveloped area had succeeded in carrying urbanisation to a degree which was rare in the rest of the world. Almost two thirds of the population of 1891 lived in towns and cities, a proportion which was not matched by other new countries, particularly the United States until 1920, and Canada until 1950.

Differences in the kind of staples, in the type of settlers, and in timing produced, in Australia, an urban society before, and partly in lieu of, an agrarian one. From 1890 onwards there has been a steady consolidation of this highly nucleated settlement pattern, with rates of national population growth varying according to fluctuations in the world market for specific staples.

Political factors further cemented these patterns. Neither the partial Canadian confederation in 1867 nor the federation of Australian states in 1901 did much in the short run to break down interstate, interprovincial, and intercity rivalries. Sydney and Melbourne competed for national dominance, much as Toronto and Montreal did in early Canada (Careless, 1954). Indeed, the establishment of secondary manufacturing industry, largely a post-World War II phenomenon, encouraged a form of competition between the states for manufacturing firms. Prior to 1900, manufacturing activities were oriented to rather small, fragmented markets encouraged by the existence of customs barriers at state borders on the one hand and by a form of artificial market protection through geographic isolation on the other. The elimination of state customs in Australia opened up possibilities of much larger markets, but the framework of state autarchy had already been set, and distance remained a major constraint. In Canada, the strengthening of customs barriers against the United States after the War of 1812, and particularly after 1867, forced an earlier closure of the hinterlands of cities in Upper (Ontario) and Lower (Quebec) Canada. Unlike Australia, this political boundary encouraged the rapid, but probably premature integration of cities into a functioning urban system.

SPECIALIZATION, CENTRALIZATION, AND CONCENTRATION

Over time, powerful centralizing trends have continued to operate in the location of secondary industry. Many enterprises based in rural or country areas have lost out in competition with their city counterparts or have been forced to relocate to those cities. The desire for economies of scale and for relief from uncertainty has greatly enhanced the attractiveness of locations at the points of greatest access to regional, statewide, or international markets (Logan, 1966; Linge, 1968; Gilmour, 1972). Given the concentration of manufacturing industries in the capital cities of Australia and in the cities of central Canada, it is not surprising that the decentralization policies pursued by state and provincial governments, but especially thos of New South Wales, Quebec, and Ontario, have focused almost exclusively on manufacturing.

It is difficult to quantify the precise cost advantages of a metropolitan as opposed to a nonmetropolitan location for firms. Higgins (1972), for example, argues strongly in favor of metropolitan concentration in the province of Quebec, because of the extensive multiplier-effects of Montreal's growth. In contrast, estimates from the report of the Committee of Commonwealth/State Officials on Decentralization (1972) in Australia suggest that the cost disadvantages of a nonmetropolitan location may be less than 1% of sales. This, according to Stilwell (1974), is so small a differential that the reasons for manufacturing concentration do not lie primarily in cost minimization but rather in the tendency for entrepreneurs to make the same locational decisions as their predecessors. Decentralization policies of the state governments, consequently, have been refined somewhat over time into strategies for "selective" decentralization, encouraging firms to agglomerate in only a small number of country towns. These, in turn, provide the background for the Australian government's new cities program (Neutze, 1974).

In Canada, most provincial governments, except Quebec, have followed a similar line of reasoning on decentralization, but, given the larger number of existing small cities and towns, they have placed less weight on new cities per se. It is interesting to contrast the decentralization policies in Australia with those in Canada, where the involvement of the national government, primarily through the Department of Regional Economic Expansion, has been much greater and of longer duration (Firestone, 1974). In Australia the states have carried the principal responsibility, and consequently interest has focused on shifting growth from metropolitan to nonmetropolitan areas. In Canada, where regional inequalities are substantially higher and more obvious (Brewis, 1970) and where regional cost differentials for manufacturing or any other secondary activity are substantial, the federal government has taken most of the formal initiatives.

Canadian interest has therefore focused on reducing regional inequalities through fiscal redistribution and economic decentralization from the national heartland to the periphery (the Maritimes, the mid-Canada agricultural belt, and the prairies).

The historical forces of economic specialization and concentration noted above have their direct spatial counterpart in the persistence of such inequalities and in the continued growth of metropolitan dominance. The spectacular expansion in the 1960s and 1970s of tertiary and quaternary activities in both countries has added new weight to these same forces. Quaternary services, such as medical and educational research and information services to management, are even more highly concentrated in larger metropolitan areas than secondary or tertiary activities. There is not even a rudimentary system of such services in most smaller cities and rural areas.

Adding to and complementing these trends, government departments and agencies for each of the two senior levels of government are heavily localized in the larger metropolitan centers, particularly in Australia. This contrasts with the situation, for example, in the United States. Since quaternary activities are the growth points in the modern economy, one can appreciate the real difficulties faced in any strategy designed to redistribute growth away from the major cities (Logan et al., 1975). Partly for this reason, one still hears of recommendations for the creation of new political units (e.g., within New South Wales and Ontario) and, thus, new capital cities. Neither are likely, although decentralization from existing capitals to regional administrative centers will almost certainly take place in certain sectors of government.

There is, therefore, considerable evidence which supports the contention that settlement patterns characterized by a high degree of concentration or primacy will tend to remain (Rose, 1966; Pred, 1973; Simmons, 1976). In a recent study of the Australian system, Stilwell (1974) argued that the cumulative advantages which accrued to the original settlements have been further increased in this century by the broadening internationalism of the capitalist system. Australia's industrial structure is increasingly dominated by large firms, and economic power generally is more concentrated than in the United States, the United Kingdom, or Canada (Vernon, 1968; Karmel and Brunt, 1962; Wheelwright, 1957). Overseas ownership and control of large industries is common in both Canada and Australia. International movements of information and capital, once again, are heavily biased toward the major metropolitan cities.

THE CURRENT SETTLEMENT PATTERN

Both Australia and Canada are among the most highly urbanized nations in the world. In Australia, according to the 1971 census, over 86% of a population of 12.7 million lived in urban centers over 1,000 population; over 58% resided in the six state capitals; and 40% resided in Sydney and Melbourne alone. In

Canada, in 1971, over 76% of a population of 21.6 millions lived in urban centers, with 30% in the three largest cities (Montreal, Toronto, Vancouver); 45% lived in these and the next five largest centers. All 22 census metropolitan areas accounted for 56% of the national population in 1971, compared to just under 47% in 1951.[1]

DOMINANCE AND GROWTH

The dominance of the capital cities, in terms of size and rates of growth, is the outstanding characteristic of the Australian settlement system. And it is increasing. Between 1954 and 1961 the six capital cities grew by 21%, compared with 19% for other urban areas. From 1961 to 1966 they grew by 15% (7% for other urban areas). It should be pointed out, however, that while the national population is highly concentrated in urban places, geographic concentration is less pronounced in that the six capital cities are widely dispersed around the eastern and southern coastlines. Although there is no single dominant city on the national scale, there is extreme primacy within each of the states. In fact, in several states the capital is the only city.[2]

There is evidence, however, in the most recent intercensal period in particular, that numerous smaller urban areas are again beginning to grow quite rapidly. The proportion of Australia's population living in nonmetropolitan urban areas has steadily increased from 13% in 1947 to almost 20% in 1971. The increase in metropolitan dominance has been largely at the expense of the rural population, which declined from 31% of the nation's population in 1947 to 14% in 1971. In Canada the rural (farm and nonfarm) population dropped in the same period from 39% to 24%.

One of the most significant demographic trends in both countries in the most recent intercensal period (1966-1971) has been the large population buildup in the urban extensions of the metropolitan areas, usually outside the present boundaries of census metropolitan areas.[3] In this period, for example, Gosford, about 50 miles north of Sydney, experienced a growth of 32% (from 42,807 to 56,373), and the Blue Mountains, about 50 miles west of Sydney, a growth of 20% (from 30,733 to 36,727). Both areas, originally recreation and vacation areas for Sydney's population, have now become dormitory suburbs. Likewise, the Gold Coast urban area immediately to the south of Brisbane has experienced a growth rate of 30% from 1966 to 1971. Medium-sized centers also grew: Wollongong (near Sydney) and Geelong (near Melbourne) experienced growth rates of 100% and 200% respectively between 1947 and 1971. Similarly, in Canada the fastest growing centers are the medium to smaller cities located within the sphere of influence of the large metropolises (see Bourne and Gad, 1972). Oshawa, Guelph, Barrie, and Kitchener around metropolitan Toronto are examples. All these centers, though not within the census definitions of Toronto, are now being enveloped in the housing market area and commuting hinterland largely because of a shortage of space and housing in the metropolis itself and because of restrictions on suburban sprawl. This phenomenon is similar to that in U.S. cities, but, unlike in the United States, however, in both Australia and Canada the pattern is largely a function of congestion (and costs) in the metropolitan cores rather than of the deterioration of those cores. If overall population growth rates slacken in the next decade, as they are already beginning to do, it will be interesting to observe what effect this has on the dispersal of urban populations and on the relatively high-quality but increasingly congested central cities.

Small towns did not necessarily stagnate. Given the large geographic areas of both countries and their low population densities, many rural centers cannot be easily bypassed in going to a metropolitan center. Also, many serve as catchment points for rural out-migrants, and others increasingly have developed as retirement locations. Still others have benefited by the growth of massive resource development projects—for oil and gas, potash, iron ore, nickel, gold, and so forth. Although comparative data here is scarce, Canadian rural towns and cities, given generally higher rural densities, have faired better than their Australian counterparts.

STABILITY AND INTEGRATION IN CITY SIZE DISTRIBUTIONS

The usual views presented of the Australian urban system are one with a sharply truncated size distribution and one that disaggregates into a set of state subsystems each centering on the capital city (Rose, 1966; Bourne, 1974b). Both conditions are true, and both have been shown to be highly stable over time. There is also little doubt that state loyalties remain strong; for example, only about 5% of the total number changing their place of residence between 1966 and 1971 migrated across state borders. Horizontal or nonhierarchical interaction among the capital cities in Australia has received very little attention (Jeffery and Webb, 1972), yet there is evidence to suggest that it may now be as important a structural attribute to the urban system as the strictly hierarchical ties which connect each metropolis to its state hinterland.

The distribution of ownership in the national economy mirrors, if not anticipates, this pattern of concentration. In one of the few empirical studies of the ownership of economic activities in Australia, Johnston (1966) was able to show the great spatial concentration of corporate directorships in the capital cities, particularly Sydney and Melbourne. In some ways, however, their dominance appears to be complementary, with Sydney a leader in financial activities (banking, investment, and insurance) and Melbourne the leader in industrial activities. In another tabulation (Australian Financial Review, 1963), of the largest 100 companies in Australia arranged in order of total shareholder funds, 50 had their headquarters in Melbourne, 41 in Sydney, 5 in Adelaide, 3 in Brisbane, and 1 in Perth. Of the top 20 mining companies, 11 were located in Melbourne, 4 in Sydney, 3 in Perth, and 1 in Adelaide. This pattern in private-sector decision making is almost matched in the public sector, where once again, both state and federal government administrative offices are heavily concentrated in the state capitals. Canberra, of course, is the exception: it is Australia's first attempt at decentralization.

In Canada the economic system is not nearly as concentrated because of its larger population base, more diverse settlement history, and proximity to and integration with the U.S. economy. Nonetheless, concentration with complementarity is extensive. Toronto and Montreal historically have competed, as Sydney and Melbourne have done, for national dominance (Kerr, 1968; Simmons, 1974). Both have achieved it, but in different sectors–Toronto in banking, insurance, and mining offices; Montreal in manufacturing and transportation–and with different cultural groups, the English- and French-speaking populations respectively. Both have reached out to truncate and eventually absorb the hinterlands of smaller centers. For example, a recent study by Storey and Bulker (1975) of airline traffic flows shows that, even in the relatively short period since 1960, Toronto and Montreal, notably the former, have reordered (with the help of long-haul aircraft) what was essentially a set of regional airline systems into a national system under their dominance.

But how "mature" are these urban systems of Australia and Canada? That is, how strong and persuasive are the links between cities in the spatial organization of society? The heavy concentration of the head offices of large private companies and of public administration in Sydney, Melbourne, and Canberra, and the strength of direct communications among them, as measured by airline and telephone connections, does imply that a nationally integrated set of large urban places has come into existence. Australian society is virtually organized by decisions made by large private and public concerns in Sydney, Melbourne, and Canberra, just as Canadian society is by Montreal, Toronto, Ottawa, and, more recently, Vancouver. The existence of strong horizontal linkages at this level of the system probably means that expanded activities in Sydney and Melbourne are more likely to stimulate the growth of Adelaide, Brisbane, and Perth than to stimulate the growth of smaller centers at lower levels of the urban hierarchy.

Further nonhierarchical integration of the urban system also implies that growth initiated at one or a set of smaller centers will be felt primarily at the location of the head offices–the metropolis. This movement of growth-effects (multipliers) upward through the urban hierarchy, as Pred (1973) suggests, follows from the increasing scale of corporate ownership in capitalistic economies. It is also a function, in Australia and Canada, of an economy dominated by unstable resource exploitation and extensive foreign control of the means of production (Caves and Reuber, 1969; English et al., 1972; Rotstein and Laxer, 1972).

Again, the essential difference between the economic concentrations in Australia and Canada is that in the former it is predominantly urban (i.e., metropolitan) while in the latter it is regional as well as urban (Cameron et al., 1974). That is, the pattern of economic (and political) power in Canada is viewed by many simply as the core (i.e., Ontario and Quebec) versus the periphery. Although this power is largely in Toronto and Montreal, the issue is one of dominance by the commercial-industrial corridor, defined as extending from Windsor to Quebec City, including Ottawa (Whebell, 1969; Yeates, 1975). This area has maintained its share of the national population and its stranglehold on industrial production, banking, and communications—all three being highly sensitive regional as well as provincial issues.

COMPONENTS OF DEMOGRAPHIC CHANGE IN THE URBAN SYSTEMS

The three basic components of urban population change—natural increase, internal migration, and immigration from abroad—are changing. The latter component is of particular significance here for two reasons: it is the only one that is subject to direct control by the national governments involved, and it has been a critical element in the economic and urban growth of both countries in the post-World War II period. It is also the major question mark in forecasting future urban population growth. About half the growth of Australia's population and one-third of Canada's since 1947 has come from overseas immigrants and the immigrants' native-born children.

The immense and uneven impact on national immigration policies in each country becomes most striking when we examine the growth processes in the respective major cities, Sydney and Melbourne, and Toronto and Montreal. In the case of Sydney about three-fourths of its population growth from 1947 to 1971 is accounted for by immigrants and their children, and in Melbourne about four-fifths. This process is described in the First Report of the National Population Inquiry (1975) in this way:

> For the most part the new settlers did not immigrate to Australia, but to the metropolis of their choice . . . a high proportion . . . coming to join relatives in the urban areas of Australia rather than in the urban areas of their own country or of other parts of Europe.

The distinctive characteristics of Australia's postwar immigration policy are not the notions of assisted passage or the stimulation of economic growth through a steady input of new labor, but the idea of extending the search for immigrants to non-British settlers and the impact that this has had on metropolitan growth and life-styles.

Both countries historically have had high rates of natural increase and immigration as well as emigration. Canada has had high net immigration ratios only from 1901 to 1911, in settling the West, and from 1951 to the present. Between 1951 and 1971 Canada absorbed 2.95 million immigrants and lost 1.05 millions. Net migration averaged 22% of total growth. While this declined somewhat in the 1961-1971 decade, it has risen since, both absolutely and proportionately, as birth rates have fallen to near the level of zero population growth. More to the point, over one-half of these immigrants have gone to the

TABLE 1

COMPONENTS OF METROPOLITAN POPULATION GROWTH, IN URBAN AREAS, AUSTRALIA 1947-1966

Metropolitan Division	Natural Increase		Net Migration of Australian-born		Net Migration of Foreign-born		Total Population Increase (=100%)
	(in 000s)	%	(in 000s)	%	(in 000s)	%	(in 000s)
Sydney	378.7	45.0	0.3	0.0	463.4	55.0	842.5
Melbourne	365.6	41.1	1.4	0.2	522.1	58.7	889.2
Brisbane	123.6	38.6	87.7	27.4	108.8	34.0	320.2
Adelaide	104.3	28.9	52.6	14.6	203.9	56.5	360.8
Perth	103.9	40.6	37.5	14.6	114.7	44.8	255.9
Hobart	28.7	50.4	9.9	17.5	18.3	32.1	57.0

three largest metropolitan areas, and over 30% to Toronto alone. Toronto has been shell-shocked and services have been strained under this growth rate, but the city remains livable. Not surprisingly, much of the recent antigrowth feeling derives from this environment.

The uneven distribution of population growth is clearly apparent when we examine the components of that growth for individual urban areas in Australia (Table 1) and Canada (Table 2). Table 1 shows the extremely small contribution of *internal* migration and the huge contribution of *overseas* migration to the growth of Sydney and Melbourne. Both actually "exported" large numbers of native-born Australians to the other capitals, especially to Brisbane and Perth, and, of course, to urban places just beyond the metropolitan boundaries.

The intermetropolitan movement of people is, however, becoming obvious for the first time and is an important indication that horizontal interaction at the highest level of the urban system is increasingly important relative to the classical vertical movement from rural to urban areas. The intermetropolitan movements, summarized in Table 3, should be considered in context, where, from 1966 to 1971, 44% of the nation's population changed its place of residence, but, as previously noted, only about 5% actually crossed state boundaries. Although 5% is a rather small figure, in absolute terms an interesting pattern of interstate migration emerged between 1966 and 1971.[4] There is some evidence of a shift of population to the north and to the west, away from the most densely settled southeastern corner of the country. Although it is too early to make any claim that this trend will lead to a substantial redistribution of the nation's population (Henderson and Holland, 1975), it may at least reduce some of the growth pressures on Sydney and Melbourne.

While *net* migration had a minor effect on the growth of Sydney and Melbourne in the most recent intercensal period, there was nevertheless a considerable volume of movement. Even so, the contribution of internal migration was much greater in the past; from 1954 to 1961, for example, which was a period of rapid urbanization, the six state capitals grew by 146,000 from internal migration, mainly from the rural sector. In contrast, from 1961 to 1966, while out-migration from rural areas actually increased, there was a substantial net movement away from Sydney and Melbourne; in this period the gain from internal migration to all capital cities was only 51,000, with almost all this gain going to Brisbane, Perth, and Adelaide. Sydney actually had a net loss through internal migration of over 10,000 people in the 1966-1971 intercensal period, a loss which was more than compensated for by immigration from overseas. Melbourne had a net gain of only 10,000 through internal migration in the same period. In contrast, Brisbane had a relatively high net gain of 17,500 people, gaining population from New South Wales, Victoria, South Australia, and Tasmania. With a surplus of 28,000, Perth had the largest gain from internal migration of any capital city in the 1966-1971 period, reflecting its pivotal role in the expanding resource development industry. It had a net in-flow of people from every part of the nation except Canberra, to which it "lost" 500 people. In

TABLE 2

COMPONENTS OF POPULATION CHANGE, METROPOLITAN AREAS, CANADA, 1966-1971

Census Metropolitan Area (CMA)	Population Increase	Net Migration (Inter-CMA)		Natural Increase*		% Population 1971 Foreign-born
		Number	% Pop. 1966	Number	% Pop. 1966	
Montreal	172,226	–7,030	–0.3	179,256	7.0	14.8
Toronto	338,143	7,565	0.3	330,578	14.4	34.0
Vancouver	149,261	41,615	4.5	107,646	11.5	26.5
Ottawa-Hull	17,976	13,000	4.6	4,976	1.7	12.5
Winnipeg	31,503	–18,230	–3.6	49,733	9.8	19.9
Hamilton	41,113	5,215	1.1	35,898	7.9	26.7
Edmonton	70,332	–3,170	–0.8	73,502	17.3	18.3
Quebec	73,736	805	0.6	72,931	13.8	2.2
Calgary	72,744	11,170	3.9	61,574	20.3	20.5
Niagara-St. Catharines	8,302	–2,225	–1.7	10,527	7.9	22.9
London	32,310	5,670	2.2	26,640	10.5	20.0
Windsor	20,320	–290	–0.1	20,610	8.7	21.5
Kitchener	34,571	5,080	2.6	29,491	15.3	21.8
Halifax	12,736	–4,675	–2.2	17,411	8.3	7.2
Victoria	20,538	8,810	5.0	11,728	6.7	24.7
Sudbury	18,685	–685	–0.5	19,370	14.7	12.4
Regina	43,584	–9,000	–2.1	52,584	12.0	13.1
Chicoutimi-Jonquiere	749	–5,015	–3.8	5,764	4.3	1.4
St. John's	14,281	–4,055	–3.5	18,336	15.6	2.9
Saskatoon	10,549	8,225	–7.1	18,774	16.2	13.9
Thunder Bay	14,058	–1,750	–1.6	15,808	14.6	21.1
St. John	2,549	–810	–0.8	3,359	3.2	4.8

*Includes other migrants (and immigrants).

TABLE 3
INTERMETROPOLITAN MIGRATION, AUSTRALIAN CAPITAL CITIES, 1966-1971

Residence in 1966	Residence in 1971						
	Sydney	Melbourne	Brisbane	Adelaide	Perth	Hobart	Canberra
Sydney		15,849	11,621	5,205	6,251	1,193	9,559
Melbourne	19,121		8,603	6,074	7,550	1,965	5,323
Brisbane	12,570	6,761		1,332	1,690	359	2,144
Adelaide	6,522	9,231	1,943		3,431	470	1,979
Perth	4,652	4,717	1,424	1,625		195	989
Hobart	1,665	2,973	557	527	617		418
Canberra	4,287	1,980	867	594	590	94	

SOURCE: Census of Population, Commonwealth of Australia, 1971.

fact, Canberra with a "surplus" of 24,000 was second to Perth in terms of net gains in exchanges; but over half of Canberra's gains came from New South Wales, the state in which it is situated.

Throughout the entire period of 1947 to 1971, Australia experienced a relatively high and extremely stable rate of natural increase of around 1.2% per year. Thus, even without immigration, cities would double their population in about 58 years. But, once again, in the most recent years there have been signs that the nation has come very close to zero population growth, which, in Australia, has been taken to mean zero *natural* growth.

The Canadian pattern is similar, but somewhat more complex, and it too is changing. In terms of net migration from 1966 to 1971 between census metropolitan areas (inter-CMA) the medium-sized centers dominated (Table 2). Vancouver gained 41,615, Ottawa-Hull 13,000, Calgary 11,170, and Victoria 8,810. For Calgary and Vancouver, growth is a response to their roles as centers of resource development as well as developers of larger trade areas at the expense of eastern centers. Ottawa-Hull's attraction is entirely that of the growth of the federal bureaucracy, while Victoria is becoming the Canadian retirement spa.

Montreal, in contrast, lost over 7,000 persons in the inter-CMA migration process, while Toronto gained only 7,600. The latter figure, however, conceals important differences in the sources of these migrants (Table 4). Toronto, in fact, had a net migration lost with other metropolitan areas taken together, excluding Montreal. It gained in total only through a very high net balance (nearly 16,000) with Montreal (25,300 in-migrants; 9,400 out-migrants). Montreal's economic troubles in the late 1960s and the quiet (and continuing) revolution in Quebec have substantially altered migration flows within the Canadian urban system. The difference lies primarily not in the outflow from Montreal, though that has increased of late, but in the decline of in-migrants from other areas of Canada and from abroad.

The Canadian interurban migration pattern also has a regional overlay which is suggestive of future urban growth paths. As expected, French-speaking cities have migration exchanges very different from those of other cities. Quebec City,

TABLE 4

INTERMETROPOLITAN MIGRATION FLOWS, SELECTED METROPOLITAN AREAS, CANADA, 1966-1971

Place of Residence, 1966		Place of Residence, 1971											Totals (All 22 CMAs)[a]
		1	2	3	4	5	6	7	8	9	10	11	
1. Calgary		0	7,575	255	320	850	890	85	680	2,300	7,565	1,620	26,030
2. Edmonton		10,605	10,325	350	405	895	1,270	50	820	2,365	9,390	1,780	31,905
3. Halifax		650	485	9,240	500	1,815	2,640	175	65	3,510	1,055	595	15,775
4. London		455	330	255	6,610	790	1,105	40	65	6,725	1,025	460	17,050
5. Montreal		2,400	1,790	2,240	1,700	269,990	12,935	9,475	255	25,315	8,970	2,010	78,885
6. Ottawa-Hull		1,100	1,300	1,100	1,305	6,005	34,740	1,125	165	8,440	3,025	1,015	30,060
7. Quebec		130	155	260	85	11,765	2,025	33,915	0	740	230	75	17,150
8. Saskatoon		2,910	2,630	45	125	350	500	20	0	920	2,590	1,345	15,010
9. Toronto		3,050	2,465	2,295	8,560	9,435	8,870	380	485	153,135	11,635	3,345	84,795
10. Vancouver		3,610	3,830	370	385	2,245	1,665	70	550	6,030	99,130	2,170	28,625
11. Winnipeg		4,590	3,445	480	655	2,440	2,515	335	995	5,670	10,435	16,325	38,075
Totals (All 22 CMAs)	a	36,105	28,430	10,750	21,560	44,920	41,480	15,260	6,655	95,335	69,230	19,835	501,115
	b	36,105	38,755	19,995	28,180	314,920	76,225	49,180	6,650	248,465	168,350	36,155	

a. Excluding main diagonal (i.e., intermunicipal moves within a metropolitan area).

b. Including main diagonal.

NOTE: Minor differences due to random rounding error.

SOURCE: Statistics Canada, preliminary figures, 1974.

for instance, appears almost isolated in Table 4 from the other metropolitan areas except for Montreal. The growth centers, in terms of urban migration, are in the West (Calgary, Edmonton, and Vancouver) and southern Ontario outside metropolitan Toronto (London, Ottawa-Hull, and Kitchener-Waterloo, the last not shown). Most other centers are being drained (Saskatoon and Winnipeg) as their regional service functions are usurped by newer growing centers (i.e., Calgary). Saskatoon, in fact, had the highest gross migration ratio, acting as a transient base for out-migrants from the province.[5] These patterns in turn are correlated with regional population movements. Between 1966 and 1971 only three of 10 provinces—Ontario, Alberta, and British Columbia—had positive net migration rates.[6] There are few if any signs that this trend will not continue, even with firm government action (Science Council of Canada, 1975).

FOREIGN IMMIGRATION

As noted, foreign immigration is increasingly the dominant demographic component in urban growth in Australia and Canada. Combined with declining birth rates, the fact that immigration increased in both countries in the late 1960s and 1970s (until controls were recently introduced) added to the pressures facing selected cities and regions. In the case of Ontario, out of a total population increase in the province of 743,000 (from 5.9 to 7.7 millions) between 1966 and 1971, 373,000 was due to natural increase and 370,000 to net migration. Of the latter, 243,000 was *net* foreign immigration. More than half of this went to Toronto. In 1971, over 34% of Toronto's population was foreign-born (Table 2)—a proportion not seen in any U.S. cities since prior to World War I. Over 26% were foreign-born in Vancouver and Hamilton, and over 20% in Calgary and Edmonton. The average for urban Canada was 17%. One consequence of this population movement, aside from the concentration issue noted earlier, is the prospect and problem of future employment growth. More persons are expected to enter the Canadian labor force by 1980, given the youthful age structure of the immigration population, than will enter the labor forces of Britain, West Germany, and Italy combined. How will this employment be distributed through the urban system? Will the need for economic growth continue to dominate over issues of the distribution of that growth in national policy making?

SUMMARY

From the discussion above we can draw a number of broad generalizations on the structure and growth of urban systems in Australia and Canada. The Australian urban system is still clearly dominated by the state capitals, more particularly by Sydney and Melbourne, and this dominance has been remarkably stable over time. However, there remain some interesting recent indications of a possible change in the nature of that system—a movement from the metropolitan areas into physically separate but accessible urban extensions, rapid growth in a few large centers with strong industrial bases, and a moderately high growth rate in a limited number of regional centers.

In Canada the situation is again more complex, but a number of similarities do exist. The dominance of Toronto and Montreal continues and, in some sectors, is expanding. But other metropolitan centers are growing more rapidly, and new centers continue to appear. The urban system is still being formed. As Simmons (1976) notes, almost 50% of Canada's urban population has been added since 1951. Over one-third of all places with over 10,000 population in 1971 were below that threshold in 1951. Sudden changes in stable markets (i.e., oil, potash, nickel) have dramatically altered the fortunes of some centers. The urban system is clearly flexible and growing, but seemingly stable overall. In this apparent paradox of stability with change lies a major challenge for national governments in formulating urban growth policies at the national level.

URBAN POLICIES AT THE NATIONAL LEVEL

We are not the only ones comparing urban problems and policies in Australia and Canada. In the last few years, both countries have initiated an evaluation of the need for, and the substance of, a strategy to shape urbanization at the national level. Both have closely followed the other's experience.

The decision to seek this national strategy was a significant political step in each case, for three obvious reasons. There was little evidence from other western countries of what a national urban strategy actually was or whether it was feasible; constitutionally, there was and still is heated disagreement over the extent to which urban issues are the responsibility of state (provincial) or federal governments; and finally, as essentially conservative capitalistic societies, neither country has been particularly effective in the past in any major effort at national planning, at least in peacetime.

THE INSTITUTIONAL RESPONSE

Nevertheless, both countries proceeded to establish urban policy-making agencies at the national level. Canada was first. Acting on one of the recommendations in an impressive set of studies by N.H. Lithwick (1970) on the state of urban Canada, the federal government established in 1971 a new Ministry of State for Urban Affairs. This ministry incorporated the existing and jealously independent national housing agency, the Central Mortgage and Housing Corporation, but otherwise it was to have primarily a policy-coordinating rather than a program-delivery role. Exactly how it was to achieve this coordination has never been made explicit, although as in any parliamentary system of government the influence of the minister responsible within the cabinet often determines the priority assigned to that policy area.

The ministry's initial organization involved both policy and research functions. The emphasis in the former was to be on intergovernmental policy cooperation, and in the latter on policy sensitive research (Miles et al., 1973; Gertler, 1974). It has not thus far been particularly successful at either (Kaplan, 1972; Lithwick, 1972b, 1972c; Blumenfeld, 1974), although there are some exceptions. Instead, it has spent most of its time and energy responding to immediate problems requiring instant solutions and, of course, justifying its existence. At the same time, and complicating its operations, some provincial governments and municipalities have become actively involved in initiating broad development strategies of their own. This, of course, is precisely what is needed. A national urban strategy is, as Swain (1972) argued, not only a federal responsibility but an integration of policies across all three levels of government. This renders the task of the new ministry considerably more difficult, and it will certainly be different from that initially conceived.

The Australian government, viewing this development with justifiable skepticism, established in 1972 a Department of Urban and Regional Development under its own minister. Like the U.S. Department of Housing and Urban Development, it was assigned substantial program functions and consequently a much larger budget (and power) than its Canadian equivalent. While details of the department's organization and objectives are described elsewhere (Australia, 1973; Logan and Wilmoth, 1974; Bourne, 1975), two divisional responsibilities warrant mention here.

The first is the Cities Commission. This agency was established out of the short-lived National Urban and Regional Development Authority created by a previous government just prior to its downfall in the 1972 national election. The Cities Commission was given principal responsibility for the government's new cities program. This commission, as well as the program itself, has many of the earmarks of a successful venture, including fiscal authority and enabling legislation similar to that of the British New Town development corporations, although its functions are soon to be absorbed within the department.[7] The second is the proposed Land Commissions, one in each state, funded by the federal and state governments. These commissions, of which only three have been established at the time of writing, are potentially very important institutions in that they are directed at tackling the core of the problem of plan implementation–land use control, acquisition, and development costs. There are also promising indications that political rivalries which have hindered their progress in the past are being overcome.

Equally important, the Australian department assumed responsibilities in regional development, whereas the Canadian ministry, because of the existence of a large well-established agency for regional development, did not. It is difficult to conveive of an effective urban strategy without a coordinated regional policy. Granted, the problem areas of urban and regional policies, as traditionally defined, are somewhat different–the former focusing on containing the growth of the "big city" and reducing its problems of affluence (i.e., congestion) and the latter focusing on welfare issues relating to peripheral distressed areas–but their roles in shaping the spatial structure of national development are inseparable. That is, regional problems are seen to be closely related to differentials in urban size and growth. Ray and Villeneuve (1974) have shown in the Canadian context that income inequalities, for example, are more a function of variations in city size and urban growth rates than of regional differences. Regional policy, in turn, has looked increasingly to the evolution of a national system of growth centers to assist in ameliorating these differences (Goddard, 1974). Such growth centers are primarily urban areas. Urban policy, on the other hand, has broadened its initial focus from a singular concern with the largest metropolitan areas to include a much larger range of the city-size hierarchy. Both Australia and Canada demonstrate these directions in policy thinking (Atcheson et al., 1974; Bourne, 1975).

In any case, the presence of these federal-level urban agencies has explicitly demanded that the ongoing activities of other federal agencies, such as in transport and manpower, make at least some effort to evaluate the impact of their programs on urban areas. Those that do not are beginning to feel the pressures of an increasingly enlightened electorate.

DEMOGRAPHIC ISSUES

Not surprisingly, in light of the figures in Tables 1 and 2 above, immigration policies in both Canada and Australia have also recently come under careful scrutiny and severe criticism, on widely diverse grounds. Some have argued that a high rate of immigration has contributed to inflationary pressures in land and housing and has aggravated unemployment. Others see it as contributing to economic growth when the important social objectives should be moving toward more equitable distribution of opportunities and quality-of-life issues. STill others (Lithwick, 1970; Australian Institute of Political Science, 1971; Burnley, 1974) have seen immigration as being one if not the main cause of the excessive growth of the major cities and, therefore, as a contributor to the assumed deterioration in the quality of urban life.

Both countries consequently have established commissions to review immigration policy. Australia established a National Population Inquiry, from which reports are now beginning to appear. Canada has a federal commission on immigration underway, and a first draft of a government policy paper is now in circulation (Information Canada, 1975). Wide and intense debates have ensued.

The results of both procedures are potentially of immense importance for their respective urban systems. As immigration restrictions tighten around the world the pressures on Canada and Australia will increase. Already both reviews have resulted in the closing of some obvious loopholes in the existing legislation and a general reduction in the rate of admissions. The Australian review is, however, somewhat broader in that it has been asked to inquire into a national population distribution policy for that country. The Canadian commission was not so asked, but it is finding that it cannot resolve immigration issues, except for loopholes, without addressing the overall population question. The prospect for Australia is almost certain to be substantially lower immigration rates, while for Canada, greater social plurality in the present population ensure that few groups will be kept out and, thus, that immigration levels will remain relatively high.

URBAN ISSUES AND POLICIES

The rationale for Australian and Canadian government involvement in urban issues lies in a number of key but traditional areas. There is, apparently, more widespread concern about quality of life issues and some critical questioning of the need for maintaining high rates of population growth. Disenchantment with economic growth, typical of the late 1960s and early 1970s, was tempered somewhat by the recession and rising unemployment of the mid-seventies. Yet, many social trends and policies which had, in varying degrees, been accepted largely without question for the two previous decades are now in open debate. The debate is manifest in the criticisms of assisted immigration policy in Australia and in the arguments in favor of zero population growth, environmental conservation, and income redistribution.

To many, economic growth is synonomous with the large city (Richardson, 1973), and it is here that inequalities in access to a large range of both public and private goods and services are most obvious and politically sensitive. Given the rapid growth of Sydney and Melbourne up to 1971 and the social welfare goals and objectives of the present Australian government (Logan and Wilmoth, 1974) it was logical that urban public policy should be directed toward those areas (Alonso, 1974). In Canada, the persistence of regional inequalities and of unemployment has overshadowed the urban policy debate. But in that debate, the growth problems of Toronto, Montreal, and Vancouver have dominated. The government has in fact now committed itself to a strategy of decentralizing the growth of these centers to at least a dozen alternative locations—largely medium-sized cities.

The essence of the present policy debate may be succinctly presented by the following quotations. The Report on New Cities for Australia prepared by the Australian Institute of Urban Studies (1972) argues that

> at themoment the biggest of Australia's metropolitan regions are in obvious trouble. The social problems are evident to everyone: they range from long journeys to work to the ills of personal alienation. It is also clear that, as they are organised at the moment, the biggest cities cost too much. . . . Now, throughout Australia, there is a wave of new aspirations for the cities and a new sense of purpose and excitement.

The Canadian government minister responsible for urban affairs argued on April 2, 1975, that if current trends and their impacts continue,

> Our largest, most rapidly growing cities would become unmanageable; and others, continuing to lose population, would wither. Regional economic disparities would be accentuated and political power would shift, perhaps to quasi city-states, but more certainly to the dominant provinces, to an even greater extent than today. [Canada, 1975:4]

The minister went on to note that these trends . . . "threaten the character of our nation, the very fabric of Confederation as we know it today" (p. 5). Clearly, politics is important in defining urban goals.

Political issues, however, are relative. By international standards, neither Australia nor Canada have serious urban problems. There are no very large cities. Most existing cities are still livable and governable. Central cities are healthy, but expensive and congested, increasingly more on the centralized European than the dispersed American model. New urban growth is more equitably distributed among several rather than a few growing centers, than is true in many other countries, and both social services and employment opportunities are improving in most disadvantaged regions. Thus it is not surprising that urban issues are still given only modest priority in the hierarchical ordering of political priorities.

Yet, and in part because of this low profile, politicians in both countries have had to contend that a national crisis prevails. No doubt, in certain sectors and in some specific cities and regions, crises do prevail. Aside from political rhetoric, such crises are, on the one hand, largely crises of anticipations: nostalgic reflections on what urban life was like in the past and overambitious expectations on what the present should have brought. On the other hand, such crises reflect the need for more effective urban management and for the coordinated delivery of social services.

Possibly the most widely quoted national urban issue in both countries is one of increasing metropolitan concentration combined with rural depopulation. In countries with small national populations spread over large geographic areas, even minor redistributions of population can have serious social consequences in rural areas. In Canada the latter concern has generated long-standing programs of regional development incentives and fiscal transfer payments, while in Australia it has lead to a century of debate on the decentralization of population from the state capitals to the interior.

Even if the absolute level of metropolitan concentration is not high in relative terms, there is concern over the results of that concentration on life-styles, political attitudes, and access to economic opportunities. The growth of metropolitan areas in Canada, as Simmons and Bourne (1974) note, has

> to an increasing degree polarized the national economic landscape . . . through their market influence and fiancial control. Politically and culturally they are creating new urban images and generating new policies which tend to obliterate or ignore the cultures, values and independence of other places.

This, they argue, is the dominant issue behind the recent discussions of a national urban strategy.

The main urban strategy followed to date in Australia has been to establish a number of designated growth centers both distant from and near the two largest metropolitan areas. This strategy, while conceived quite independently, is in general accord with recommendations made by the Australian Institute of Urban Studies in its August 1972 report entitled *New Cities for Australia.* It based its arguments on projections of an Australian population by the year 2000 of almost 22 millions (i.e., an increase of almost 9 million over the 1972 population). In July 1973, when the Cities Commission made its recommendations to the Australian government for a range of new city and growth center projects, the situation was largely unchanged: for the year 2000 the Australian Bureau of Statistics projected a low population of 17.8 millions based solely on natural increase and a high of 23.7 millions assuming an average immigrant intake of 140,000 per year. Official projections prepared by the bureau were revised downward for the World Population Conference in Bucharest in April 1974 to 16.9 millions by the year 2000, based on a somewhat slow rate of natural increase and eliminating any net immigration. The upper limit, allowing for immigration at the rate of 80,000 net immigrants per annum from 1973 to 2000, gave a total of 19.8 millions.

It was on the basis of what now appeared to be rather exaggerated projections, and on a fairly widespread assumption that both Sydney and Melbourne would reach four million by 2000, that "growth centers" with varying degrees of Australian and state government commitment were designated at Canberra, Albury-Wodonga, Bathurst-Orange, Monarto, Geelong, and Campbelltown-Holsworthy.[8] Work on Albury-Wodonga and Bathurst-Orange has already begun. Clearly, the new cities program amounts to a major restructuring of the urban system, not simply a mechanism for slowing down the growth of Sydney and Melbourne. It is likely to involve a huge allocation of public resources and therefore will no doubt be scaled down considerably over time from the initial overly ambitious recommendations for 24 new towns and growth centers by the year 2000 (Australian Institute of Urban Studies, 1972).

The conditions (and expectations) under which these proposals were made have of late changed rather dramatically. That is, problems arising from rapid population growth, and the need for metropolitan containment and new cities, may be sharply reduced by demographic forces outside the jurisdiction of national planners. The same is true in Canada. Projected population growth rates have declined consistently since the 1960s. The Economic Council's (1968) estimate of the most likely total national population in the year 2001 was 35 millions. Lithwick's (1970) median projection was 33.8 millions. More recent estimates have dropped these figures to 31 millions (Bourne et al., 1974) and 29 millions (Statistics Canada, 1974). Current estimates are as low as 28 millions (Science Council of Canada, 1975), and even these may be exaggerated. But perceptions of the problem in Canada do not seem to have changed accordingly.

What, for example, are the effects of a substantially lower rate of population growth, even zero growth, on patterns of urbanization and on the issues created by those patterns? How will internal migration patterns change? Will Toronto and Vancouver (and Edmonton, Calgary, and Ottawa) still grow at the expense of the nation?

The realization has also grown in Australia that far from experiencing quite high rates of population growth through natural increase and immigration, the country may come to face a situation in which population itself is a scarce resource, to be allocated among competing developmental opportunities. The First Report of the National Population Inquiry (1975) makes three population projections to the year 2001, the "most preferred" resting on assumptions that the net reproduction rate declines to 1.0 by 1975-1976 and remains constant thereafter and that immigration runs at a net gain of only 50,000 per year. The immigration intake will be permanently restricted by economic circumstances within Australia and by a dwindling supply of immigrants from Europe. These assumptions yield a population of 17.6 millions by 2000, the lowest projection yet made. The report admits that growth to at least 15.5 millions is assured without any immigration simply because of the present age and sex composition and because of rates of family formation. There is a third assumption, that present patterns of internal migration will continue.

All three assumptions, of course, are open to question, especially the third, where the thrust of government policy is precisely to alter past trends. Even if the assumptions are correct, the anticipated population of 17.6 millions will still permit the major part of the growth center program to continue, but it suggests that "competition" among places for people is going to be increasingly strong. Given the strength of centralizing pressures and of interaction among the metropolitan areas, strong instruments of public policy will be required to steer any significant proportion of new population away from the larger metropolitan areas and into the new urban places.

SYNOPSIS

Urbanization in Australia and Canada appears to have entered a new period. Growth rates are beginning to decline, and their distribution across the two urban systems is changing. Internal migration streams have shifted away from the two dominant major metropolitan areas in each country toward medium-sized cities and to smaller centers just outside the metropolitan regions. The shift to smaller centers is a continuation of the process of urban population dispersal common throughout this century and may result in higher degrees of regional population concentration through the creation of megalopolitan urban forms (Bourne, 1974b; Yeates, 1975). The drift to medium-sized cities appears to be new in the late postwar era and is in part a response to the problems of rapid growth in the larger metropolises. Foreign immigration is now the key policy variable in containing the population growth of the metropolises.

Over time, the urban systems of both countries have been shown to be remarkably stable. They have grown rapidly; new member cities are continually added at the frontier and within; and they have adapted their structures (notably in Canada) to a wave of differing demands generated by changes in the staple products produced in the economy. Both systems remain relatively open to foreign influence and to fluctuations in international events. Both countries remain peripheral to global economies, and both continue their preoccupation with filling in national territory, through extensions to the urban system (i.e., new cities). It is increasingly clear that, as population growth rates decline and pressures for a more equitable distribution of opportunities within the present urban system grow, these extensions will not be necessary of appropriate. Long-term policy making at the margin of the world economy, and in such a turbulent environment, is a hazardous exercise.

The policy question still under debate in both countries is not whether the federal governments have a role to play in the urban system, but what that role is to be. These governments already have a major and unequal investment in urban areas (in employment, land, housing, and income). It is now recognized that even their explicitly nonurban policies and programs have immense impacts on urban areas and that these often conflict among themselves and with the actions of state (provincial) and local governments. Although the most recent tendency in both countries is toward a greater decentralization of policy-making responsibility, particularly in taxing powers, to the states (provinces), new regional governments and municipalities, federal involvement will continue to be significant. But that involvement will have to be more explicit and imaginative than it has been thus far, and be more closely coordinated with the actions of lower levels of government, if future growth is to be effectively managed.

The first and most difficult prerequisite in planning urbanization at the national level is to achieve a consensus on objectives, including the desired future state of the urban system in question. However, a national consensus on urban goals, in such pluralistic, federated states as Australia and Canada, is not likely to be attainable in the foreseeable future. Regional, social, and political differences assure that an agreement on what the overall problems are, or how the urban system should be changed, will be slow in coming.[9] This is not a phenomenon unique to Canada or Australia.

The result is likely to be a preservation of the status quo in terms of urbanization. Interestingly, what seems to concern politicians most is a sharp departure from the current state (i.e., greater population concentration). Thus political pressures in both countries will add to the pressures of the economic system to maintain population and city-size distributions in their present form. Stability is self-reinforcing, and deliberate change is, therefore, all the more difficult to achieve. What is encouraging, however, is that these issues are being openly debated among an interested and growing audience in both countries. Debate, of course, will not solve the problems, but it should help to clarify future alternatives and the possible consequences of government actions and encourage more sensitive policies in the future.

NOTES

1. Such figures should be interpreted with caution. Urban population changes can occur through population growth of urban areas, as defined at a previous point in time, or through redefinition of their boundaries. Stone (1967) estimated the redefinition effect on population growth between 1961 and 1966 at 15%.

2. In South Australia, for example, Adelaide is 25 times the size of Whyalla, the second city; and in Western Australia, Perth is 31 times the size of Kalgoorlie-Boulder, the second city.

3. Part of this pattern is attributable to that fact that metropolitan areas in both Canada and Australia tend to be underbounded, relative at least to the definition of Standard Metropolitan Statistical Areas and the statistical consolidated regions used in the U.S. census. This has the effect of underestimating geographic population concentrations.

4. Tasmania experienced a loss of 9,000, South Australia 17,000, Victoria 32,000, and New South Wales 21,000. In contrast, Queensland gained 19,000, the Northern Territory 13,000, Western Australia 28,000, and the Australian Capital Territory 24,000.

5. Interregional population changes may still be substantial, but annual migration rates are highly erratic. For example, the population of the province of Saskatchewan dropped from a peak of 965,000 in 1969 to below 906,000 in early 1974. Taking into account average rates of natural increase and allowing for some migration into the province, the population loss in just five years is over 100,000. During this period the two cities in the province, Regina and Saskatoon, continued to grow, but at a slower rate. Since 1974, however, this trend appears to have been reversed, with the provincial population growing in absolute terms.

6. The respective totals were 369,000 for Ontario, 222,400 for British Columbia, and 59,400 for Alberta (including net foreign immigration). In contrast, Saskatchewan lost 80,000, Quebec 42,000, Manitoba 24,000, Newfoundland 20,000, and New Brunswick 17,500.

7. After this paper was prepared, the Labor government was defeated in the national election of December 13, 1975, and the Department of Urban and Regional Development was abolished, and its functions were incorporated into a new Department of Environment, Housing, and Community Development.

8. Anticipated year 2000 populations for these proposed centers varied, but reasonably firm estimates were made for Albury-Wodonga of 300,000, Bathurst-Orange of 200,000, and Canberra of 700,000.

9. A slow rather than hasty response to urban problems of course has its advantages (see Goldberg and Seelig, 1975). Initial mistakes are likely to be less severe and costly, and the inevitable learning process inherent in national planning will have time to mature.

REFERENCES

ALONSO, W. (1974). "A report on Australian urban development issues." In Urban and regional development: Overseas experts reports. Canberra: Cities Commission.

ANDREWS, J. (1964). Australia's resources and their utilisation. Sydney: Department of Adult Education, University of Sydney.

ATCHESON, J., CAMERON, D., and VARDY, D. (1974). "Regional and urban policy in Canada." Discussion paper no. 12, Economic Council of Canada, Ottawa.

Australia, Department of Urban and Regional Development (1973). First annual report. Canberra: Author.

Australian Financial Review (1963). Directory of the top 800 Australian companies. Sydney: Author.
Australian Institute of Political Science (1971). How many Australians: Immigration and growth. Proceedings of the 37th summer school. Sydney: Author.
Australian Institute of Urban Studies (1972). First report of the Task Force on New Cities for Australia. Canberra: Author.
BENSUSAN-BUTT, D.M. (1965). "A case for new cities." Economic Papers, 19:16-27.
BERRY, B.J.L. (1973). The human consequences of urbanization. Divergent paths in the urban experience of the twentieth century. Basingstoke: Macmillan.
BLAINEY, G. (1971). The tyranny of distance. Melbourne: Sun Books.
BLUMENFELD, H. (1966). "The role of the federal government in urban affairs." Journal of Liberal Thought, 2(2).
--- (1974). "The effects of public policy on the urban system." In L.S. Bourne et al. (eds.), Urban futures for central Canada: Perspectives on forecasting urban growth and form. Toronto: University of Toronto Press.
BOOTHROYD, P., and MARLYN, F. (1972). "National urban policy: A phrase in search of a meaning." Plan Canada, 12(1):4-11.
BOURNE, L.S. (1974a). "Forecasting urban systems: Research design, alternative methodologies and urbanization trends, with Canadian examples." Regional Studies, 8:197-210.
--- (1974b). "Urban systems in Australia and Canada: Comparative notes and research questions." Australian Geographic Studies, 12:152-172.
--- (1975). Urban systems: Strategies for regulation–A comparison of policies in Britain, Sweden, Australia and Canada. Oxford: Clarendon Press.
BOURNE, L.S., and GAD, G. (1972). "Urbanization and urban growth in Ontario and Quebec: An overview." Pp. 7-35 in L.S. Bourne and R.D. MacKinnon (eds.), Urban systems development in central Canada: Selected papers. Toronto: University of Toronto Press.
BOURNE, L.S., and MacKINNON, R.D. (eds., 1972). Urban systems development in central Canada: Selected papers. Toronto: University of Toronto Press.
BOURNE, L.S. et al. (1974). "Population forecasts for cities in central Canada." In L.S. Bourne et al. (eds.), Urban futures for central Canada: Perspectives on forecasting urban growth and form. Toronto: University of Toronto Press.
BREWIS, T.N. (1968). Growth and the Canadian economy. Toronto: McClelland and Stewart.
--- (1970). "Regional economic disparities and politics." Pp. 335-351 in L.H. Officer and L.B. Smith (eds.), Canadian economic problems and policies. Toronto: McGraw-Hill.
BUCKLEY, K. (1975). "Primary accumulation: The genesis of Australian capitalism." In E.L. Wheelwright and K. Buckley (eds.), Essays in the political economy of Australian capitalism (vol. 1). Sydney: Australian and New Zealand Book Co.
BURNLEY, I.H. (ed., 1974). Urbanization in Australia: The post-war experience. Cambridge: University Press.
BUTLIN, N.G. (1964). Investment in Australian economic development. Cambridge: University Press.
CAMERON, D., EMERSON, D.L., and LITHWICK, N.H. (1974). "The foundations of Canadian regionalism." Discussion paper no. 11, Economic Council of Canada, Ottawa.
Canada, Department of Regional Economic Expansion (1973). Report on regional development incentives. Ottawa: Author.
Canada, Ministry of State for Urban Affairs (1975). Interim national report for habitat: UN Conference on Human Settlements, Vancouver, 1976. Ottawa: Author.
CARELESS, J.M.S. (1954). "Fronterism, metropolitanism and Canadian history." Canadian Historical Review, 25:1-21.
CAVES, R.E., and REUBER, G.L. (1969). Canadian economic policy and the impact of international monetary flows. Toronto: University of Toronto Press.

Committee of Commonwealth/State Officials on Decentralization (1972). Report. Canberra: Australian Government Publishing Service.

CHOI, C.Y., and BURNLEY, I.H. (1974). "Population components in the growth of cities." In I.H. Burnley (ed.), Urbanization in Australia: The postwar experience. Cambridge: University Press.

DRUMMOND, I.M. (1972). The Canadian economy. Georgetown, Ont.: Irwin-Dorsey.

Economic Council of Canada (1964). First annual review: Economic goals for Canada to 1970. Ottawa: Queen's Printer.

--- (1968). Fifth annual review: The problem of poverty (chap. 6). Ottawa: Queen's Printer.

ENGLISH, H.E., WILKINSON, B.W., and EASTMAN, H.C. (1972). Canada in a wider economic community. Toronto: University of Toronto Press.

Feldman, L., and Associates (1971). A survey of alternative urban policies. Research monograph no. 6. Ottawa: Central Mortgage and Housing Corporation.

FIRESTONE, O.J. (ed., 1974). Regional economic development. Ottawa: University of Ottawa Press.

First Report of the National Population Inquiry (1975). Population and Australia (vol. 1). Canberra: Australian Government Publishing Service.

FITZPATRICK, B. (1949). The British Empire in Australia: An economic history, 1834-1939. Melbourne: Melbourne University Press.

FRIEDLY, P.H. (1975). National policy responses to urban growth. London: Saxon House.

FRIEDMANN, J. (1972). "The spatial organization of power in the development of urban systems." Comparative Urban Research, 1:5-42.

--- (1973). Urbanization, planning and national development. Beverly Hills, Calif.: Sage.

GERTLER, L.O. (ed., 1968). Planning the Canadian environment. Montreal: Harvest House.

--- (1974). "The research strategy of the Ministry of State for Urban Affairs." Unpublished paper, Ottawa.

GILMOUR, J.M. (1972). The spatial evolution of manufacturing southern Ontario, 1851-1891. Department of Geography research paper no. 10. Toronto: University of Toronto Press.

GODDARD, J. (1974). "The national system of cities as a framework for urban and regional planning." In M. Sant (ed.), Regional policy and planning for Europe. London: Saxon House.

GOLDBERG, M.A., and SEELIG, M.Y. (1975). "Canadian cities: The right deed for the wrong reason." Planning: The A.S.P.O. Magazine, 41(3):8-13.

GREEN, A.G. (1971). Regional aspects of Canada's economic growth. Toronto: University of Toronto Press.

HARTWICK, J.M., and CROWLEY, R.M. (1973). "Urban economic growth: The Canadian case." Working paper A.73.5. Ottawa: Ministry of State for Urban Affairs.

HELLYER, P. et al. (1969). Report of the Federal Government Task Force on Housing and Urban Development. Ottawa: Queen's Printer.

HENDERSON, A.T., and HOLLAND, R.P. (1975). "Internal migration in Australia: Implications for urban policy." Paper presented to the ANZAAS Conference, Canberra.

HIGGINS, B. (1972). "Growth pole policy in Canada." In M. Hansen (ed.), Growth centres in regional economic development. New York: Free Press.

Information Canada (1975). Immigration policy perspectives (vol. 1). A report of the Canadian Immigration and Population Study. Ottawa: Manpower and Immigration.

INNIS, H. (1957). Essays in Canadian economic history. Toronto: University of Toronto Press.

JACKSON, J.N. (1974). "The relevance to Canada of British policies for controlling urban growth." Paper presented to a conference on the Management of Land for Urban Development, Toronto, April 5-6.

JEANS, D.N. (1965). "Town planning in N.S.W., 1829-1842." Australian Planning Institute Journal, 3:191-195.

JEFFREY, D., and WEBB, D.J. (1972). "Economic fluctuations in the Australian regional system." Australian Geographic Studies, 10:141-160.

JOHNSTON, R.J. (1966). "Commercial leadership in Australia." Australian Geographer, 10:49-52.

KAPLAN, H. (1972). "Controlling urban growth." Discussion paper B.72.20. Ottawa: Ministry of State for Urban Affairs.

KARMEL, P., and BRUNT, M. (1962). The structure of the Australian economy. Melbourne: Cheshire.

KERR, D. (1968). "Metropolitan dominance in Canada." Pp. 551-555 in J. Warkentin (ed.), Canada: A geographical interpretation. Toronto: Methuen.

KING, L.J. (1966). "Cross-sectional analyses of Canadian urban dimensions, 1951 and 1961." Canadian Geographer, 10(4):205-224.

LANG, R. (1972). "Oh Canada, A national urban policy." Plan Canada, 12(1):15-32.

LINGE, G.J.R. (1968). "Secondary industry in Australia." In G.H. Dury and M.I. Logan (eds.), Studies in Australian geography. Melbourne: Heinemann.

LITHWICK, N.H. (1970). Urban Canada: Problems and prospects. Report prepared for the Minister responsible for the Central Mortgage and Housing Corporation. Ottawa: Central Mortgage and Housing Corporation.

--- (1972a). "An economic interpretation of the urban crisis." Journal of Canadian Studies, 7(3):36-49.

--- (1972b). "Political innovation: A case study." Plan Canada, 12(1):45-56.

--- (1972c). "Urban policy-making: Shortcomings in political technology." Canadian Public Administration, 15(4):571-584.

LOGAN, M.I. (1966). "Capital city manufacturing in Australia." Economic Geography, 42:139-151.

--- (1968). "Capital city development in Australia." In G.H. Dury and M.I. Logan (eds.), Studies in Australian Geography. Melbourne: Heinemann.

LOGAN, M.I., MAHER, C.A., McKAY, J., and HUMPHREYS, J.S. (1975). Urban and regional Australia: Analysis and policy issues. Melbourne: Sorrett.

LOGAN, M.I., and WILMOTH, D. (1974). "Australian initiatives in urban and regional development." Paper presented at Conference on National Settlement Systems and Strategies, Vienna: IIASA. Also pp. 119-178 in H. Swain (ed.), National urban strategies, East and West: Proceedings of the Conference on National Settlement Systems and Strategies, Vienna. Laxenburg, Austria: IIASA, 1975.

MacNEILL, J.W. (1972). "Environmental management constitutional study prepared for the Government of Canada." Ottawa: Information Canada.

McCARTY, J.W. (1973). "A general approach to Australian economic development." Unpublished paper. Melbourne: Monash University Faculty of Economics and Politics.

MEEKISON, J.P. (ed., 1971). Canadian federalism: Myth or reality (2nd ed.). Toronto: Methuen.

MILES, S. (1972). "Developing a Canadian urban policy: Lessons from abroad." Plan Canada, 12(1):88-106.

MILES, S., COHEN, S., and De KONING, G. (1973). Developing a Canadian policy. Toronto: Intermet.

NADER, G. (1975). The cities of Canada. Toronto: Macmillan.

NEUTZE, G.M. (1974). "The case for new cities in Australia." Urban Studies, 11:259-275.

POWELL, A.J. (1974). "National urban strategy." Royal Australian Planning Institute Journal, 12(1):8-10.

PRED, A.R. (1966). The spatial dynamics of U.S. urban-industrial growth, 1800-1914. Interpretive and theoretical essays. Cambridge, Mass.: Massachusetts Institute of Technology Press.

--- (1973). "The growth and ddevelopment of systems of cities in advanced economies." In A.R. Pred and G. Törnqvist (eds.), Systems of cities and information flows. Lund, Sweden: Gleerup.

PRESSMAN, N. (1974). Planning new communities in Canada. Waterloo: University of Waterloo.

RAY, D.M. (1971). Dimensions of Canadian regionalism. Ottawa: Information Canada.

--- (1972). "The allometry of urban and regional growth." Discussion paper B.72.10. Ottawa: Ministry of State for Urban Affairs.

RAY, D.M., and VILLENEUVE, P.Y. (1974). "Population growth and distribution in Canada: Problems, processes and policies." Paper delivered to a Conference on the Management of Land for Urban Development, Toronto, April 8.

RAYNAULD, A., MARTIN, F., and HIGGINS, B. (1970). Les orientations du développement économique régional dans la province du Québec. Ottawa: Ministére de l'Expansion Economique Régional.

RICHARDSON, H.W. (1973). "Optimality in city size, system of cities and urban policy: A sceptic's view." In G. Cameron and L. Wingo (eds.), Cities, regions and public policy. Edinburgh: Oliver and Boyd.

ROBINSON, A.J. (1974). Economics and New Towns: A comparative study of the U.S., U.K. and Australia. New York: Praeger.

ROBINSON, K.W. (1962). "Processes and patterns of urbanisation in Australia and New Zealand." New Zealand Geographer, 18:32-49.

ROSE, A.J. (1966). "Dissent from downunder: Metropolitan primacy as a normal state." Pacific Viewpoint, 7:1-27.

ROTSTEIN, A., and LAXER, G. (eds., 1972). Independence: The Canadian challenge. Toronto: Committee for an Independent Canada.

Science Council of Canada (1971). Cities for tomorrow: Some applications of science and technology to urban development. Ottawa: Queen's Printer.

--- (1974). Population growth and urban problems: Perceptions 1. Study on Population and Technology. Ottawa: Author.

--- (1975). Population distribution and technology. Preliminary report: Study on population and technology. Ottawa: Author.

SCOTT, P. (1964). "The hierarchy of central places in Tasmania." Australian Geographer, 9:134-147.

SIMMONS, J.W. (1970). "Inter-provincial interaction patterns." Canadian Geographer, 14(4):372-376.

--- (1972). "Interaction among the cities in Ontario and Quebec." Pp. 198-219 in L.S. Bourne and R.D. MacKinnon (eds.), Urban systems development in central Canada: Selected papers. Toronto: University of Toronto Press.

--- (1974). "The growth of the Canadian urban system." Research paper 65. Toronto: University of Toronto, Centre for Urban and Community Studies.

--- (1975). "Canada: Choices in a national settlement strategy." Research paper 70. Toronto: University of Toronto, Centre for Urban and Community Studies.

--- (1976). The Canadian urban system. Toronto: University of Toronto Press.

SIMMONS, J.W., and BOURNE, L.S. (1974). "Defining the future urban system." In L.S. Bourne et al. (eds.), Urban futures for central Canada: Perspectives on forecasting urban growth and form. Toronto: University of Toronto Press.

SIMMONS, J.W., and SIMMONS, R. (1974). Urban Canada. Toronto: Copp Clark.

SMAILES, P.J. (1969). "Some aspects of the South Australian urban system." Australian Geographer, 11:29-51.

Statistics Canada (1974). Population projections: Canada and the provinces. Ottawa: Author.

STILWELL, F.J.B. (1974). Australian urban and regional development. Sydney: Australia and New Zealand Book Co.

STONE, L.O. (1967). Urban development in Canada. 1961 Census monograph. Ottawa: Dominion Bureau of Statistics.

STOREY, K., and BULKER, F. (1975). "Changing patterns of spatial integration: Transaction flow analysis of Canadian air passenger traffic, 1960-1973." Paper presented at the annual meeting of the Canadian Association of Geographers, Vancouver.

SWAIN, H. (1972). "Research for the urban future." Working paper B-72-13. Ottawa: Ministry of State for Urban Affairs.

--- (ed., 1975). National urban strategies. East and West. Proceedings of the Conference on National Settlement Systems and Strategies, Vienna. Laxenburg, Austria: IIASA.

SWAIN, H., and MacKINNON, R. (eds., 1975). Issues in the management of urban systems. Laxenburg, Austria: IIASA.

SWAN, N.M. (1972). "Differences in the response of the demand for labour to changes in output among Canadian regions." Canadian Journal of Economics, 5:373-386.

Systems Research Group (1970). Urban Canada 2000. Population projections. Toronto: Author.

THOMPSON, W. (1972). "The national urban system as an object of public policy." Urban Studies, 9:99-116.

VERNON, J. (1965). Report of the Committee of Economic Enquiry. Canberra: Australian Government Publishing Service.

WARKENTIN, J. (ed., 1968). Canada: A geographical interpretation. Toronto: Methuen.

WATKINS, M. (1963). "A staple theory of economic growth." Canadian Journal of Economics and Political Science, 24:141-158.

WHEBELL, C.F.J. (1969). "Corridors: A theory of urban systems." Annals, Association of American Geographers, 59:1-26.

WHEELWRIGHT, E.L. (1957). Ownership and control of Australian companies. Sydney: Law Book.

WILLIAMS, M. (1965). "A note on the influence of Adelaide on rural shopping habits." Australian Geographer, 9:312-314.

--- (1968). "Two studies in the historical geography of South Australia." In G.H. Dury and M.I. Logan (eds.), Studies in Australian geography. Melbourne: Heinemann.

WONNACOTT, R.J., and WONNACOTT, P. (1967). Free trade between the United States and Canada. Cambridge: Harvard University Press.

YEATES, M. (1975). Main Street: Windsor to Quebec City. Toronto: Macmillan.

6

Constrained Urbanization: White South Africa and Black Africa Compared

T.J.D. FAIR
R.J. DAVIES

□ URBANIZATION IS DESCRIBED as "one of the most compelling societal forces operating in history" and city growth "as an irrepressible accompaniment of modern development experience" in underdeveloped countries (Friedmann, 1973:65, 25). To some, the role that cities have played throughout history is the role of major generators and diffusers of change (e.g., Soja, 1968:49), and they are seen as "beach-heads" or "centres of modernization which act as catalysts for economic growth" (McGee, 1971:13). Others see urbanization as no more than "a symptom of processes operating at a societal level" (McGee, 1971:31) and even perceive overconcentration in cities as "destructive, dehumanising and corrupting" (as reported in Berry, 1973:106). But whatever interpretations are attempted and however the societal benefits or costs of urbanization are perceived, "on one issue there is little debate. The cities of the Third World are growing rapidly" (McGee, 1971:14), and the trend is "irreversible in the immediate and intermediate future" (Ford Foundation, 1972:1).

Thus, while many students of Western urbanization, reinforced in their views by current trends in Third World countries, regard the process as a largely unconstrained and inevitable one, many governments around the world view with alarm the urban tide as it engulfs more of mankind each year. The latter seek to constrain urbanization by strategies of varying kind and degree. Laws of 1931 and 1939 in Italy were introduced in order to half the growth of cities by "forbidding rural workers to leave the land for cities or towns . . . unless already assured of steady employment" (Fried, 1967:509). Friedmann (1973:244) has drawn attention to "direct controls over migration . . . attempted in several socialist countries." Berry (1973:99) refers to the fear of urban "gigantism" on the part of public officials in Third World countries who try "to gain control over the urbanization process and to restrict population growth in the primate cities." Moreover, land-use planning in West European countries has been concerned with indirectly containing metropolitan growth by strategies designed to decentralize the urban system.

The objective of this chapter is to examine processes of urbanization and urban containment: first, in black or tropical Africa, where governments at least in the past 20 years have attempted, generally unsuccessfully, to employ various strategies to regulate and slow down the rate of cityward movement of rural peoples, and, second, in white South Africa, where the flow of blacks to cities is being contained, largely for political reasons, by legislative measures associated with strong authority-dependency relationships existing in that country.

BLACK AFRICA

In the decade of the 1970s "the world balance of city dwellers will for the first time in modern history swing from Europe and North America to the other continents" (Ford Foundation, 1972:2). Between 1970 and 1980, the increase in the urban population of the developing countries will be twice that in the developed countries. Africa, mainly black Africa, has registered an annual rate of increase of the urban population of 5.4% as against 3.2% for the world (United Nations Economic Commission for Africa, 1972:28). Moreover, with the coming of political independence in the 1950s and 1960s, with the remarkable decrease in urban death rates and the increase in birth rates as medical conditions have improved, and with undiminished economic dependence on Western countries, black Africa is currently experiencing rates of growth in its large cities that are outstripping the growth rates of colonial times. In the 1960s, Lusaka experienced a growth rate of 11-12% per year, Kinshasa 11.8%, Nairobi 9.6%, and metropolitan Lagos 11.5% (compared with only 3% in the 1930s). Clarke (1972b:451) indicates that a majority of the countries of tropical Africa exhibit marked urban primacy with two-city indices of 2 to 7 and that this condition is increasing, although it is not yet of the order of the "hypercephalism" of Latin American and some Southeast Asian cities.

While any assessment of the causes of growth of African cities depends upon such variables as fertility, mortality, migration, and population structure, "there can be little doubt that [migration] has been the major factor in the early stage of urban expansion" and "perhaps has been the most significant demographic event in Africa so far this century" (Clarke, 1972a:69). The percentage of urban inhabitants born outside the towns varies from 10% to 90%, while 50% and more of the annual growth of a number of tropical Africa's largest cities is attributable to in-migration (Clarke, 1972a:68-69). Moreover, in Africa, where generally the proportion of urban to rural inhabitants is comparatively low, "even a relatively small migration of rural population can have a dramatic effect on the rate of growth of urban population" (International Bank, 1972:11).

Whether Africa's cities are orthogenetic or heterogenetic (Redfield and Singer, 1954), generative or parasitic (Hoselitz, 1955), overurbanized (McGee, 1971:115), or in the process of being "ruralized" (Mangin, 1970:xiii) is of less consequence to the theme of this chapter than the reality of their "alarming" rates of growth and the "great social and economic concern" that all African governments feel about this trend (United Nations, 1972:28)—all the more so since "the impact of urbanization will press most heavily upon those societies which at present are most deficient in the economic, technological and managerial resources required to maintain and improve a complex urban environment" (Ford Foundation, 1972:2). The evidence of this deficiency is clear. Green and Milone (1971:14-15) write of metropolitan Lagos that

> chaotic traffic conditions have become endemic; demands on the water supply system have begun to outstrip its maximum capacity; power cuts have become chronic as industrial and domestic requirements have both escalated; factories have been compelled to bore their own wells and to set up stand-by electricity plants; public transport has been inundated; port facilities stretched to their limits, the congestion of housing and land uses has visibly worsened and living conditions have degenerated over extensive areas within and beyond the city's limits, in spite of slum clearance schemes; and city government has threatened to seize up amidst charges of corruption, mismanagement and financial incompetence. Moreover, although employment has multiplied in industry, commerce and public administration, there is no doubt that thousands of in-migrants have been unable to find work, and the potential for civil disturbances has increased.

The magnitude of the problem—i.e., the intensity of the squeeze between the limitation of overall resources and the high cost of urban services (International Bank, 1972:18)—cannot be overestimated. In the countries of Africa, Asia, and Latin America, if only 80% of urban households are provided with low-cost housing, it will necessitate an investment of about 5% per year of their aggregate national income (United Nations, 1971:112).

These conditions and these views elicit, on the part of the public officials and ruling elites in Third World countries, rightly or wrongly, a real fear of a revolutionary, ill-housed, poor, and unemployed urban mob (Nelson, 1970:413). Strategies have accordingly been devised and implemented in certain countries to slow down, contain, or halt the flow of migrants to cities. Such strategies are expressions of three main types of policy that governments have adopted in the hope that the problem may be eased: namely, rural development, urban dispersal, and direct curbs on rural-urban migration.

RURALIZATION

With the vast majority of the population of black Africa still concerned with farming, forestry, and fishing, it is logical to promote policies of rural and agricultural development. "Unless there are improved incomes in this sector," states Hance (1972:658), "it is difficult to see how there can be a healthy urban growth." Moreover, in parts of Africa high rates of out-migration of people have had a depressing effect upon rural population growth rates and upon the rural economy (Hanna and Hanna, 1971:41).

Attempts to alleviate and rehabilitate rural areas in tropical Africa include agricultural and pastoral programs, resettlement schemes, and villagization projects as well as educational and publicity campaigns aimed at deterring young people from moving to cities. While critics accept the need for such policies in rural areas, they regard their efficacy in slowing down the flow of migrants to urban centers as a fallacy. Hirsch (1969) shows that higher agricultural incomes in rural areas, far from stabilizing the population, encourage a shift to nonagricultural activities and so to towns. Moreover, evidence is strong that it is mainly the educated, not the uneducated, who migrate to urban areas in Africa (Caldwell, 1969:62; Lea, 1974:676). Attempts to force people back to the land or to keep them there could today be perceived as a revival of colonial policies. For example, the East African Royal Commission (1955:200-201) reported that "for many years Africans were regarded as temporary inhabitants of the towns" and that "the theory of indirect rule . . . led to a concentration on the development of rural tribal societies rather than the training of an educated urban elite, and also to the view that the town was not a suitable habitat for a permanent African society." By contrast, the overwhelming evidence is that enterprising rural people migrate to cities primarily in order to enjoy the greater opportunities of employment, education, and upward mobility that are offered there (Ford Foundation, 1972:7). This is the "revolution of values" described by Hanna and Hanna (1971:32) as the basic spontaneous cause of urban in-migration in contemporary black Africa.

URBAN DISPERSAL

The development of small and medium-sized towns of between 10,000 and 100,000 inhabitants has been advocated as a means to reduce primacy in African urban systems, to curb metropolitan growth, and to diffuse modernization to rural areas (Miner, 1960:77; Lewis, 1967:16; Clarke, 1972b:451; Harrison Church, 1972:663). Hunter (1967:66) states that "the future urbanization of Africa, if it is not to be atrociously expensive, may well have to follow much more closely the pattern of Ibadan or Kumasi than of Nairobi or of Lagos. The urbanization of Calcutta should be the constant nightmare of Africa; the growth of market towns should be its dream."

However, while the policy of promoting growth in smaller centers may be desirable, it is appreciated that it is extremely difficult to turn off migration to major centers (United Nations, 1968:85) or to create costly new towns which have little chance of offsetting growth in major centers (Ford Foundation, 1972:9). Moreover, as Berry (1973:99) points out, contrary to Western notions,

> There is no incentive for growth to decentralise. Modern enterprise remains concentrated in the major cities. Modernising influences reach the migrants but in the hinterlands traditional ways of life remain in the small towns and villages. Increasing primacy is, in turn, a sign that economic growth is taking place and affecting more people.

The International Bank for Reconstruction and Development (1972:4) argues that the outcome of strategies emphasizing rural against urban development and smaller towns against large cities is inconclusive and that a significant reduction in the urban growth rate or the problems of the larger cities in the next 20 years is unlikely. In fact, "rural development . . . may even accentuate migration to the towns, and from the smaller towns to the larger."

DIRECT URBAN CONTAINMENT

The inadequacy of curbing the growth of major cities in Africa by indirect means has encouraged many governments to employ more direct means. With an African urban labor force increasing by 3.75% per year and employment opportunities by only 2.5% (Hanna and Hanna, 1971:76), attempts have been made to hold the rate of in-migration to cities to that at which employment needs are generated. Hance (1972:656-657), drawing on various sources, lists a number of attempts made by black African governments to contain the population growth of their cities:

> In the early 1950s, Congo (Brazzaville) offered free transport, limited exemption from taxes, and loans of seed and equipment as inducements for urban idlers and vagabonds to return to their villages, and in 1954 the government started a paysannat in the Niari Valley for unemployed urban youth. In January 1959 authorities in Bangui more or less forcibly dispatched to a 2,000 hectare area . . . an initial contingent of 100 unemployed youths recruited in the capital city, and in 1964 a decree directed that all unsalaried citizens living on their wits in urban centers be rounded up and put to work rehabilitating old plantations. It was stated that this was not a violation of international work conventions but an economic necessity. Niger enacted a law in 1962 stipulating that all unemployed youth must either perform some public service or return to their villages; in 1963 the police in Niamey were reported as having rounded up all such youths and shipped them back to the country, and in June 1964 the government ordered all unemployed men to take up agricultural work on pain of being prosecuted as vagabonds. In 1964 Kenyatta announced that his government was taking immediate steps to return to the land all able-bodied people living in towns but without jobs.

> In the same year, Tanzania attempted to restrict travel to Dar es Salaam and ordered the police to pick up unemployed people and issue them warrants to return home. In January 1966 the Interior Minister of Congo (Kinshasa) [now Zaire] ordered unemployed persons in the cities to return home by the end of the month; a similar order made in March 1968 also called for the deportation of all foreigners without sufficient means of support. In 1969 Ghana announced that it planned to develop controls over internal migration to prevent the excessive movement of persons to the cities. . . . Other measures . . . include raising urban taxes with the aim of discouraging residence in the cities, and, in a few cases, levelling of bidonvilles or shanty towns and restrictions on construction of new dwellings.

Other than legislation enacted to repatriate aliens to their countries of origin, Hance comments that none of these direct measures has been successful. The deliberate withholding of urban services, too, has had no effect upon the growth of squatter settlements whose inhabitants still perceive the urban life as superior to that in the country (International Bank, 1962:10). The view of many is that all these measures are radical, negative, authoritarian, immoral, unpopular, expensive, and ineffective and are a form of neocolonialism. Latin American countries, faced with the massive growth of primate cities over the past 25 years, have never applied policies restricting the residence of national citizens in urban areas. According to Hauser (1961:294), few authorities in the countries of Latin America would consider such measures either feasible or desirable.

Consequently, the Ford Foundation (1972:2) sees policies of ruralization, urban dispersal, and urban containment as forms of "escapism." The irreversibility of cityward migration, instead, must be accepted as hard reality, and policies must be adopted that will convert it inot a positive development factor instead of an impediment (United Nations, 1971:3). Friedmann (1968:366) suggests that "it may be easier to guide development along its normal course than to reverse its general direction" for "one cannot regard the city as a Pandora's box of social evils." This theme runs through much of the findings of modern research in the Third World. Berry (1973:82-83) reports that

> rapid migration has not produced the alienation, anomie, psychological maladjustments and other symptoms of disorganisation held in the Wirthian model to be hallmarks of rapid urbanisation. . . . Far from a "detribalising" process, much of the rich associational life of African cities is based upon common interests, mutual aid, and the need for fellowship of people in the towns who are members of the same tribe or ethnic group, speak the same language, or have come from the same region.

Moreover, Hanna and Hanna (1971:102) indicate that "squatting offers immediate relief from the burden of rent and the threat of eviction, and a long-run prospect of a modicum of comfort and respectability" since some "shanty towns evolve over ten or fifteen years into acceptable working-class neighbourhoods." Nelson's (1970:394) findings show that "contrary to widespread speculation by both foreign observers and elites in the countries concerned, neither new migrants nor the urban poor are likely to play a direct destabilising role." Finally, McGee (1971) shows that the remarkably absorptive quality of the tertiary sector in Third World cities (his theory of involution) offsets the findings of those who view problems of urban unemployment and overemployment in terms of "Western-centered preconceptions" of the urban economy.

It is against this general background of a largely unconstrained urbanization in independent black Africa, and the generally ineffective attempts to constrain it, even partially, that the more deliberate, politically motivated policies of the white South African government with regard to the cityward flow of blacks and their absorption within the urban fabric of South African society are now examined.

WHITE SOUTH AFRICA

POLICY

The policy of apartheid or separate development in South Africa, as practiced by the present government of the Afrikaner National Party, aims, on the one hand, to prevent blacks (collectively, Africans, Asians, and Coloreds of mixed descent) from gaining access to positions of power and authority in white areas and, on the other, to transform the African peripheral regions, the former Native Reserves, now termed Bantu Homelands, into politically viable national systems (Schmidt, 1973). The proposed distribution of land between blacks and whites is shown in Figure 1. No provision is made for homelands for Coloreds or Asians. The policy in its present form was initiated by the present white government in 1948, but its roots lie much deeper—in the physical separation of black and white in Natal in 1846, when the British government established the Native Reserves, and in the ideology of the Constitution of 1856 of the South African Republic (Transvaal), which held that there shall be no equality of black and white in Church and State.

The policy underlies the politics of white survival in southern Africa and, "at the national level, [it] provides a means of transferring the political power of the African majority from the metropolis to areas where it poses less of an immediate threat" (Smith, 1973:99). The policy aims explicitly at controlling the flow of Africans to the cities of white South Africa and at discouraging the integration of all black groups into the fabric of white urban society. As in black Africa, strategies to control the flow of rural people to cities are both indirect and direct, but in South Africa the motivation behind them and the effectiveness of the methods and their impact upon the way of life of Africans are very different.

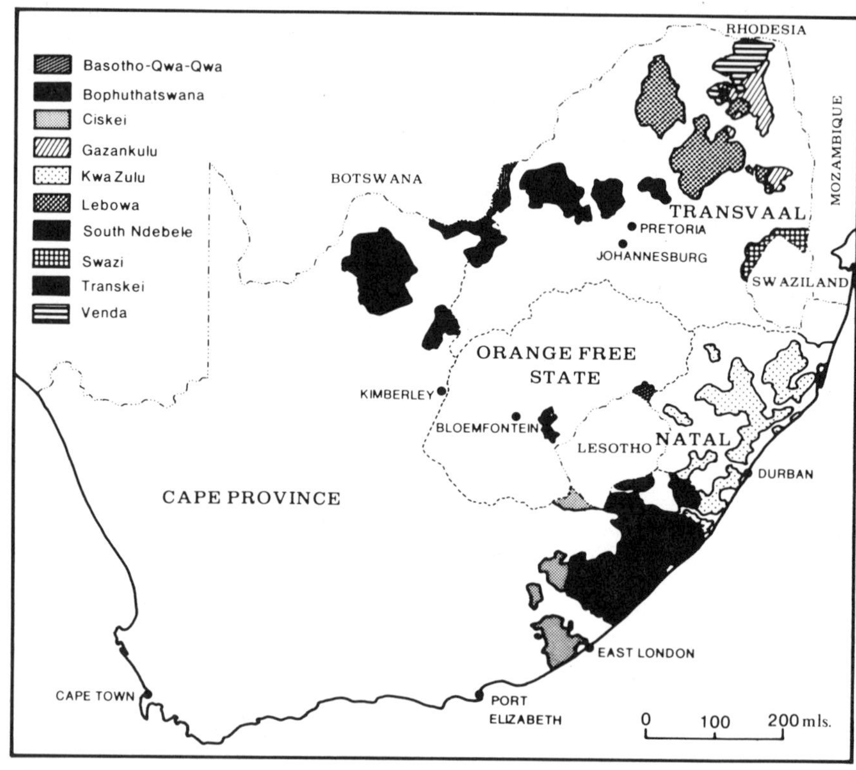

Figure 1: SOUTH AFRICA: BLACK HOMELANDS

DECENTRALIZATION

In 1970, of a total African population of 15,058,000 in South Africa, 46% were resident in the Bantu Homelands, but the overwhelming number of African wage earners were employed in the white areas and particularly in the major metropolitan and urban centers. The government's indirect strategy of urban containment is associated with that part of its policy which envisages not only ultimate political independence for the Homelands but also their greater socioeconomic development. The latter, it is considered, will help not only to contain the bulk of South Africa's African population within the Homelands as permanent citizens but also, by implication, to slow down the flow of Africans to the white cities. This has meant considerable investment in the advancement of agriculture and in industrial growth centers both within and on the borders of the Homelands, along with the provision of housing and other infrastructural facilities.

However, according to the *Financial Gazette* (Johannesburg), despite an investment of $900 million between 1960 and 1974, only 115,000 industrial jobs were created in these peripheral areas, 92,000 of them for Africans. With the demand in the Homelands running at 60,000 jobs per year, efforts to decentralize or to divert industry to peripheral black areas has had no material impact upon the continuing growth of African employment in metropolitan areas. Many of the towns developed in the Homelands are instead becoming residential areas for the families of migrants working in white cities–not economically based urban centers in their own right.

The reasons for the failure to disperse African employment in new peripheral centers on any substantial scale are fundamentally similar to those in black Africa. Polarization is still the dominant spatial process in South Africa in its current phase of industrialization (Fair, 1965). The flow of African labor from the periphery to the core areas is more powerful than the flow of capital and industrial investment in the opposite direction. Moreover, in order to counter these centripetal forces, the incentives provided by government to industry are inadequate to overcome the physical and economic distances between Homeland and market, and the inferior economic and social environments of poor and unattractive peripheral locations. The government has implicitly abandoned its attempts to contain the growth of the black population of the country's major cities mainly by such means and, instead, has focused its efforts more firmly upon long-practiced methods of direct urban containment, the extent and supporting legislation of which has no parallel in any country in black Africa.

DIRECT URBAN CONTAINMENT

The Urbanization Process. South Africa is a rapidly industrializing and urbanizing country. In the period 1960 to 1973 its gross domestic product (at constant 1963 prices) doubled to $14,300 million. The contribution of the primary sector in this time declined from 24% to 18% of the total and that of the secondary sector rose from 26% to 33%.

The total population in 1970 was 21,400,000, of which 10,230,000 or 48% lived in urban areas. Of this latter number only 32% were whites, and the remainder comprised Africans (49%), Asians and Coloreds (19%). Similarly, whites accounted for only 27% of the country's nonagricultural and therefore mainly urban employment, and blacks accounted for 73%. By contrast, patterns of education, occupation, and earnings reflected quite different levels of urbanization. In 1970, the majority of black groups accounted for only 36% of white-collar jobs; whites accounted for 65%. Only 21% of those with an education equivalent to two years in an American high school and beyond were black; 79% were white (Fair and Schmidt, 1974:163). In the major sectors of the economy, whites earned two to three times more than Coloreds and Asians, and over five times more than Africans (Smith, 1973:93). These contrasts emphasize the respective roles of whites and blacks in the urbanization process in South Africa. This process displays two main forms (Friedmann, 1973:65):

Urbanization$_1$: The geographic concentration of population and nonagricultural activities in urban environments of varying size and form.

Urbanization$_2$: The geographic diffusion of urban values, behavior, organizations, and institutions.

The former includes the cityward migration of people and their physical presence in towns; the latter includes urbanism as a way of life.

Constraining Urbanization$_1$. As the migration of rural Africans to cities gathered pace, especially from the 1920s on, the problems that beset white urban inhabitants centered essentially on how to segregate the Africans residentially, how to restrict their geographic and social mobility in towns, how to reduce the burden of a large low-paid African population upon the white taxpayer, how to ensure sanitary living conditions in African residential areas, and how to avoid an urban influx of Africans on a scale that might threaten the security of whites (Reader, 1961:10). While no constraints on the mobility of whites has ever been applied, the notion of regulating the flow of blacks was explicitly stated as early as 1922 in the report of the Transvaal Local Government Commission (1922, para. 42):

> The native should only be allowed to enter the urban areas, which are essentially the white man's creation, when he is willing to enter and to minister to the needs of the white man, and should depart therefrom when he ceases so to minister.

That this fundamental view on the part of the ruling white elites has not changed is seen in an official pronouncement made in 1972 which states:

> The Africans in the White area, whether they were born here [i.e., in the white area] or whether they were allowed to come here [from African homelands or foreign territories] under our [i.e., white] control laws are here for the labour they are being allowed to perform. . . . Every African person in South Africa, wherever he may find himself, is a member of his specific nation [but not of the white nation] . . . and . . . fundamental citizenship rights may only be enjoyed by an African person within his own ethnic context, attached to his own homeland. [Van der Merwe, 1972:82]

The policy of governments over the years, therefore, has been to ensure an adequate supply of African labor to the cities of white South Africa but to control the flow of that labor only to meet employment demand and not to allow, as far as possible, the growth and the establishment of a permanent African urban population.

As for the minority black groups, there are legally no constraints upon the geographic mobility of Coloreds, who are domiciled mainly in the Western Cape Province. There are some constraints upon the movement of Indians between the four provinces of South Africa, but none upon their movement within each province, excepting the Orange Free State, from which Indians are barred from permanent residence. (Most Indians live in Natal.) The control of the flow of blacks to cities applies, therefore, overwhelmingly to the majority African group, which forms the bulk of the South African labor force.

The chief measure employed to regulate the flow of black labor to cities is termed "influx control," and it is enforced through a considerable body of supporting legislation and administrative machinery (Davenport, 1969; Horrell, 1971). In terms of the Natives (Urban Areas) Consolidation Act of 1945, and as subsequently amended, only those Africans who have resided in a town continuously since birth or have been employed continuously for 10 years by the same employer or for 15 years by different employers are accorded some degree of permanency. The term "temporarily permanent" has been applied to them (Wilson, 1972:35). However, they may forfeit this status if deemed "idle or undesirable," are considered a threat to peace and order, or are declared unfit for employment. Those who do not have this status are generally migrants. They must first register at labor bureaus in the Homelands, receive permits to enter urban employment, and sign on for one-year contracts with a specific employer. They may not bring their families to town with them, and they must return home for at least one month before being allowed to return to the same or other place of work for another one-year stint.

It is estimated that migrant workers comprise one-half of the African males working in Johannesburg, the country's major urban center (Wilson, 1972:40). Moreover, according to the Johannesburg *Star,* of the 137,000 Africans who made use of government aid centers in the larger cities in 1973, only 1% were assisted to find employment while 67% were repatriated to their respective Homelands.

Since 1970, in the large Witwatersrand region, the government will not generally allow the expansion of existing industries or the location of new ones which employ more than two Africans to one white unless the industry is defined as "locality bound" (South African Department of Industries, 1970:7). Thus, as far as possible, black labor-intensive industries are encouraged to seek rural or Homeland locations, though the limited success of the decentralization program indicates some relaxation of this condition.

A major consequence of these measures is that African population growth rates in those South African cities which are distant from Homelands are becoming related more closely to, and even falling below, those of employment. In Johannesburg, for instance, the African population growth rate fell from 5% per year between 1946 and 1951 to 2% between 1960 and 1970 (Urban and Regional Research Unit, 1973:23), while the African employment growth rate between 1960 and 1970 was 3.4% per year. In Cape Town, African labor is permitted only when Colored labor is unable to meet the demand. In that city the growth rate of the African population fell from 11.2% per year between 1946 and 1951 to 2.6% between 1951 and 1960, but rose, owing to employment demand, to 3.7% between 1960 and 1970. In Durban, East London, and Pretoria, circumstances are different in that portions of the Kwa

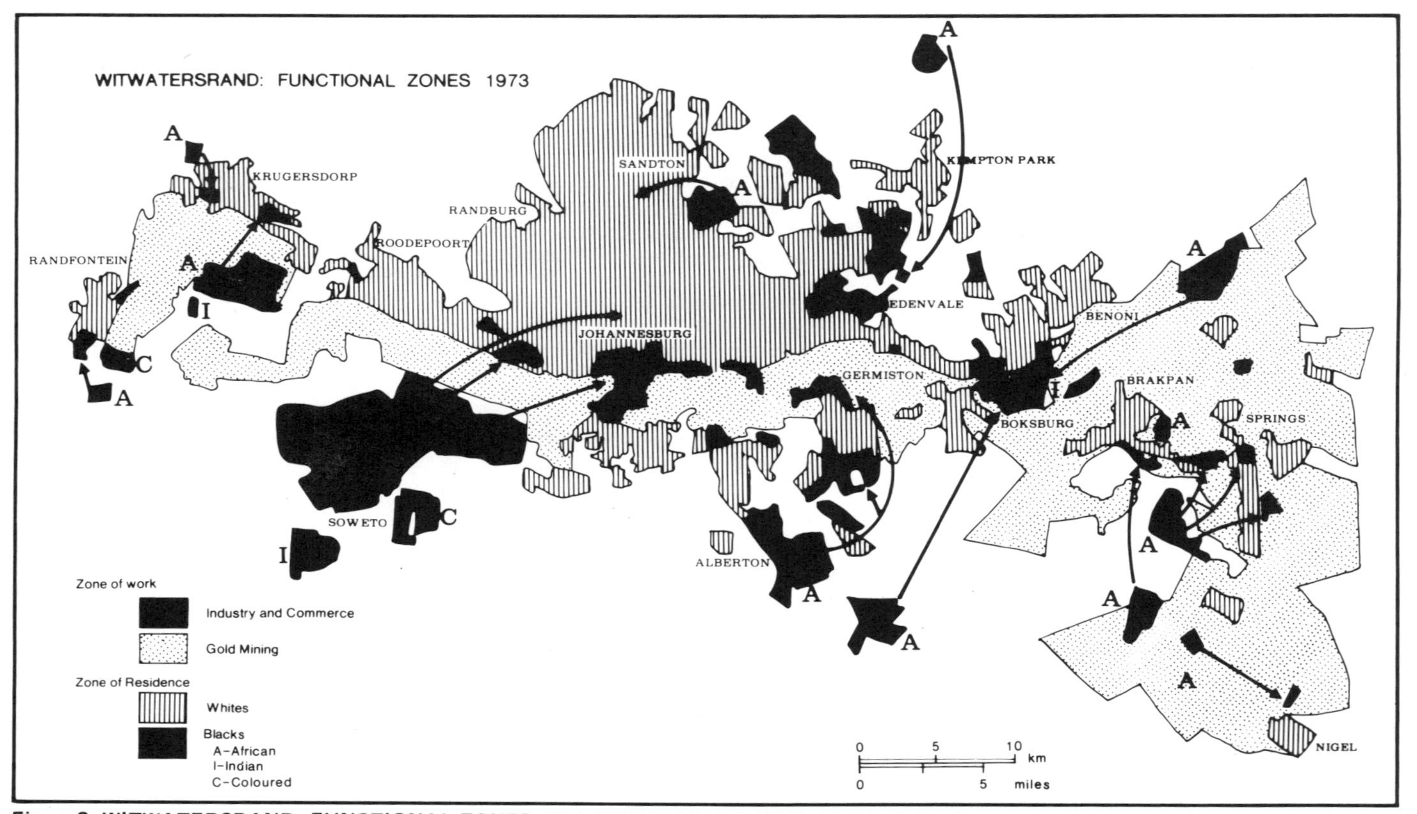

Figure 2: WITWATERSRAND: FUNCTIONAL ZONES AND GROUP AREAS 1973. ARROWS INDICATE THE MAIN MOVEMENTS OF AFRICANS BETWEEN PLACES OF RESIDENCE AND PLACES OF WORK.

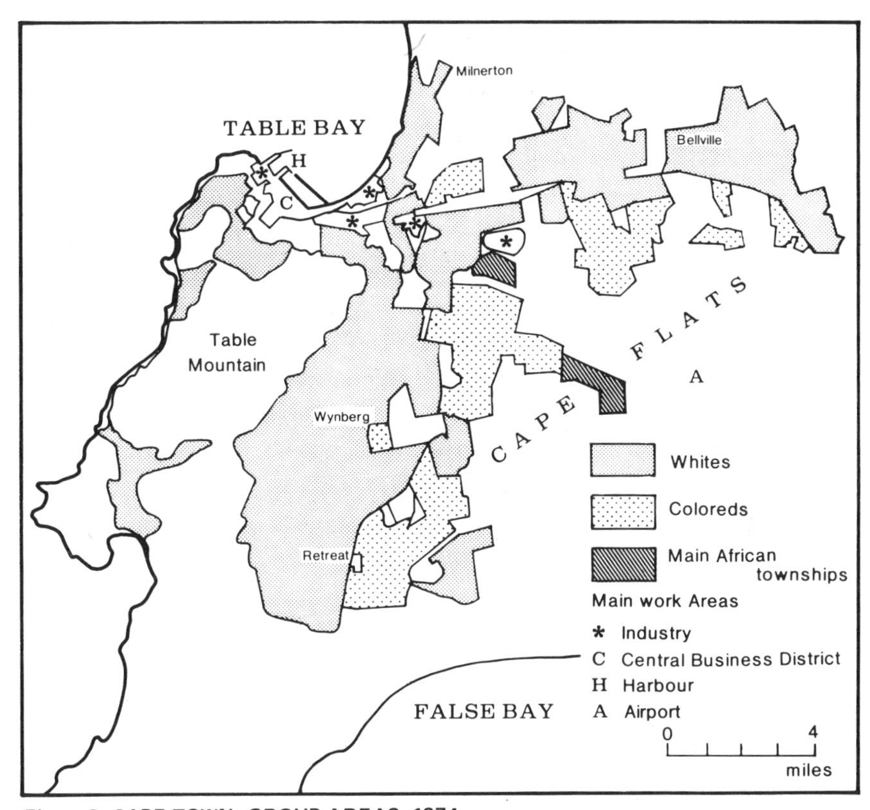

Figure 3: CAPE TOWN: GROUP AREAS, 1974

Zulu, Ciskei, and Bophutatswana Homelands, respectively, border immediately upon those cities and so provide permanent residence for many of the workers (and their families) who commute daily to their places of work in those cities.

Constraining Urbanization$_2$. "The condition of urbanization," stated Mayer (1961:5), "refers not just to the length but to the quality of the life that is lived in town" and to being "urban cultured." Urban life for blacks in white South Africa in this sense is restrictive and incomplete. The measures outlined go far beyond an attempt merely to contain the flow of Africans to white cities. Implicitly and explicitly they are also aimed, along with a mass of associated laws and regulations, at reducing the perception of opportunity and of permanence on the part of Africans in urban areas and at inhibiting urbanization as a modernizing process (Fair and Schmidt, 1974:159).

The Group Areas Act of 1950 segregates the residential areas of whites from those of Coloreds and Asians, and property transactions between these groups are forbidden. Africans are segregated in urban townships and former rights to ownership of homes and land or to leases of long duration were swept away with the Bantu Laws Amendment Act of 1964. African urban dwellers henceforth

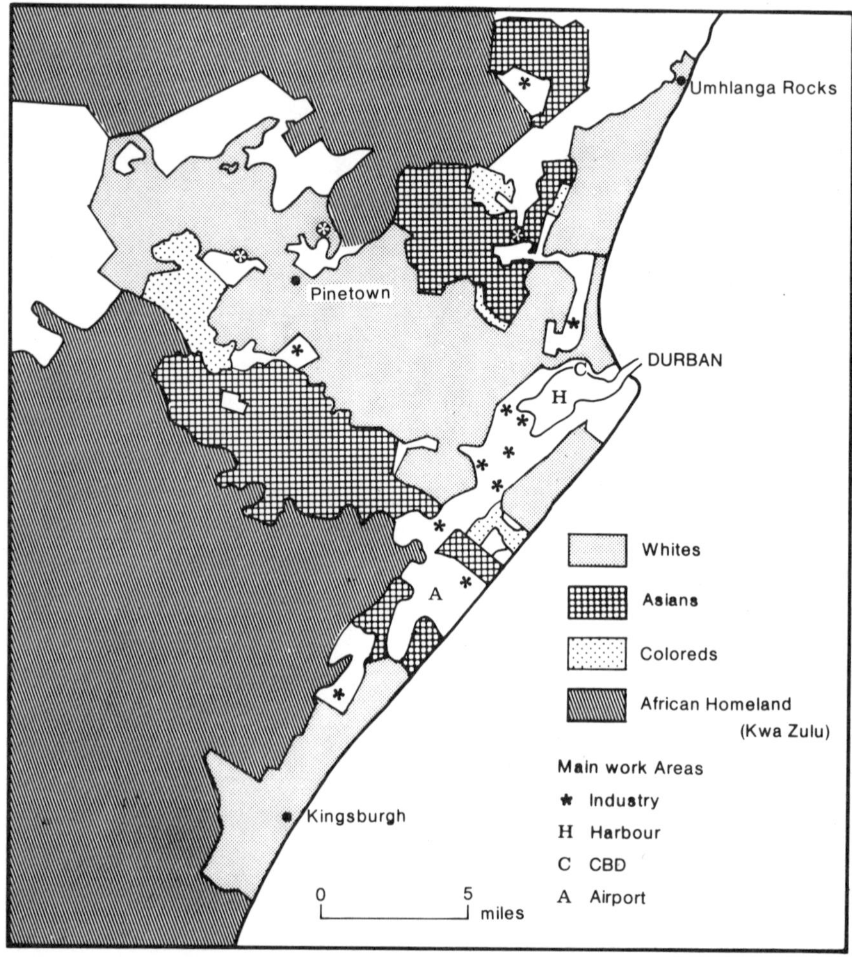

Figure 4: DURBAN: GROUP AREAS, 1974

were regarded as "temporary sojourners" whose permanent homes lay in the Homelands to which they could at any time be compelled to return if their residence in town was disqualified by their being unemployed or deemed superfluous or undesirable. Residential impermanence is being fostered by the government's efforts to increasingly migratize "African labor so that the ideal labor unit is the single, male contract laborer" who oscillates back and forth between Homeland and city (Randall, 1973:36). Along with this policy goes a discouragement of the further construction of family houses in urban areas and the implementation, instead, of a substantial housing program in the Homelands both to accommodate the families of migrants and to entice others from the city to seek new homes there.

"Civilized labor," "job reservation," and "closed-shop" policies reserve the more skilled urban jobs for whites, while universities and most new high schools for Africans are established, not in the cities, but in the rural Homelands. African trade unions are not recognized, and the activities of African traders, entrepreneurs, and professional people, few as they are, are severely circumscribed in white urban areas although encouraged in the Homelands. All nonwhite groups are barred from sharing fully with whites in civic amenities of all kinds. In addition, life in the townships displays many of the "symptoms of social deprivation" (Smith, 1973:93). Duncan (1970) indicates that two-thirds of African families in Johannesburg live in poverty, suffer from inadequate educational, health, and recreational facilities, are subjected to a high rate of violent crime and long journeys to work, and experience high rates of alcoholism, low ownership of household appliances, together with social pressures, anxieties, and instabilities generated in part by the disruption and breakup of family life that compliance with official regulations promotes.

These measures and conditions are also aimed at placing great physical and social distance between whites and blacks in general. Social intercourse is drastically reduced other than on a largely employer-employee basis, and those normal processes that entitle people to be called urbanized "in some profounder sense than the mere physical change of abode" (Mayer, 1961:4) are severely inhibited.

Present government policy clearly "signifies a reluctance to accept the necessary concomitants of the growth of an industrial economy and an attempt to arrest, and even reverse, normal processes of urbanization accompanying economic development" (Van der Horst, 1960:35). Thus, although the oscillations associated with rural-urban migration normally tend to diminish as economic development proceeds and more people settle permanently in town, in South Africa "despite almost a century of spectacular economic growth" (Wilson, 1972:152), the amplitude of migrant oscillations on the part of Africans is undiminished. Political forces are clearly maintaining this circulatory system. "So long as South Africa continues to pursue the twin goals of Separate Development and economic growth, so long will the system of migratory labor remain a central feature of the economy" (Wilson, 1972:165). This implies an ambivalence in the policy-making process which obviously encourages economic and urban development on the part of the white entrepreneur and denies the full fruits of that development to black working groups.

The Double Constraint. Arising out of this ambivalence, policy-making and planning styles as they affect whites and blacks are remarkably dissimilar. The urban planning process is doubly constrained. On the one hand, there are the constraints imposed by the governing elites of the white National Party government upon the white entrepreneur in a white-dominated market economy, and, on the other hand, there are the constraints imposed by a white ruling elite generally upon the geographic and social mobility and the economic well-being of a subordinate nonwhite, mainly African, group. Berry (1973:178-179) has listed four main policy-making and planning styles. In any context there will be a combination of styles, but "predominant value systems so determine the preferred policy-making style that significantly different processes assume key roles in determining the future in different societies." In South Africa these differences are intrasocietal as well.

Especially since the mineral discoveries of the late 1800s "the ownership of the means of production has remained largely in private hands, and the economic generator has enjoyed a relative autonomy which is unlikely to be seriously breached in the foreseeable future" (Urban and Regional Research Unit, 1974:8). Consequently, an exploitive opportunity-seeking style, assisted by government, has dominated the approach of private enterprise to economic development in South Africa. In turn, urban development as it affects whites has been largely subject to no more than the usual health, amenity, and town planning controls. In fact, within the white social system, urban planning by government has not been allocative trend-modifying with its future-oriented emphasis, but rather ameliorative problem-solving in approach, displaying a natural tendency to do nothing until undesirable dysfunctions are seen to demand corrective action. Over the past 40 years, urban planning has generally been viewed "rather as a negative regulator of physical development than as a positive generator of growth and change in human society" (Urban and Regional Research Unit, 1974:51). In consequence, urban planning as it affects whites has not been a highly constraining process. Rather has it been adaptive and past-oriented with the implied goal of preserving the "mainstream values of the past" (Berry, 1973:178).

Insofar as the urban white entrepreneur is an employer of African labor, however, his economic activities impinge upon the very different and highly constrained world of the urban black and the normative goal-oriented planning style which marks the regulation of the latter's life in the cities of white South Africa (Fair, 1976). Such a style, as Berry points out, first sets future ideological goals, in this case separate development; policies are then formulated to achieve these goals, and plans are implemented as direct instruments of public will. A high degree of "closure of means and ends" has been achieved (Berry, 1973:179). Moreover, the degree to which government is committed to "its particular value propositions" is seen in the continuing strength of the "supporting political [white] consensus" (Friedmann, 1973:36) despite nearly 30 years of worldwide opposition to its goals. Some explanation for these political commitments are discussed in the section which follows.

Meanwhile, the spatial manifestations of the differing planning styles adopted toward white and nonwhite are a reflection of the differing worlds in which each lives. Politically, economically, and socially, South African cities are perceived as white men's creations and domains. Municipal councils are white-controlled, and nonwhites have no vote in municipal affairs. White-owned shops and offices dominate central business districts, while the Group Areas Act has either eliminated or severely reduced and contained black business sectors. The rate of land consumption for white residential areas and their low gross densities and generally high quality display all the characteristics of a virtually unconstrained suburbanization typical of North American cities, as their inhabitants go in search of space and mobility. By contrast, for blacks the choice of residential location and housing is severely limited, even nonexistent. African urban townships are subeconomic housing estates; they exhibit gross densities four times that found in white residential areas, and these densities are rising as the government decrees that growing numbers must be accommodated within existing township boundaries. Whereas whites display largely unconstrained and complex journey-to-work patterns made possible mainly by private transport, Africans in particular move in less complicated patterns between designated places of residence and designated places of work, and mainly by public transport. That "the rich and relatively resourceful can reap great benefits while the poor and necessarily immobile have only restricted opportunities" (Harvey, 1973:64) is representative of most capitalist urban societies. In South Africa the contrasts in policy-making and planning styles ordain that it shall be so.

TOWARD EXPLANATION

Explanations of the constraints which operate upon the functional and spatial relations of population groups in South African urban society are difficult to articulate simply and objectively. Like constraints in other societies, they arise from complex value systems which govern the norms of the society and control the manner in which it is functionally and spatially organized. Dahrendorf (1959) recognized two basic forms of social organization which Harvey (1973) has defined as integrative and coercive respectively. In Dahrendorf's words, an integrative society is one in which "social order results from a general agreement of values, a *consensus omnimum* which outweighs all possible or actual differences of opinion and interest." In contradistinction, a coercive society is one in which "coherence and order in society are founded on force and constraint, on the domination of some and the subjection of others."

To a substantial degree, the form of social organization and the nature of social constraints within societies arise from processes of social stratification and from the level of social mobility that they permit. It is within the context of the two theories of social organization and of social stratification in particular that the constraints of the type described earlier in this paper should be seen. It is profitable also to view the constraints within a comparative framework and against the backdrop of stratification in advanced urban industrial societies elsewhere.

Interrace or ethnic group relations have for centuries and in many societies been a cause of social stress to which processes of social behavior and spatial organization have responded in varying ways from place to place and from time to time. As a phenomenon, social constraints which arise from stress are not unique to contemporary urban society, nor indeed are they necessarily confined to urban society, though propinquity brought about by concentration in urban places is likely to intensify their impact.

Problems associated with intergroup relations may arise from the process of assimilation and adjustment of racially, ethnically, or culturally distinct immigrant groups within a larger society. They may also arise from the dominance of an immigrant settler group over a technologically backward and racially distinct indigenous society. Whatever their source, however, they are brought about by the operation of interrelated normative processes which include the development of class divisions and cultural distance, the growth of sets of stereotyped attitudes toward people of a different racial, ethnic, or cultural group and the institution of a plethora of discriminatory practices which arise from such attitudes and values. Discriminatory practices may be either unstructured and informal, stemming from customary social behavior, or structured and formal, expressed in legislative or administrative sanctions and regulations. They are expressed in concrete terms in the functional behavior patterns of population groups toward one another, in varying degrees of social distance, in social and economic inequities, in patterns of spatial segregation and interaction within cities and territories and in social policy and development strategies.

The stratification of society into class or classlike groups is a fundamental divisive process within which relationships between race or ethnic groups may be caught up, and it is a process particularly relevant to the discussion which follows.

Class is perhaps most satisfactorily defined in terms of the distribution of authority (or legitimate power) in the political, economic, and social associations of society (Dahrendorf, 1959). By definition, the distribution of authority is dichotomous and can produce only two classes in any given association—a dominant or ruling class and a dependent or subject class. Competing dominant and subject groups may arise in as many cases as there are authority associations (either political or economic) in the society. A class-structured society is normally characterized also by relatively rigidly defined class boundaries determined by ascription and concomitantly by low levels of economic and social mobility. By definition, a class-structured society is coercive in organization. Typical examples exist in many preindustrial and early capitalist societies.

While there may be powerful arguments to the contrary (see, for example, Harvey, 1973), processes that underlie social stratification in advanced urban industrial societies, including divisions between races or ethnic groups, are, in general, decreasingly to be sought within a strictly defined class context. Though class divisions may continue to exist in varying degrees, there is evidence to show that their boundaries have become increasingly blurred, their identity less evident, and their impact upon the functional organization of society substantially reduced. The blurring of class boundaries and the creation of greater freedom of participation in the organization of society in a political, economic, and social sense have been brought about, Dahrendorf (1959) suggests, by several interrelated factors. These include the diffusion of authority roles within society, the spread of mass education, expanding employment opportunities, an increase in economic and social mobility (particularly intragenerational mobility), and the resolution of conflict through the growth of institutions within which conflict situations may be contained.

Ascriptive processes that maintained class-structured societies in the past have given way to a process of competitive achievement within a market exchange economy and within an open class system. Thus, although society remains stratified, its strata are not determined so much by class as by socioeconomic status determined by the degree of success that an individual or group achieves in gaining access to scarce resources in the competitive market exchange economy. Status is measured by such variables as income, occupation, and educational attainment.

A free economy furthermore is characterized by the functional and spatial mobility of capital, resources, and labor and by the freedom to contract regardless of race, economic sector, or area (Kuper, 1965). Ideally, also, the process of competitive achievement in an open class system takes no cognizance of racial or ethnic origins, and recruitment in the economy is determined only by ability and training. Members of different races or ethnic groups theoretically have equal opportunities to participate in political and economic associations commensurate with their abilities to compete and be economically and spatially mobile. Policy-governing relationships and contact between groups thus becomes increasingly based upon a race-cooperation ideology which maintains that common interests and harmony develop from contact and shared experience.

In practice, inequities arise, not only from vestigial elements of class consciousness which may remain, but also from other causes. These include, on the one hand, the ineffective and discriminatory operation of the wealth redistribution process. The poor are as a result placed at a disadvantage in their ability to compete successfully. On the other hand, discriminatory practices may arise, for example, from deep-seated racial or color prejudice and generate undesirable levels of social (as in housing), economic (as in job opportunities and income differentials), and spatial (as in racial segregation) distance between groups. When discrimination is identified or perceived to be identified with particular races or ethnic groups and when discrimination results in absolute rather than relative deprivation, the free operation of the society becomes severely constrained. The case of black ghetto formation and its associated economic, racial, and spatial inequities in American, British, European, and other cities of advanced urban industrial society is a particularly severe example.

However undesirable and reprehensible constraints of that kind may be, and they are not uncommon, it is important in the context of the discussion to note that, in advanced urban industrial societies, they are brought about mainly by unstructured and informal social forces and, more often than not, result in relative rather than absolute deprivation. These forces in practice oppose stated social policy goals, and they are among the basic reasons for the development of social and economic development programs designed to redress functional and spatial inequities in society. Thus, rather than defining them as indicators of a coercive form of organization, one should perhaps see them as evidence of manipulations of the freedom of society or as elements of social disorganization against which society may act.

Processes of social stratification in South Africa urban society are similar in nature to those which have been outlined in the model of advanced urban industrial society. Fundamental differences exist, however, in the emphasis which is placed upon the counterpoised but interrelated stratification processes which arise from racial differences, class structuring, and competitive achievement and socioeconomic status in a market exchange economy. It is from these differences that explanations of the social constraints which operate within the society may be drawn.

Deep social divisions in South African society stem from a colonial past in which a white, technologically advanced, immigrant minority group was juxtaposed with a traditional and culturally distinct black, technologically backward, indigenous majority group. The identification of ideas of dominance and dependency with race, coupled with the development of paternalistic attitudes toward blacks, created the means for the evolution of a social system in which racial differences were emphasized and progressively projected into a closed race-class-like structure. The race-class structure, with severely restricted levels of mobility, has enabled the dominant or ruling white class in varying degrees to decide the dynamics of association in political, economic, and social spheres of society functionally and spatially.

At the same time, socioeconomic stratification arising from the process of competitive achievement in an expanding urban, industrial market-exchange economy has developed strongly and coexists with the race-class structure. It is a process, however, which has been secondary to, and constrained by, the evolution of the race-class system. To a very high degree, socioeconomic strata are ascribed to race-class groups to establish an integrated occupational hierarchy, and occupational and income mobility is constrained to upward and downward movements within race-class groups rather than between race-class groups.

Controls over the levels of political and economic participation and mobility of blacks arose at an early date, and, although they may have varied in intensity from territory to territory in the period before Union in 1910, they were characteristic of both the former British colonies of the Cape and Natal and the Afrikaner republics of the Transvaal and Orange Free State. Since Union and particularly since 1948, when the present governing Nationalist Party came to power, and notwithstanding the growth of an urban industrial economy, the gradual spread of education, and the progressive acculturation of urban blacks, the race-class structure has tended to intensify.

Minority black groups, particularly the Coloreds, who derive from miscegenation between whites and nonwhites, and the Asians, principally Indians, who were added to the population as immigrant workers between 1860 and 1911, have been equally caught up in the process.

In the past, the Colored population traditionally identified itself with the whites and enjoyed a measure of political and economic freedom greater than that of any other black group. They tended, Kuper (1965) suggests, "to accept their position as appendages to the Whites and to guard their privileges by maintaining an extreme social distance in relation to Africans."

The relations of the Indian population, on the other hand, were governed by a complex interplay of social and economic constraints comprised, in part, of voluntary processes of separation and, in part, of imposed regulations which affected trade and residence. To a degree, the Indian population in particular —unlike the Africans—came into open economic competition with the whites in trade, property ownership, and, to a lesser degree, employment.

Prior to 1948, therefore, the race-class system was governed by structured controls which represented periodic and pragmatic responses to changing political, economic, and social conditions. The controls lacked an overall conceptual framework of functional and spatial organization, were incomplete, were differentially applied to the various racial groups, and were permissive to a degree in practice. Since that date, the racialization of society on highly structured lines has become the dominant and pervasive force affecting practically every facet of political, economic, and social life in the country.

The intensification of stratification based on racial criteria has been a concomitant, and perhaps partly a result, of increasing levels of urbanization of all population groups and the growth of an urban industrial economy with a widening spectrum of opportunity. Increasing propinquity has heightened race consciousness and prejudice and the perceived need on the part of the white ruling class to preserve and strengthen the race-class structure as a means of maintaining racial privilege. In contrast to the situation in other countries with advanced urban industrial societies, policy-governing relationships between the races in South Africa have become increasingly framed within a conflict ideology that postulates that physical, social, and cultural differences between peoples of different races are incompatible and that harmonious relations between racial groups can be secured only by reducing points of contact to a minimum. It follows that the racialized society can be maintained only through coercive organization.

CONCLUSION

In South Africa, racial legislation provides for the comprehensive, systematic, and highly structured control of a wide range of human relationships in order to underpin the race-class-like system. A policy of urban containment of the black group is merely one facet of this system of control aimed at maintaining existing authority-dependency relationships. Clearly, the urbanization of blacks is constrained, since the industrial-urbanization process not only promotes but also demands a mobility and a flexibility in the socioeconomic system of modern societies.

However, in South Africa the policy of separate development, generally, and the policy of black urban containment in white areas, specifically, are inevitably tied to the politics of white survival. For an issue as visceral as this, a majority consensus among whites is not unexpected. Justification for these policies, however, is also sought in the wider realm of black-white relationships in South Africa today. Thus, the supporting philosophy, it is claimed, is one rather of racial differentiation than of discrimination. The African, for example, in the city is Homeland-based, not urban-based. For every denial of political rights to Africans in white areas, compensatory rights are granted to them in the Homelands. Influx control, it is claimed, is necessary not only to protect rural Africans against the effects of too rapid and uncontrolled change (Brandel0Syrier, 1971:xvi) but also to protect African urban workers against loss of their jobs through competition. Influx control prevents large-scale urban unemployment and the growth of uncontrolled shantytowns. It permits the government to house the African population in urban areas at standards greatly superior to those generally found in the cities of black Africa. Moreover, a small African elite has arisen in the urban townships of South African cities because of opportunities which the policy of separate development has afforded and which the Africans would otherwise not have had.

REFERENCES

BERRY, B.J.L. (1973). The human consequences of urbanisation. London: Macmillan.

BRANDEL-SYRIER, M. (1971). Reeftown elite. London: Routledge and Kegan Paul.

CALDWELL, J.C. (1969). African rural-urban migration: The movement to Ghana's towns. Canberra: Australian National University Press.

CLARKE, J.I. (1972a). "Demographic growth of cities in black Africa and Madagascar: The mechanism of growth and general characteristics of demographic structures." Pp. 65-76 in La croissance urbaine ẹn Afrique noire et à Madagascar. Paris: Centre National de la Recherche Scientifique.

--- (1972b). "Urban primacy in tropical Africa." Pp. 447-453 in La croissance urbaine en Afrique noire et à Madagascar. Paris: Centre National de la Recherche Scientifique.

DAHRENDORF, R. (1959). Class conflict in industrial society. Stanford, Calif.: Stanford University Press.

DAVENPORT, R. (1969). "African townsmen? South African natives (urban areas) legislation through the years." African Affairs, 68(271):95-109.

DUNCAN, S. (1970). "The plight of the urban African." Topical talks no. 23. Johannesburg: South African Institute of Race Relations.

East Africa Royal Commission (1955). Report. London: Her Majesty's Stationery Office.

FAIR, T.J.D. (1965). "The core-periphery concept and population growth in South Africa, 1911-1960." South African Geographical Journal, 47:59-71.

--- (1976). "Metropolitan planning and the policy of separate development with special reference to the Witwatersrand region." In D.M. Smith (ed.), Separation in South Africa (Occasional paper no. 7). London: Queen Mary College, University of London.

FAIR, T.J.D., and SCHMIDT, C.F. (1974). "Contained urbanization: A case study." South African Geographical Journal, 56(2):155-166.

Ford Foundation (1972). "International urbanization survey: Findings and recommendations." Unpublished paper. New York: Author.

FRIED, R.C. (1967). "Urbanization and Italian politics." Journal of Politics, 29:505-534.

FRIEDMANN, J. (1968). "The strategy of deliberate urbanization." Journal of the American Institute of Planners, 34(6):364-373.

--- (1973). Urbanization, planning, and national development. Beverly Hills, Calif.: Sage.

GREEN, L.P., and MILONE, V.M. (1971). "International urbanization survey: Urbanization in Nigeria, a planning commentary." Unpublished paper. New York: Author.

HANCE, W.J. (1972). "Controlling city size in Africa." Pp. 653-658 in La croissance urbaine en Afrique noire et à Madagascar. Paris: Centre National de la Recherche Scientifique.

HANNA, W.J., and HANNA, J.L. (1971). Urban dynamics in black Africa. New York: Aldine-Atherton.

HARRISON CHURCH, R.J. (1972). "The case for industrial and general development of the smaller towns of West Africa." Pp. 659-665 in La croissance urbaine en Afrique noire et à Madagascar. Paris: Centre National de la Recherche Scientifique.

HARVEY, D. (1973). Social justice and the city. London: Arnold.

HAUSER, P.M. (ed., 1961). Urbanization in Latin America. Paris: United Nations Educational, Scientific, and Cultural Organization.

HIRSH, G.P. (1969). "Planning for development of rural areas–A criticism of present practice." Pp. 55-58 in Town and Country Planning Summer School Report, United Kingdom.

HORRELL, M. (1971). Legislation and race relations. Johannesburg: South African Institute of Race Relations.

HOSELITZ, B.F. (1955). "Generative and parasitic cities." Economic Development and Cultural Change, 3:278-294.

HUNTER, G. (1967). The best of both worlds? London: Oxford University Press.

International Bank for Reconstruction and Development (1972). Urbanization. Washington, D.C.: Author.

KUPER, L. (1965). An African bourgeoisie: Race, class and politics in South Africa. New Haven, Conn.: Yale University Press.

LEA, J.P. (1974). "Population mobility in rural Swaziland: A research note." Journal of Modern African Studies, 12(4):673-679.

LEWIS, W.A. (1967). "Unemployment in developing countries." World To-day, 23:13-22.

MANGIN, W. (ed., 1970). Peasants in cities: Readings in the anthropology of urbanization. Boston: Houghton Mifflin.

MAYER, P. (1961). Townsmen or tribesmen. Cape Town: Oxford University Press.

McGEE, T.G. (1971). The urbanization process in the Third World. London: Bell.

MINER, H. (ed., 1960). The city in modern Africa. New York: Praeger.

NELSON, J. (1970). "The urban poor: Disruption or political integration in Third World cities?" World Politics, 22(April):393-414.

RANDALL, P. (ed., 1973). South Africa's political alternatives. Johannesburg: Spro-Cas.

READER, D.H. (1961). The black man's portion. Cape Town: Oxford University Press.

REDFIELD, R., and SINGER, M. (1954). "The cultural role of cities." Economic Development and Cultural Change, 3(October):53-73.

SCHMIDT, C.F. (1973). "The South African regional system: Political independence in an interacting space economy." Unpublished Ph.D. dissertation, University of South Africa, Pretoria.

SMITH, D.M. (1973). "An introduction to welfare geography." Occasional paper no. 11. Johannesburg: Department of Geography, University of the Witwatersrand.

SOJA, E. (1968). The geography of modernization in Kenya. Syracuse, N.Y.: Syracuse University Press.

South African Department of Industries (1970). White paper on the report by the Inter-Departmental Committee on the Decentralisation of Industries. Pretoria: Government Printer.

Transvaal Local Government Commission (1922). Report. Pretoria: Government Printer.

United Nations (1968). "Urbanization: Development policies and planning." International Social Development Review, 1:79-88.

--- (1971). Improvement of slums and uncontrolled settlements. New York: Author.

United Nations, Economic Commission for Africa (1972). Survey of economic conditions in Africa, 1971 (Part 1). New York: Author.

Urban and Regional Research Unit (1973). Population and land use. Johannesburg: Witwatersrand University.

--- (1974). Implications for strategy. Johannesburg: Witwatersrand University.

Van Der HORST, S. (1960). "The economic implications of political democracy." Optima, 10(suppl.).

Van Der MERWE, P.J. (1972). "Manpower in South Africa." Finance and Trade Review, 10:73-113.

WILSON, F. (1972). Migrant labour in South Africa. Johannesburg: South African Council of Churches and Spro-Cas.

7

Urban Growth and Economic Development in Brazil in the 1960s

SPERIDIAO FAISSOL

□ IN THIS PAPER, the relationship of Brazil's urban growth and structure to the nation's economic development is explored for the decade of the 1960s. While some strong relationships are found, in other respects the findings are inconclusive, and a follow-up analysis for the entire period 1950-1975 is deemed desirable. There are two reasons:

(1) The beginning of structural economic transformations might clearly be better understood by going back to 1950. In particular, the relationship between "functional size" of urban centers, development levels, and urban primacy is likely to be clearer in 1950 than 1960.

(2) Since structural transformations are being impelled by industrialization, inclusion of data from the 1975 Economic Census might well show yet another phase of an accelerating process: the move from import substitution of consumer goods toward capital goods, the attendant sectoral economic shifts, and their impacts on specific parts of the urban system.

The focus of this paper is upon urban growth and structure. Related topics, such as tendencies toward regional concentration, are not discussed in more detail, but they underlie most of the analysis and are present even in some of the differences between segments of the system in the two main regions of the country–North and South–along the classical line of core-periphery differences. Likewise, another critical issue–whether the gap between the developed and the underdeveloped parts of Brazil was narrowed in the period–cannot be answered in this study in a positive way. The introduction of relevant data on relative and absolute increases in income would require lengthy discussion of inflation-correcting indexes and the controversial use of particular income classes for the computation of mean income. The fact is, however, that differences between the two parts of the nation are still a serious handicap to the expansion of the productive capacity of the developed part, to the extent that a very clear line of economic policy in the second half of the seventies is turning toward a correction of income distribution patterns, both vertically in the society and spatially between regions. The creation of a Social Development Council and the implementation of a system of social indicators, along with economic indicators, is sufficient evidence of the effort being made to reduce there inequalities.

GENERAL FEATURES OF THE BRAZILIAN URBAN SYSTEM AND THEIR RELATIONSHIPS TO THE ECONOMIC DEVELOPMENT PROCESS: 1960-1970

Brazil is one of the few countries in the world which has two metropolitan areas of about the same size and importance so close together. As a result, its urban system has peculiar characteristics. This statement is, however, time-specific—applicable only to the period 1940-1960. By 1970 São Paulo was already outgrowing Rio, becoming the biggest Brazilian metropolis, whereas up to 1940, it was Rio de Janeiro that was first. The Brazilian urban system, having only one national metropolis up to the thirties, changes as a consequence of São Paulo turning itself into the nucleus of Brazilian industrialization. Everything points to São Paulo being, in the eighties, the biggest Brazilian metropolis.

This implies that attempts to reverse this trend might need and call for solutions of a political rather than an economic nature, at the expense of associated high social costs. Throughout this paper we shall have the opportunity to consider the relative positions of these two cities with respect to different dimensions of the Brazilian urban system. These positions seem to indicate the irreversibility of the process, for the trend points to the gradual fading of Rio de Janeiro.

The evolution is very much related to the whole process of Brazilian development: the concentration of economic power in an important urban-industrial focus, which is established in São Paulo and spilling out for about 200 kilometers around it, reminding one of a statement made about the U.S.A.: "A great heartland nucleation of industry and the national market, the focus of large scale national serving industry, the seed bed of new industry responding to the dynamic structure of national final demand, and the center of high levels of per capita income" (Perloff and Wingo, 1963).

Some facts about Brazilian growth should be presented to provide a framework for understanding this process of urbanization, the nature of which will become clearer as we analyze a set of cities that we have assumed to be an adequate sample of the Brazilian urban system in the sixties. The first of these facts relates to urban population increase. Only 28 million in 1960, by 1970 the urban population had already surpassed 47 million people. In the same period, the Brazilian population increased from 70 to 93 million, so that urban population growth was 66%, against 30% for the total population, which shows the intensity of the urbanization process.

The second point to be made is the growth of the GNP, which was, at the beginning of the period, around 15 billion dollars and, by 1970, was already almost 40 billion dollars. This growth increased per capita income from $170 in 1960 to about $400 in 1970.

Both demographic and economic growth were accompanied by heavy concentration, following the lines of the classical economic model for developing countries in which heavy concentration is held to occur in the early periods of economic evolution, followed by a shift toward a state of equilibrium, or tendency to it after some generations.

Table 1 presents the number of Brazilian cities in the 1960-1970 period, classified according to population size. The number of cities increased by 43%, whereas the urban population increase was 66%, which shows the growing concentration of people in towns. If we look at the first three classes in the table (up to 100,000 inhabitants), we see that the population increase was around 50%, relatively similar to the national index. The smallest increase occurred in the class with less than 10,000 inhabitants. On the other hand, the 100,000-300,000 class increased by 100% (population increased by 140%), indicating the emergence of medium-sized towns, set between the large national and regional metropolises and the small hinterland towns. In 1960 there were 22 cities in this class, against 46 in 1970. The southeastern and southern regions claimed 12 and 33 of these cities, respectively. Of the latter 33 towns, 13 are located in the São Paulo subsystem. In 1960, the ratio "population of the metropolitan region/population living in cities over 100,000 inhabitants" was 0.929; in 1970, the same ratio was 0.834.

There were only two cities with populations over a million in 1960: the two national metropolises, Rio de Janeiro and São Paulo. By 1970 there were six of that size (Rio and São Paulo, plus the regional metropolises of Recife, Salvador, Belo Horizonte, and Porto Alegre). The two national metropolises experienced a population increase of 5 million (from 8 to 13 million) in the period under consideration, and it is estimated that São Paulo will have, in 1975, a population of around 10 million inhabitants.

This mushrooming of the biggest places is aggravated by the fact that one of the most general assumptions of demographers, that of low natural increase of the urban population, does not hold in Brazil. The Demographic Census data for 1970 show that in the metropolitan region of Recife (in the Northeast) the general fertility rate was 260%, whereas for São Paulo it was 180%.

The concentration of income is even greater than the urban concentration, producing a strong asymmetry in the system. This asymmetry may be detected

TABLE 1

BRAZILIAN CITIES: NUMBER AND POPULATION 1960-1970

POPULATION SIZE	CITIES				CHANGE 60/70	
	1960		1970		%	
	NUMBER	POPULATION	NUMBER	POPULATION	NUMBER	POPULATION
Less than 10,000	2,392	6,563,635	3,361	9,024,936	40,51	37,50
10,001 to 50,000	303	6,064,047	476	9,837,238	57,10	62,22
50,001 to 100,000	37	2,602,218	55	3,765,951	48,65	44,72
100,001 to 300,000	22	3,181,773	46	7,605,768	109,09	139,04
300,001 to 1,000,000	7	3,739,478	9	4,288,772	28,57	14,69
more than 1,000,000	2	6,388,212	5	12,771,015	150,00	99,92
TOTAL	2,763	28,539,363	3,952	47,293,680	43,03	65,71

SOURCE: Synopsis of the Demographic Census, Brazil, 1960 and 1970.

by comparing data on the income distribution of Brazil as a whole with two of its more asymmetric regional units, São Paulo and the Northeast. Table 2 shows data for these two regions and for Brazil as a whole in 1972, in units of minimum wages. In the Northeast almost 90% of the employed population had a montly salary below two minimum wages, while for São Paulo the corresponding figure was around 60%. If we consider the lowest level (0.25 of a minimum wage), we find that in this category were only 3% of the employed population of São Paulo, against 26% of the employed population of the Northeast. The figures show great discrepancies between these two regions, even though income levels are quite low in both cases. If urban-rural inequalities in both regions are taken into account, the former proportion drops from 26% to 15%, while in São Paulo it remains around 3%.

Another useful index of the differences between these two regions concerns the ownership of durable goods such as radio and television sets, refrigerators, and automobiles. After 1970, car ownership surpasses the other listed items, whereas the other goods presented higher totals before this period. This implies increasing diffusion of these goods hand in hand with the growing output of the factories associated with personal income levels. The high rates of TV and car ownership lead one to suppose that a "consumer society" attitude may be leading Brazilians into buying goods that have more weight in a social status value system, rather than the goods with greater utility (television sets versus refrigerators). In the Northeast, out of a total of 250,000 cars in 1972, just a little over 63,000 were bought before 1970, whereas 41,000 were bought in 1970 and 42,000 between 1971 and 1972. Some 90,000 of these cars were paid for in cash; the remainder were bought on installments. The proportion of new and used cars was relatively even. Of the 920,000 cars bought in São Paulo state, only 240,000 were paid for in cash. It is easy to see that in São Paulo less than 30% of the cars were paid for in cash, against almost 40% in the Northeast. This seems to imply that the structure of the income distribution in the Northeast is such that when a person can afford to buy a car, he can buy it in cash. The proportion of cars, relative to the year of acquisition, is quite similar in São Paulo and in the Northeast, but not the number of cars per unit of population.

These data highlight two important things:

(1) The consumer society is inducing the lower income classes into buying more consumption items than they can really afford, possibly at the expense of other more important items like food. This tendency accelerated after 1970, an accompaniment of the rapid development of the Brazilian economy in this period.

(2) This penetration of durable goods has not been associated with any improvement in the income distribution, which continues to be heavily concentrated at the regional level in São Paulo and the Northeast, as well as at the intraregional level between the urban and rural sectors in the underdeveloped areas. Rural Brazil has the greatest concentration of absolute poverty. In the urban Northeast, about 15%

TABLE 2
INDIVIDUAL MONETARY INCOME

INCOME CLASSES (MINIMUM WAGES) (a)	BRAZIL		NORTHEAST		SÃO PAULO	
	NUMBER OF PERSONS	%	NUMBER OF PERSONS	%	NUMBER OF PERSONS	%
Less than 0.25	3,214,991	12	2,114,116	26	200,418	3
0.25 to 2.00	16,592,198	63	5,206,064	65	3,849,723	58
more than 2.00	6,540,445	25	703,801	9	2,618,508	39
TOTAL	26,470,008	100	8,070,237	100	6,691,428	100

(a) The minimum wage was about U.S. $70, that is, about U.S. $800 yearly.

SOURCE: PNAD/1972: National Household Survey.

of the employed population earns only 0.25 of a minimum wage, while in the rural sector this percentage exceeds 30%.

THE BRAZILIAN URBAN SYSTEM: ITS STRUCTURE IN 1970 COMPARED TO THAT IN 1960

We now turn to the specifics of our analysis, which consists of factor analyses of a matrix of 28 variables (listed in Appendix 1) and 104 cities (listed in Appendix 2). The set of variables was selected with specific hypotheses about the dimensions of the Brazilian urban system and the process of Brazilian development in mind. In this way, the functional size, level of development, occupational structure, and accessibility in the system (reflected by the density in the urban network)–as determined by several previous studies made by the author–were assumed to be important dimensions which would emerge from the two analyses. Separate factor analyses were completed for 1960 and 1970; the results are compared in the sections that follow.

Table 3 shows some aspects of the change process in the 1960s, through a simple analysis of means and standard deviations of some of the variables used in the analyses. One important point to be observed relates to the variable "number of cars per a thousand people." The mean for 1960 is clearly much lower than the one for 1970, but the coefficient of variation is much higher, revealing the extent to which cars have penetrated in the society. This is consistent with the data we presented in the previous section, as well as with the fact that Brazil's automobile industry is a fundamental segment of the industrialization process.

Another variable that it is interesting to compare is the one relating to the labor force in industry, for it also shows the diffusion of industrial innovations. There is a significant decrease in the coefficient of variation associated with higher mean values, but the same phenomenon is not observed if one looks at the variable "added value in manufacturing" per capita, which increased substantially, but with an even greater increase in the coefficient of variation. Increasing concentration is clearly shown by the increase in this coefficient.

Table 4 shows the factor structures in the two years and indicates an essential persistence of the same factors, although some changes are to be noted, especially in the way the factors describe structures that are beginning to differentiate.

AGGREGATE FUNCTIONAL SIZE

The first differentiating dimension–aggregate functional size–allows comparisons between the two structures which are quite revealing of the nature of the change process during this period, even though in both periods this factor explains about 25% of the variance contained in the 28 variables.

In 1960 functional size was related to variables which also describe the level

TABLE 3

MEANS, STANDARD DEVIATIONS, AND COEFFICIENTS OF VARIATION 1960-1970

VARIABLES	1960			1970		
	MEAN ($\bar{x}$)	STANDARD DEVIATION (s)	COEFFICIENT OF VARIATION ($\frac{s}{\bar{x}} \times 100$)	MEAN ($\bar{x}$)	STANDARD DEVIATION (s)	COEFFICIENT OF VARIATION ($\frac{s}{\bar{x}} \times 100$)
Cars per 1,000 inhabitants	18.22	13.60	74.64	30.48	18.23	59.81
Labor force in industry/urban labor force	0.50	0.20	40.00	0.47	0.12	25.53
Labor force in commerce/urban labor force	0.32	0.13	40.63	0.21	0.05	23.81
Labor force in services/urban labor force	0.18	0.08	44.44	0.32	0.08	25.00
Added value in manufacturing industry/labor force in industry	0.29	0.16	55.17	10.91	17.37	159.21
Urban population/total population	0.74	0.18	24.32	0.84	0.13	15.48

TABLE 4

FACTOR LOADINGS: 1960-1970

EIGENVALUES	1960						1970				
	23.98	15.76	13.42	11.34	7.06		25.04	17.82	13.92	13.90	
VARIABLES	FUNCTIONAL SIZE	FUNCTIONAL STRUCTURE	DEVELOPMENT LEVEL	DENSITY OF THE URBAN NETWORK	INDUSTRIAL EFFICIENCY	COMMUNALITIES	FUNCTIONAL SIZE	DEVELOPMENT LEVEL	DENSITY OF THE URBAN NETWORK	FUNCTIONAL STRUCTURE	COMMUNALITIES
1	-0.67	-0.65	0.14	-0.23	0.07	96.02	0.88	0.17	-0.19	-0.27	92.83
2	0.94	-0.01	0.15	-0.04	-0.04	93.15	0.96	0.08	-0.11	0.11	95.94
3	0.94	0.03	0.15	-0.14	0.03	93.68	0.95	0.11	-0.18	0.16	97.06
4	0.74	-0.25	0.39	-0.35	0.15	91.86	0.79	0.47	-0.28	-0.10	93.76
5	0.92	-0.08	-0.03	-0.11	0.05	94.59	0.97	0.07	-0.13	0.01	97.59
6	0.77	-0.03	0.35	-0.01	-0.21	75.81	0.78	0.30	-0.10	0.19	74.89
7	0.34	-0.32	0.65	-0.22	0.22	73.80	0.26	0.79	-0.34	-0.23	86.79
8	0.47	-0.00	0.60	-0.08	0.07	65.97	0.33	0.71	-0.23	0.06	69.66
9	0.57	0.20	0.59	0.02	-0.07	76.28	0.37	0.52	-0.17	0.43	62.30
10	-0.09	-0.16	0.61	-0.38	0.09	71.64	-0.11	0.53	-0.51	-0.16	58.27
11	0.19	-0.21	0.61	-0.22	-0.05	68.44	0.22	0.60	-0.56	-0.15	74.94
12	0.05	-0.46	-0.04	-0.25	0.61	78.62	0.10	0.29	-0.30	-0.63	79.66
13	0.06	0.36	0.06	-0.22	-0.47	73.95	-0.14	-0.01	-0.15	-0.06	87.83
14	-0.06	-0.94	0.00	-0.25	0.09	96.71	-0.17	0.11	-0.15	-0.93	92.58
15	0.08	0.87	-0.01	0.34	-0.13	88.97	0.15	-0.18	0.32	0.75	71.93
16	0.03	0.88	0.00	0.07	-0.03	80.19	0.15	-0.04	0.01	0.92	87.40
17	-0.00	-0.06	-0.01	-0.08	0.88	78.57	0.34	0.36	0.05	-0.31	34.07
18	0.06	-0.64	0.15	0.08	0.54	74.25	0.40	0.75	-0.02	-0.32	88.95
19	0.48	-0.26	0.60	0.00	0.16	68.09	0.24	0.82	0.15	-0.21	81.07
20	0.57	-0.19	0.51	0.35	0.18	79.39	0.29	0.74	-0.02	-0.38	79.22
21	0.36	-0.11	0.27	-0.18	-0.04	83.58	0.50	0.17	-0.58	-0.08	67.63
22	-0.32	0.04	0.02	0.54	0.10	65.11	-0.30	-0.13	0.62	-0.09	50.06
23	-0.25	0.27	0.63	-0.12	-0.14	61.14	-0.35	0.75	-0.02	0.28	58.07
24	0.21	0.10	0.56	-0.14	-0.20	42.89	-0.19	0.44	-0.22	0.10	31.94
25	0.90	0.28	-0.02	-0.04	0.01	88.77	0.88	-0.01	0.00	0.31	86.95
26	0.05	-0.23	0.25	-0.88	0.08	89.32	0.07	0.06	-0.89	-0.14	83.24
27	0.07	-0.18	0.22	-0.91	0.12	92.03	0.04	0.05	-0.89	-0.12	82.66
28	-0.24	0.43	-0.03	0.59	-0.04	62.24	-0.28	-0.17	0.57	0.37	58.12

TABLE 5
FUNCTIONAL SIZE AND POPULATION: SELECTED CITIES, 1960-1970

CITIES	FUNCTIONAL SIZE 1960	FUNCTIONAL SIZE 1970	URBAN POPULATION 1960	URBAN POPULATION 1970
1. São Paulo	29.06	29.85	3,300,218	5,869,966
2. Rio de Janeiro	28.98	29.99	2,912,923	4,252,009
3. Porto Alegre	19.53	18.60	625,957	869,795
4. Belo Horizonte	17.95	17.15	663,215	1,228,295
5. Santos	15.38	13.36	263,054	343,890
6. Recife	14.34	14.85	788,569	1,046,454
7. Salvador	13.53	13.87	638,592	1,005,206
8. Curitiba	13.50	13.79	351,259	583,857
9. Campinas	11.44	10.40	184,529	336,279
10. Niterói	11.32	12.63	229,025	292,255
11. Fortaleza	10.04	10.43	470,778	828,763
12. Belém	7.57	8.07	380,667	609,699
13. Ribeirão Preto	7.34	4.88	119,429	195,680
14. Petrópolis	5.22	2.51	120,113	154,602
15. Vitória	5.10	4.49	83,900	132,132
16. Goiânia	4.55	7.69	133,462	363,304
17. Santo André	4.54	9.21	231,705	417,023
18. Juiz de Fora	4.11	3.81	128,364	220,282
19. Bauru	3.98	1.08	85,881	120,878
20. Pelotas	3.81	0.68	129,517	154,949
21. Maceió	3.33	3.46	170,134	251,622
22. Blumenau	3.25	-1.07	48,014	86,517
23. Manaus	3.02	3.99	154,040	284,118
24. Sorocaba	2.90	3.10	119,477	169,799
25. São Luís	2.65	0.28	139,075	205,248
26. Natal	2.11	3.30	155,860	257,673
27. João Pessoa	2.10	3.89	137,788	213,495
28. Londrina	2.08	2.37	77,382	163,871
29. São José do Rio Preto	2.06	1.30	67,921	109,695
30. Aracaju	1.92	0.79	112,516	178,512
31. Jundiaí	1.62	1.81	84,010	145,785
32. Ponta Grossa	1.34	-0.14	78,557	113,144
33. Rio Grande	1.29	-1.99	87,528	104,156
34. Campos	1.08	-1.39	131,974	175,889
35. Santa Maria	0.62	-1.74	84,014	124,288
36. Nova Iguaçu	0.49	6.61	257,516	724,862
37. São Caetano do Sul	0.32	3.62	114,039	150,171
38. Uberlândia	0.09	-0.34	71,717	111,640
39. Piracicaba	0.01	1.45	82,303	127,914
40. Campina Grande	-0.02	-0.05	126,274	168,045
41. Taubaté	-0.03	-2.36	65,911	100,031
42. Araraquara	-0.07	-1.59	60,591	84,582
43. Presidente Prudente	-0.15	-1.74	54,980	92,420
44. Campo Grande	-0.16	0.48	64,934	131,282
45. Volta Redonda	-0.21	1.07	83,973	120,645
46. Mogi das Cruzes	-0.27	-1.10	71,335	110,156
47. Uberaba	-0.49	-1.29	72,053	108,605
48. São Gonçalo	-0.51	4.32	195,872	430,349
49. Duque de Caxias	-0.74	4.17	176,306	404,380
50. Caxias do Sul	-0.81	-0.84	69,269	113,404

of development. This relationship reveals a stage of urban-economic development in which primacy is dominant in the urban system, together with concentration of development in the big cities.

The main difference between the structures of the functional size dimension in 1960 and 1970 is that, in the 1970 factor, size appears much more clearly because the variables related to development practically disappear. Further variables related to size of industry appear in 1970, while in 1960 they also loaded on a functional structure factor which separates industrial cities from the others, regardless of size.

The first aspect indicates that development filtered, to a certain degree, to smaller cities or to cities of intermediate size; i.e., diffusion took place down the urban hierarchy. The variable "telephones per 1,000 inhabitants" (representing purchasing power), which had a correlation of 0.47 in 1960, almost disappears in 1970; at the same time the variable "physicians per 10,000 inhabitants" changes its correlation from 0.57 to 0.37, and the "wages in commerce" and "wages in services" correlations go down from 0.48 and 0.57 to 0.24 and 0.29, respectively.

On the other hand, besides the changes in the structure of the factors, we may observe changes in the relative position of the cities in the two analyses under consideration. At the level of national or regional metropolises these changes were very small, as expected, because the diffusion process moved from the higher to the lowest levels of the urban hierarchy. But, as it also moved at the interregional level from the center to the periphery some metropolises of the Northeast, particularly Recife and Salvador, which in 1960, for example, had appeared with a score smaller than the one for Santos, in 1970 had scores slightly above it. Another pertinent aspect of this comparison concerns what happened to the cities in the periphery of the main metropolises. Nova Iguaça, for example, in the Rio de Janeiro metropolitan area, underwent quite a change in its relative position, going from the 36th place in 1960 to the 15th in 1970–a good example of metropolitan spread. Notice, besides, that it also had an almost threefold population growth (from 260,000 inhabitants to more than 700,000). Other details of the relative changes, at the level of the cities of intermediate size, will be shown later on.

Table 5 presents the scores of the selected cities, in 1960 and 1970, and their urban population in those years. There are some significant examples like São Luís and Teresina which changed their relative positions considerably. The first went down and the second went up in size, which seems to correspond to a greater dynamism during the period, which caused a faster growth of Teresina. The fact that in 1970 some variables related to development do not weigh as much in the factor composition as in 1960 makes it difficult to compare cities in areas that are unequally developed, but this is not the case here. The same occurs with João Pessoa and Campina Grande, the first ascending and the second stationary in this period. Other northeastern cities changed very much, like Natal, which went from 2.1 to 3.3 which, even if we consider the change in the composition of the factor, is a significant change. Feira de Santana is another similar example, which indicates that the growth process of the cities in the Northeast is not even (this is not surprising) and that the relative position of the cities is changing. The cities undergoing this change should be looked at according to their positions on the other factors, not only because it may explain the changes, but also because the other factors give a better description of the economic process underlying their growth.

Observing what is happening in the southern and southeastern regions, one can see that many cities are changing their relative positions, but the type of cities that are changing is different. It is not the state capitals that are getting bigger, but the majority of the centers in the metropolises' peripheries, and certain other specific areas. This implies a diffusion of metropolitan growth out to the periphery. While Duque de Caxias and Nova Iguaça had scores of −0.7 and 0.5 in 1960, they changed to 4.2 and 6.6 in 1970. The radical nature of this change becomes clear if one considers that Petrópolis went from 5.2 to 2.5 in the same period and Campos had changed from 1.0 to −1.4. The same problem occurred in São Paulo, where Ribeirão Preto and Bauru went from −1.3 and 0.3 to 3.0 and 3.6 respectively and Guarulhos went from −2.6 to 3.4. Campinas and Santos did not lose very much, as they changed from 11.4 and 15.4 to 10.4 and 13.4. This confirms, once again, the process of more accelerated expansion in the different rings of the metropolitan periphery of São Paulo than in its hinterland.

This is an example of what is happening in the metropolian regions of Rio de Janeiro and São Paulo. Meanwhile in the Northeast, Natal and João Pessoa moved from the 26th and 27th positions, respectively, to the 21st and 20th, reversing their regional positions and rising in status, in opposition to what happened to São Luís or Aracaju (also in the Northeast), or even to Ponta Grossa, in Paraná.

Another illustrative example of one of the important aspects of the Brazilian urban growth process involves cities in the second developing nucleus, the Porto Alegre area. Caxias do Sul, the area's most important industrial center presented a score of −0.81 in 1960 and −0.84 in 1970, keeping its relative position. Meanwhile Santa Maria, Pelotas, and Rio Grande, being out of the main axis of industrial expansion, fell to much lower relative positions, as can be seen in Table 5.

These numbers reveal two important aspects of the Brazilian urban growth process: (1) concentration in the metropolitan regions, with some cities on their periphery showing evidence of the beginning of metropolitan spread effects, and (2) concentration in cities which are foci of the regional development process, like Natal and João Pessoa in the Northeast, Goiânia (near Brasilia) and beyond the periphery of the São Paulo metropolis, in its industrial ring.

FUNCTIONAL STRUCTURE

This factor is bipolar, differentiating centers of industrial activity from those concentrating in commerce and services. This was the second most important factor in 1960, explaining almost 16% of the variance, but in 1970, as may be seen in Table 4, the general structure is different. The variable "labor force in industry" no longer loads on the factor, a change which indicates dispersion of industry to smaller centers, but the variable "labor force in modern industry" is added to the factor structure, indicating an association between industrial dispersion and the modernization of industry.

Table 6 shows the cities' scores on the two 1960 factors and on the one for 1970. Notice, for example, the position of Rio de Janeiro and São Paulo in the two periods, reinforcing the idea of industrial intensification in São Paulo, although, as far as the factor indicating "industrial efficiency" is concerned, in 1960 the difference is not substantial. Substantial differences appear in the metropolitan peripheries of the two big centers. Towns like São Bernardo, Santo André, and São Caetano, around São Paulo, were among the centers of higher industrial specialization in 1960 and continued to be so in 1970, while the ones in the metropolitan area of Rio de Janeiro lost their relative position.

Another important aspect that we should notice is that the metropolitan areas as a whole changed from centers with negative signs, i.e., approaching the more industrial cities in relative terms, to positive signs, consequently approaching the centers of commerce and services in 1970. Recife changes from −2.43 to 2.10 (this shows that industrial growth localized in its metropolitan periphery) and the same happens to Belo Horizonte, Porto Alegre, etc. Meanwhile, Juiz de Fora loses its relative position, Campinas remains in the same position, and Jundiaí grows considerably, the same occurring with Americana, Rio Claro, Franca, etc., indicating an expansion of São Paulo's industrial ring, as opposed to a relative concentration of Rio's industrial ring.

DEVELOPMENT LEVEL AND DENSITY OF THE URBAN NETWORK

A "development level" factor emerged as the third in importance in the 1960 analysis and moved up to second in 1970, as shown in Table 4. In 1960 the second most important factor was functional structure, but by 1970 it has slipped to fourth place. This change shows that by 1970 the development level of cities had become more important as a differentiating factor than functional structure, even though the relationship between these two factors and the industrialization process is quite obvious.

Comparison of Tables 6 and 7 clearly reveals that the factor is made up of variables which describe the city's development level through such indicators as telephones, cars, water supply, electricity, and salaries. The difference in the composition of the factor in 1970 is that the first two indicators (cars and telephones—more clearly representative of purchasing power) have higher

TABLE 6

INDUSTRIAL CITIES DEFINED BY FUNCTIONAL STRUCTURE AND INDUSTRIAL EFFICIENCY, 1960-1970

CITIES	1960		1970
	FUNCTIONAL STRUCTURE	INDUSTRIAL EFFICIENCY	FUNCTIONAL STRUCTURE
1. São Bernardo do Campo (1)	-12.87	6.99	-11.83
2. Volta Redonda	-12.84	11.25	- 9.17
3. São Paulo (1)	-11.80	4.59	- 6.42
4. São Caetano do Sul (1)	-11.80	5.53	-12.34
5. Santo André (1)	-11.43	2.92	-10.95
6. Jundiaí	- 8.75	1.84	- 7.87
7. Guarulhos (1)	- 7.95	3.55	- 7.67
8. Mogi das Cruzes (1)	- 7.52	2.52	- 5.32
9. Novo Hamburgo	- 6.73	1.20	- 6.54
10. Barra Mansa	- 6.73	7.02	- 4.14
11. São José dos Campos	- 6.52	1.56	- 6.51
12. Campinas	- 6.31	3.72	- 3.56
13. São Leopoldo	- 6.00	1.27	- 4.88
14. Limeira	- 5.88	2.66	- 5.07
15. Joinvile	- 5.72	1.97	- 7.21
16. Rio de Janeiro (2)	- 5.71	3.40	- 2.37
17. Sorocaba	- 5.59	- 0.36	- 3.61
18. Piracicaba	- 4.90	1.65	- 3.70
19. Caxias do Sul	- 4.78	1.61	- 5.63
20. São Carlos	- 4.72	1.83	- 3.85
21. Blumenau	- 4.65	- 0.12	- 4.58
22. Duque de Caxias (2)	- 4.42	3.95	- 1.95
23. Porto Alegre	- 4.38	2.44	1.51
24. Americana	- 3.98	0.21	- 7.81
25. Niterói (2)	- 3.89	1.93	2.07
26. Taubaté	- 3.64	- 0.42	- 2.60
27. Petrópolis (2)	- 3.41	- 0.30	- 1.81
28. Nova Iguaçu (2)	- 2.77	2.73	- 0.89
29. Itu	- 2.55	- 1.69	- 3.90
30. Recife	- 2.43	- 1.04	2.10
31. São Gonçalo (2)	- 2.19	1.51	- 0.23
32. Araraquara	- 2.14	0.72	- 0.22
33. Canoas	- 1.99	1.32	- 3.55
34. Juiz de Fora	- 1.88	- 1.08	1.35
35. Macapá	- 1.77	2.94	1.74
36. Curitiba	- 1.53	- 0.61	0.20
37. Rio Claro	- 1.51	- 0.40	- 2.87
38. Belo Horizonte	- 1.45	0.92	2.72
39. Franca	- 1.94	- 0.51	- 2.52
40. Barra do Piraí	- 1.89	1.83	- 2.21
41. Nova Friburgo	- 1.84	- 0.79	- 2.19
42. Guaratinguetá	- 1.14	- 1.23	0.16
43. Rio Grande	- 1.11	0.45	- 0.35
44. Bauru	- 1.00	0.34	- 1.38
45. Santos	- 0.35	1.77	- 1.11
46. Ribeirão Preto	- 0.24	- 0.91	0.55
47. Botucatu	- 0.17	0.46	1.56
48. Barretos	- 0.07	- 0.78	3.21

(1) "Municípios" within the Metropolitan Área of São Paulo
(2) "Municípios" within the Metropolitan Área of Rio de Janeiro

TABLE 7
DEVELOPMENT LEVEL AND DENSITY OF THE URBAN NETWORK RELATIVE TO THE SELECTED CITIES, 1960-1970

CITIES	DEVELOPMENT LEVEL		DENSITY OF THE URBAN NETWORK	
	1960	1970	1960	1970
1. Rio de Janeiro	13.63	19.03	- 2.62	- 6.77
2. São Paulo	12.38	17.01	- 8.54	-10.83
3. Santos	11.65	15.54	- 2.70	- 6.76
4. Porto Alegre	10.69	12.08	- 1.96	- 4.39
5. Campinas	9.65	9.80	- 5.66	- 6.17
6. Niterói	9.09	14.18	- 2.32	- 4.35
7. Belo Horizonte	8.81	7.45	- 2.83	- 4.34
8. Bauru	8.23	4.56	- 3.52	- 3.99
9. Vitória	6.94	8.51	0.46	- 1.25
10. Curitiba	6.72	10.36	- 2.15	- 4.25
11. Ribeirão Preto	6.21	5.40	- 3.12	- 5.41
12. Goiânia	5.68	1.28	1.64	0.33
13. Blumenau	4.86	1.84	- 1.01	- 1.31
14) Jundiaí	4.63	5.28	- 6.19	- 5.44
15) Florianópolis	3.92	3.39	1.25	- 0.86
16) São José do Rio Preto	3.90	5.61	- 1.79	- 4.71
17) Araçatuba	3.87	0.96	- 2.58	- 0.83
18) Lins	3.54	3.32	- 1.71	- 3.23
19) Petrópolis	3.47	1.52	- 1.82	- 1.38
20) Juiz de Fora	3.43	1.33	- 2.86	- 2.94
21) Salvador	3.24	3.55	0.29	- 2.66
22) Volta Redonda	3.15	4.21	- 3.24	- 3.98
23) Presidente Prudente	2.91	1.58	- 1.39	- 1.17
24) Araraquara	2.69	2.50	- 2.73	- 2.36
25) São Bernardo do Campo	2.24	9.19	- 4.68	- 5.21
26) Uberaba	2.15	- 2.04	- 0.66	0.76
27) Recife	2.07	2.86	- 4.99	- 5.96
28) Botucatu	1.98	3.24	- 1.55	- 1.69
29) Fortaleza	1.84	- 3.50	0.20	1.04
30) São Carlos	1.72	4.41	- 3.86	- 4.69
31) Santo André	1.57	9.16	- 5.28	- 7.47
32) Piracicaba	1.55	4.61	- 4.24	- 4.81
33) São José dos Campos	1.52	3.15	- 3.58	- 3.25
34) Limeira	1.44	1.48	- 3.82	- 3.85
35) Varginha	1.39	- 1.31	- 1.47	- 1.23
36) Taubaté	1.31	1.44	- 3.40	- 2.87
37) Ponta Grossa	1.25	- 1.78	1.12	0.24
38) São Caetano do Sul	1.16	12.96	- 5.73	- 7.63
39) Pelotas	1.06	- 0.99	2.11	3.26
40) Aracaju	1.00	- 3.37	- 1.15	- 1.07
41) João Pessoa	0.87	- 1.98	- 0.88	- 2.49
42) Londrina	0.84	0.52	0.45	0.62
43) Sorocaba	0.80	4.88	- 4.74	- 5.90
44) Jaú	0.73	- 0.50	- 1.74	- 1.72
45) Novo Hamburgo	0.70	- 1.10	0.07	- 0.85
46) Belém	0.64	- 1.91	2.20	1.73
47) Santa Maria	0.56	- 2.39	3.75	3.56
48) Uberlândia	0.55	0.76	1.61	0.86
49) Americana	0.47	1.16	- 3.14	- 4.62
50) Ourinhos	0.31	- 1.51	- 0.74	0.01
51) Marília	0.23	0.13	- 0.88	- 1.23
52) Rio Claro	0.23	2.50	- 3.58	- 3.92
53) Caxias do Sul	0.19	2.62	0.10	- 1.15
54) Franca	0.15	0.42	- 2.26	- 2.42
55) Catanduva	0.14	0.79	- 1.08	- 1.72

correlations. At the same time, the variables "wages in industry" and "added value in manufacturing industry," which appeared in the "industrial efficiency" factor in 1960, moved to the "development level" factor in 1970. This factor appears in both analyses somewhat related to the factor "density of the urban network," which also contains the variable "distance to the nearest metropolis" (São Paulo or Recife, depending on the case).

Table 7 indicates the relative positions of the selected cities on these two factors for the two periods. Notice that Rio de Janeiro has a position relatively higher than São Paulo's in the two periods on the development factor but a lower position on the "density of the urban network" factor. However, this does not apply to the cities on the metropolitan peripheries: the values for the main São Paulo nucleus are higher than those for Rio de Janeiro's. The diffusion process had already filtered more in São Paulo, not only because of São Paulo's greater dynamism, but also because Rio's metropolitan area is composed of towns of the old state of Rio de Janeiro (joined with Guanabara to form a new state in 1975) that did not receive the benefits of expansion of the central city (Rio de Janeiro). Some differences between the 1960 and 1970 values also are significant, revealing the process of concentration of wealth. For example, the values of São Paulo and Rio de Janeiro were 12.4 and 13.6 in 1960, respectively, on the development factor, while in 1970 they changed to 17.0 and 19.0, showing an equal relative position above the average in the second period. However, fundamental differences, which already existed in 1960 at the periphery of the two metropolitan areas, became more marked in 1970. Nova Iguaçu, São Gonçalo, and São João de Meriti, which had scores of –5.9, –4.3, and 6.0 in 1960, kept almost the same position in 1970 with scores of –5.7, –5.0, and –6.4; while (in São Paulo) Santo André, São Bernardo do Campo, and São Caetano do Sul changed from 1.6, 2.2, and 1.2 to 9.1, 9.2, and 12.9, evidencing a greater change in the São Paulo region than in the Rio de Janeiro region.

In the Northeast, the phenomenon of increasing regional differences as a whole is very clear, for out of all cities in the two analysis only Recife and Salvador kept the same relative positions (changing from the values 2.1 and 3.2 to 2.9 and 3.5), while all the others, including the remaining state capitals, declined in relative position.

In 1960, Belém in the North region–like Fortaleza, Aracaju, João Pessoa, Natal, Maceió, Campina Grande, etc., in the Northeast–either had positive values, i.e., above the average of the Brazilian cities, or had negative values near zero. In 1970 they were quite different relative to the other cities. None of them were above average, and their values became very low. Fortaleza's change was very marked, going from 1.8 in 1960 to –3.5 in 1970, revealing the strong effect of the intense migration to the city, unaccompanied by the provision of urban facilities or by an increase in the purchasing power of the population. The other northeastern state capitals underwent similar processes.

It has already been mentioned that the "development level" and "density of the urban network" factors complement each other, although, as in the case of the "functional structure," this is not revealed by the structure of the interrelations, but by the position of the cities on the factor. Nevertheless, we must observe that variables like water and electricity supply appear on this factor with low correlations, which to some extent explains this similarity. It is important to compare the values in this factor at the two periods, to observe the evolution of the urban network in the neighborhood of each city. Thus, while Recife changed from a value of –5.0 to –6.0, São Paulo changed from –8.5 to –10.8 (both negative, as the correlation with this factor is negative). In the Northeast, while Natal and João Pessoa changed from 0.5 and –0.9 to –1.2 and –2.5, i.e., increased the density of their urban networks in a radius of 100 and 200 kilometers, Maceió, Aracaju, Teresina, and their state capitals maintained similar relative positions, and Fortaleza, Campina Grande, and São Luís changed for the worse. In the South and Southeast, Ribeirão Preto, São José do Rio Preto, and some others improved their positions, indicating an expansion of the urban network in this area. The same happened to Sorocaba and to cities in the metropolitan area of Porto Alegre (including the metropolis), including Caxias do Sul.

This factor is of singular importance in describing the dynamics of urbanization, because it offers powerful support for the correlation of industrialization with the expansion of the urban network. Thus, combining the two factors (development level and density of urban network), one is able to characterize the core region and the periphery of Brazil quite well (Faissol, 1972). What is shown is that the core region is not uniformly developed, containing underdeveloped pockets, some of which are evident in both 1960 and 1970 and some of which were much more marked in 1970, particularly around the metropolitan peripheries. Also shown is the emergence of a secondary core around Porto Alegre, tending to coalesce with the main core in the same way that the metropolitan areas of Rio de Janeiro and São Paulo are coming to form a great São Paulo-Rio de Janeiro megalopolis.

CONCLUSION

Possibly the most important events that took place in the Brazilian urban system in the 1960-1970 period were the inversion of the relative positions of São Paulo and Rio de Janeiro and the consolidation of Porto Alegre's and Belo Horizonte's positions with respect to Recife and Salvador in the Northeast. These events are related to the Brazilian economic evolution in that a stage has been reached when gains through scale economies are increasing the level of concentration of population and income simultaneously. This stage is producing an accelerating "deviation amplifying process," which seems to be leading to a system with a single national metropolis, a system characterized by equilibrium of the rank-size type and characterized by the emergence of a subsystem of intermediate-sized cities, between the major metropolitan areas and the lower levels of service centers related to the rural economy. Leading these trends, the São Paulo urban subsystem has a much more balanced hierarchy and higher levels of development, with growth rates clearly dissociated from the size of the cities, and shows well-established filtering mechanisms. On the other hand, the lag of the northeastern system is equally evident, with great concentration in the bigger cities, with increasing intraregional differentiation, and with diffusion processes not able to filter growth down the urban hierarchy, mainly due to strong migratory flows in the direction of the regional metropolises with no compensating capacity for generating jobs in the same proportion. This lag presents most critical problems, for to be competitive, new development has to be capital-intensive, but for this reason it does not generate enough jobs; in order to support the massive migration flows, development has to be labor-intensive, but such activities are not competitive. The resulting dilemma is profound.

Other features of the development process revealed by the analysis are these:

(1) The great metropolitan agglomerations became much more diversified between 1960 and 1970. Both São Paulo and Rio de Janeiro had their industrial specialization rates reduced in this period, and even medium-sized cities like Campinas and Ribeirão Preto in São Paulo developed a more diversified structure, for they are no longer predominantly industrial centers. Even the northeastern metropolis (Recife) shows this trend.

(2) There is clear evidence of the existence of a "developed Brazil" and an "underdeveloped Brazil." This dualism is reinforced by the fact that the primacy of cities like Recife within underdeveloped Brazil is accompanied by income concentration in the regional metropolis.

These changes in the system's structure in the 1960-1970 period clearly reflect the Brazilian development process, in which São Paulo and its urban system dominate the spatial organization of the Brazilian economy. The problem posed by this situation is that, inasmuch as the amplification of inequalities as a process does not have any control mechanism or any decentralization capacity, it can turn into a dangerous bottleneck as development proceeds. This is the reason why the process is the object of careful research by the planners of Brazilian urban policies.

REFERENCES

FAISSOL, S. (1972). "A estrutura urbana Brasileira: Uma visao do processo Brasileiro de desenvolvimento economico." Revista Brasileira de Geografia, 34(3):19-123.

PERLOFF, H.S., and WINGO, L. (1963). Natural resource endowment and regional economic growth. Washington, D.C.: Resources for the Future, Inc.

APPENDIX 1

LIST OF VARIABLES USED IN THE FACTOR ANALYSES OF 104 BRAZILIAN CITIES, 1960-1970

1. Labor force in industry
2. Labor force in commerce
3. Labor force in personal services
4. Number of cars
5. Urban population
6. Number of hospital beds
7. Cars per 1,000 inhabitants
8. Telephones per 1,000 inhabitants
9. Physicians per 10,000 inhabitants
10. Buildings with water supply per 10,000 inhabitants
11. Electric connections per 10,000 inhabitants
12. Labor force in modern industry/labor force in industry
13. Labor force in traditional industry/labor force in industry
14. Labor force in industry/urban labor force
15. Labor force in commerce/urban labor force
16. Labor force in services/urban labor force
17. Added value in manufacutring industry/labor force in industry
18. Wages in industry/labor force in industry
19. Wages in commerce/labor force in commerce
20. Wages in services/labor force in services
21. Urban population/total population
22. Students registered in elementary school at the beginning of the school year/urban population
23. Students registered in high school at the beginning of the school year/urban population
24. Students registered in college at the beginning of the school year/urban population
25. Population of the tributary area
26. Number of centers within a range of 100 kilometers
27. Number of centers within a range of 200 kilometers
28. Distance to the nearest metropolis (Recife or São Paulo)

APPENDIX 2
THE 104 BRAZILIAN CITIES STUDIED USING FACTOR ANALYSIS
(Listed by State)

Rodônia
- Porto Velho

Acre
- Rio Branco

Amazonas
- Manaus

Roraima
- Boa Vista

Pará
- Belém

Amapa
- Macapá

Maranhão
- São Luís

Piauí
- Teresina

Ceará
- Fortaleza

Rio Grande do Norte
- Mossoró
- Natal

Paraíba
- Campina Grande
- João Pessoa

Pernambuco
- Caruaru
- Recife

Alagôas
- Maceió

Sergipe
- Aracuju

Bahia
- Feira de Santana
- Ilhéus
- Itabuna
- Salvador
- Vitória da Conquista

Minas Gerais
- Barbacena
- Belo Horizonte
- Governador Valadares
- Juiz de Fora
- Montes Claros
- Uberaba
- Uberlândia
- Varginha

Espírito Sano
- Cachoeiro do Itapemirim
- Colatina
- Vitória

Rio de Janeiro
- Barra do Piraí
- Barra Mansa
- Campos
- Duque de Caxias
- Nova Friburgo
- Nova Iguacu
- Nilópolis
- Niterói
- Petrópolis
- Rio de Janeiro
- São Gonçalo
- São João de Meriti
- Volta Redonda

São Paulo
- Americana
- Araçatuba
- Araranquara
- Barretos
- Bauru
- Botucatu
- Campinas
- Catanduva
- Franca
- Guaratinguetá
- Guarulhos
- Itu
- Jaú
- Jundiaí
- Limeira
- Lins
- Marília
- Mogi das Cruzes
- Ourinhos
- Piracicaba
- Presidente Prudente
- Ribeirão Preto
- Rio Claro
- Santo André
- Santos
- São Bernardo do Campo
- São Caetano do Sul
- São Carlos
- São José do Rio Preto
- São José dos Campos
- São Paulo
- Sorocaba
- Taubaté

Paraná
- Curitiba
- Londrina
- Maringá
- Ponta Grossa

Santa Catarina
- Blumenau
- Florianópolis
- Joinville
- Lajes

Rio Grande do Sul
- Bagé
- Campo Grande
- Canoas
- Caxias do Sul
- Cuiabá
- Erechim
- Ijuí
- Novo Hamburgo
- Passo Fundo
- Pelotas
- Porto Alegre
- Rio Grande
- Santa Cruz do Sul
- Santa Maria
- São Leopoldo

Goías
- Anápolis
- Goiânia

Part III

THE THIRD AND FOURTH WORLDS

8

Developments in North African Urbanism: The Process of Decolonization

JANET L. ABU-LUGHOD

THE HISTORIC CONTEXT OF NORTH AFRICAN URBANIZATION

□ A GLANCE AT A MAP of North Africa showing the distribution of population and urban places would reveal a dense but intermittent fringe of settlements along the Mediterranean coastline, broken only on the peripheries by two pairs of north-south extensions. One, on the east, follows the Nile upstream into the heart of Egypt and a parallel path along the Suez Canal; the other, on the extreme west, follows the Atlantic coast and a parallel interior route in Morocco. While to some extent this pattern can be accounted for by the location of fertile lands, over time there has been a marked movement of settlement toward the coastal regions, impelled as much by the economic and political fact of European domination as by any "natural" dessication of the interior (Issawi, 1969).

During the period of Arab-Islamic hegemony, between the 9th and 15th centuries, in the zone that Braudel (1972, 1:103-138) has distinguished as the western basin (roughly that half of the Mediterranean between the Straits of Gibraltar and the bridge between Italy-Sicily-Tunisia-Libya), cities were as likely to be *away* from the coasts as they were to be on it, for the critical trade routes were land routes. In A.D. 1000, according to the data assembled by Chandler and Fox (1974:49-57), of the towns with 20,000 or more inhabitants in North Africa proper, excluding eastern Libya and Egypt, some six were set back from the coast—Sijilmasa, Meknes, Fez, Tlemcen, Constantine, and Qairuwan—while only three were at the water's edge—Salé (not an active port then), Tunis, and Mahdiya. By 1200, reflecting the increased Arab control over the "Middle Sea," several new seaports had been added or revived. But there were still seven inland towns—the original six plus Marrakesh—matched by seven coastal cities—Rabat-Salé (by then a port), Ceuta, Oran, Bougie, and Tripoli, as well as Tunis and Mahdiya.

This situation prevailed up to the 15th and 16th centuries, when the critical battle for the Mediterranean (and the future) was fought between the waning Arabs-Ottomans, on the one hand, and the rising maritime powers of Christian Europe, Spain, and Portugal. By the 16th century Morocco had partially retreated from the contest. Reconquista Spain was entrenched in scattered enclaves along Morocco's coast, and Morocco had withdrawn inland, compensating for the loss of her traditional linkages to Muslim Spain by an intensification of trans-Saharan trade links to Ghana and Mali. By then, also, the Portuguese had wrested from the Egyptians control over the Red Sea path to the Indian Ocean. On the Mediterranean side, the so-called corsairs of Barbary (many of them of European origin) functioned actively out of the older ports, but the balance was clearly tipping.

By the 17th century, Arab-Ottoman-North African power had clearly been superseded by Europe's. The number of towns decreased, and the size of existing cities shrank, along with overall population, wealth, and debased currency. By 1800, across all the northern tier of Africa including Egypt, there remained only a handful of modest-sized urban centers: the four royal cities of Meknes, Fez, Marrakesh, and Rabat, which served as rotating capitals, and the other capital cities of Algiers, Tunis, Tripoli, and Cairo.

When the 19th century ushered in the modern era of urbanism, the entire dynamics of the urban system had been altered. The region became increasingly a passive, albeit far from compliant, recipient of forces generating from imperial Europe. That century witnessed first the French conquest of Algeria in 1830 and then the division of the spoils of Tunisia and Egypt between France and Great Britain by the early 1880s; finally, after a period of standoff among competing European interests for the richest prize, Morocco, that country was partitioned between the two chief contenders—Spain, whose claims were oldest but declining, and France, which took the lion's share. Protectorates over the two zones of Morocco were made official by 1912. By this time, Italy had invaded Libya, although it was not until the 1930s that she had finally subjugated that then poorest prize.

The colonial extractive economy, which was well-established in the 19th century even before political control was consolidated and which, to some extent, still persists today long after official decolonization, dictated a regional pattern of urbanization and a system of differential urban growth in North Africa which had much in common with other colonized nations of the Third World. Port cities—either old revived ones, such as Algiers, Tunis, and Alexandria, or new ones, such as Casablanca on the site of a small fishing village (for its history, see Adam, 1968) or the cities of Ismailiya, Suez, and Port Said along the Suez Canal—began to dominate the hierarchy.

To these centers flocked European military and administrative personnel often linked to commercial ventures, entrepreneurs, and banking representatives from European firms and financial houses, quite a few imaginative charlatans, followed by southern European workers (and even peasants) seeking a better life, and finally, by a motley assortment of Maltese, Greeks, and other "Levantines" whose assimilation to the European elite often went no deeper than consular protection and the privileges that that bestowed. These outsiders formed commercial alliances wherever they went with locally resident minority groups (Copts in Egypt, Jews in Tunisia and Morocco) and with small sectors of the elite, whether royal or aspirant bourgeois. The foreign population concentrated in and dominated the major cities. Only in Cairo (and only because it was the largest of the premodern capitals) did foreigners not constitute a substantial proportion of the population, as they did in Casablanca, Rabat, Tunis, Algiers, and Tripoli.

Not only did this increase in foreign population stimulate the development of entire new appendages to the existing port cities, appendages which in time came to dwarf the original nuclei, but the same extractive economy led to the creation of new towns whose only function was to extract a resource (e.g., the mining town of Khouribda in Morocco) or to ship it out (e.g., Kenitra). Interior towns of the developing urban hierarchy were more and more resource sites, military control points, and, only gradually, central places for an expanding European agriculture. This was particularly true in Algeria and later Libya, where foreigners became farmers, and it was also true in Morocco and Tunisia, if less markedly; only in Egypt, where foreign control over agricultural land and produce took place in the absence of foreign labor, was this not the case.

Interior towns of historical and religious importance often retained their integrity but at the price of decline. Such illustrious centers of the Islamic period as Tlemcen, Qairuwan, and even Fez were more and more bypassed by the imposed economy, while their economic bases in handicraft production and trade were systematically undermined by the European manufactured substitutes which passed through the new commercial firms at the ports (see Miège, 1961, for full documentation on Morocco). Rural and nomadic people of the interior, who had provided the raw materials for the traditional handicraft products and even an insignificant market for them, were increasingly displaced from the more fertile lands which, in the Maghreb particularly, were given over to foreign cash crops (such as viniculture) relevant only to external trade and foreign consumption. Some wandered southward to even more marginal land, making them especially vulnerable to the periodic disasters and droughts which eventually drove them to the foreign centers at the coasts. Others, ill-adapted to nomadism, had little choice but to head immediately for the foreign-dominated centers which were growing at extremely rapid rates. They provided a cheap labor pool used to construct the housing and infrastructure of the new cities, to man the factories capitalized from abroad and oriented toward Europe, and to serve the imported "elite" as domestics, gardeners, and porters and, as their numbers proliferated, more and more to serve one another in a widening chain of intermediary transactions, each of which made only a small marginal contribution and commanded but a minor marginal return (Montagne, 1952).

It was, however, this movement to the coastal and capital cities of displaced farmers and herders as well as craftsmen and *commerçants* from the declining interior cities that intensified the growth of the former and prevented these cities from becoming disembodied foreign enclaves with exclusive focuses on the metropole. Unlike the Portuguese "factories" along the African coast, these cities were rooted in the countries. But the nature of the colonial economy, the fact that development decisions were made from the standpoint of the European capitalists, and the fact that internal fluctuations sensitively reflected the ups and downs of the Paris bourse meant that the welfare of these new urbanites depended less upon their own enterprise and initiative than upon factors beyond their ken and control.

The final crucial phase of urban development in North Africa began in the 1950s with official decolonization. Sovereignty was restored first to Libya in 1951 and then to Tunisia and Morocco in 1956. It was not until 1962 that Algeria, the first to be subdued, was the last to be freed. (While Egypt had gained paper self-government much earlier, British troops exercised extraterritorial control in the Canal Zone until its nationalization in 1956, when Egypt became fully sovereign.) Decolonization, however, is proving to be a lengthy process in which political independence is certainly a necessary but scarcely sufficient cause (Amin, 1970). Just as the second half of the 19th century was a period of growing dependency for North Africa, even though official colonization may have lagged behind, so a reversal of this system and an undoing of its effects remains the task of the second half of the 20th century, even though official independence has been a fait accompli for more than a decade.

While this article on contemporary trends in North African urbanization has dwelt perhaps too long on the historic roots, it is our contention that the basic outlines of present-day urban patterns were established during the colonial period. Many of the contemporary problems and issues of change and policy derive from this "false start," and each of the countries must, to some extent, reorder its urban system and its cities to reflect the new goals of independence, distributive justice, and cultural integrity.

Among the direct effects of the earlier colonial pattern of urbanization might be listed:

(1) The geographic concentration of major cities within relatively shallow coastal portions of the country or, put conversely, the ezistence of wide discrepancies between overurbanized subregions at the coast and the underdeveloped backlands.

(2) The persistence of an external pattern of trade relationships and an internal pattern of investment priorities which has perpetuated the vulnerability of development schemes and which has often placed greater emphasis upon gross "returns" to a limited number of

investors than upon long-term development, the use of local manpower, and the "distributive" income effects of growth.

(3) The pattern of primate city dominance, in which the largest city not only contains an extremely high percentage of the total and urban populations but, even worse, disproportionately controls the web of decisions, communications, and "modern" facilities in the country.

(4) As a consequence of the first three, the continued migration from the "backward" regions to the already developed coastal centers (allometric growth rates) of large numbers of people for whom jobs may be scarce in the city but for whom conditions were even more precarious at their points of origin.

(5) The coexistence within the major cities of a fragmented set of ecologically distinctive (and often disjunctive) units, within which the different classes lead lives considerably encapsulated from those of other classes, a phenomenon which interferes with rather than aids the process of social and cultural integration.

(6) Disproportionate investment in urban infrastructure and services in the "Westernized" quarters of the major cities, to the neglect of both the oldest quarters and the vast zones of self-constructed housing to which the increasing local population had recourse as a solution to the shortage of housing and the preemption of centrally located land for the colonial city.

(7) These preexistent problems have recently been compounded by a rapidly increasing rate of natural increase which currently accounts for a substantial portion of the total urban growth rate in each of the countries and which, due to sustained high birth rates in the cities coupled with sharply declining death rates, is currently higher in urban than rural areas.

It is these "givens" which, taken together, define the issues of contemporary urbanization in North Africa. Policies designed to alter these realities or to change the bases for their perpetuation are at the core of governmental deliberations everywhere. Within each of the North African countries, however, the parameters of such policies are affected differentially by (a) the concrete process whereby the decolonization actually took place and the degree of real autonomy that has been achieved thus far, (b) the resources newly available for development along patterns not set by earlier colonial designs, and (c) the political ideology of action adopted by each country. We propose to explore each of these issues in greater detail in the sections that follow, to diagnose the urban situations in each country, and to suggest directions for future trends and policies.

TABLE 1

POPULATION AND PERCENTAGE IN LARGE CITIES (100,000 plus), NORTH AFRICAN COUNTRIES, 1950-1975

PART I. TOTAL POPULATION

Datum and Year	Algeria	Egypt	Libya	Morocco	Tunisia
1950 est.[1]	8,920,000	20,000,000	999,000	8,959,000	3,555,000
1951				8,004,000[2]	
1954	9,529,726[3]				
1956					**3,441,696**[4]
1958				**11,095,000**[5]	
1960 est.[1]	10,784,000		1,349,000	11,640,000	4,157,000
Census		26,062,000		**11,626,470**	
1962				11,870,000	
1963	**10,670,000**[6]				
1964				12,629,000	
Census			**1,564,000**		
1966	11,833,000[7]				
Census	12,101,994	30,000,000			4,533,351[4]
Estimate	12,096,347		**1,728,059**[8]		
1968			1,808,000[9]		
1970 est.[1]	13,663,000		1,850,000	15,519,000	4,882,000
Off. est.	13,547,000	33,000,000			
1971 census	**14,643,700**			**15,379,259**	
1972 est.			2,084,000[10]		
1973			**2,257,035**[8]		
1975[11]	16,500,000	37,000,000	2,500,000	17,388,633[11]	5,700,000[12]

PART II. PERCENTAGE OF POPULATION IN CITIES 100,000 OR MORE, 1950-1975

Datum and Year	Algeria	Egypt	Libya	Morocco	Tunisia
1950 est.[1]	11.9	19.0 (1947)	11.4	15.8	17.3
1951[2]				17.1	
1956					25.8[4]
					19.6
1960 est.[1]	16.4		21.3		
Census		27.0	25.5 (1964)	18.9	
1966 est.			17.0		
Census	15.0				24.0
1970 est.[1]	13.6	31.0	26.1	23.4	22.2
1971 census				24.1	
1973			36.2		
1975					23.0[12]
	20.0	36.0	38.0	28.0	27.0[11]

1. Kingsley Davis, *World Urbanization, 1950-1970* (Berkeley: University of California Institute of International Studies, 1969), Vol. 1, pp. 57-58.

2. Official Moroccan Census of 1951 for the French zone only. Davis' 1950 estimate evidently includes the Spanish zone also, but results are not trustworthy.

3. Amor Benyoussef, *Populations du Maghreb et communauté économique à quatre: Esquisse d'une théorie démographique de l'integration* (Paris, 1967) cites this figure on p. 78 from the Algerian *Annuaire Statistique.* It includes almost one million non-Algerians.

TABLE 1 (continued)

4. Various figures appear in the sources for the Tunisian population in 1956 according to the census of that year. We have used the "official" retrospective number presented by Groupe Huit, *Les Villes en Tunisie* (Tunis: Tunisienne Ministère de l'Economie Nationale, 1971).

5. Retrospective official estimate for the unified and newly independent French and Spanish zones.

6. The figures for 1954, 1960, and 1963 come from Benyoussef, 1967. Note the decline in total population between 1960 and 1963. In the former year there were more than one million foreigners, chiefly French, in Algeria. After independence, in 1963, only 140,000 remained.

7. Although the Algerian Census was held in 1966, various sources disagree as to the official total. The low figure is reported by Gerald Blake, "Urbanisation in North Africa: Its Nature and Consequences," in D.J. Dwyer (ed.), *The City in the Third World* (New York: Barnes and Noble, 1974), p. 70; the middle figure by the *Demographic Yearbook, 1972;* and the high by K. Sutton, "Algeria: Changes in Population Distribution, 1954-66," in J.I. Clarke and W.B. Fisher (eds.), *Populations of the Middle East and North Africa: A Geographical Approach* (New York: Africana Publishing Corporation, 1972), p. 375. The discrepancy is due to Algerian nationals living temporarily in France, who are included in the larger figure but excluded from the smaller ones. While the census estimated the Algerians abroad at less than 300,000 at the time of the census, Sutton believes the true number to have exceeded 500,000.

8. Salah El-Shakhs, "Spatial Development and the Future System of Settlements," unpublished paper prepared for the Ministry of Planning, Libyan Arab Republic (Rome: Italconsult, 1974), is the source for Libyan population and urbanization estimates for 1966 and 1973. His semiofficial figures are in great detail.

9. Borham Atallah and Mona Fikry, "Le phénomène urbain en Libye," in *Annuaire de l'Afrique du Nord,* 11 (Paris, 1972), pp. 79-103.

10. *Demographic Yearbook 1972* estimate.

11. These projections are my own, but in the case of Morocco I have accepted the projection done by the Centre de Recherches et d'Etudes Démographiques, *Projections de la Population Marocaine* (Direction de la Statistique, Royaume du Maroc, January 1975), p. 6.

12. Provisional returns of the census of Tunisia taken in 1975.

RECENT URBANIZATION LEVELS

North Africa, in the world context, is only moderately urbanized, but, like other parts of the developing world, it is changing very rapidly. In the region as a whole, about one-fourth of the population in 1970 lived in cities with 100,000 or more inhabitants (Davis, 1969) but by 1975 this has increased to over 30% (our estimate). Egypt has traditionally deviated on the high side and Algeria on the low, but the similarities are stronger than the differences. Table 1 summarizes data on the levels of urbanization for the five countries of North Africa between 1950 and 1975. Since census dates are not uniform and sources vary by stability and reliability, many data points and individual entries are presented, the most trustworthy being shown in bold face. By our best estimates, the five countries now contain a combined population of some 80 million persons, of whom probably 25 million (over 30%) are living in places with populations of at least 100,000. Perhaps another 10% are in towns of 20,000 to 100,000, this relatively small number being due to the unbalanced urban hierarchies in each of the countries, hierarchies which are most unbalanced in the tiny countries of Libya and Tunisia and the giant of Egypt and somewhat less so in Algeria and Morocco.

In each of the countries, the move to the cities has been dramatic but not smooth. While all have followed an upward trend, roughly doubling population and big city population proportions over the past quarter of a century, each had its trends modified by nondemographic factors, either episodic or endemic. At one extreme was Egypt, a country whose early emancipation from direct colonial rule and whose developing economy promised steady urbanization along with economic growth (Abu-Lughod, 1965). It was a late-in-the-day colonial-type invasion in 1967 which brought that trend to a sudden halt. Israeli bombing destroyed large portions of the major cities along the Suez Canal, and occupation of the Sinai side of the canal reduced the former million residents of Suez, Port Said, and Ismailiya to a small skeleton crew. Most of the refugees from these cities have settled in Cairo, along with other refugees from Arish and the Gaza Strip. Instead of a projected 1975 population of some 6.5 million, Cairo now houses 8.3 million persons in its metropolitan area (1975 news release, *Al-Ahram*). Because of the catastrophic character of this "urban redistribution," it would be inane to discuss the dynamics of urbanization in Egypt as if it were predictive either of Egypt's future or of other countries in North Africa. The cities in the canal zone are now being reconstructed and will, in the absence of another war, soon yield not only a redistribution of the urban population within the hierarchy but probably encourage even greater movement out of the rural areas.

At the other extreme is Libya, the smallest and, until the late 1950s, the least developed of all the North African countries. She has experienced the most marked shift in her level of urbanization, as well as in all other economic and social indicators (El-Shakhs, 1974), during the decade and a half since independence; but this remarkable change is due first to the extremely low point at which she began and, second, to the discovery and subsequent exploitation of oil. Libya's "urban revolution" since 1960 cannot be understood without reference to this "new" resource. In contrast to the usual stereotype of an Arab country, Libya is essentially underpopulated, and the movement out of agriculture and nomadism into the cities of the coastal fringe has been caused not by oversaturation of rural areas (indeed, agricultural production has suffered from the depopulation) but by the labor shortage in oil-linked processing and the consequent wide wage differential. The population of Tripoli has doubled in less than five years, Benghazi's size has increased by two-thirds, and many of the small towns along the coast have been entering the urban population range above 20,000. In the interior of the country, in desert zones previously only the site of caravan stops, towns are beginning to form.

The remaining three countries, the real Maghreb of Morocco, Algeria, and

Tunisia, show remarkably similar trends. In each a foreign-impelled urbanization was grafted upon a declining preindustrial one, and in each foreigners constituted a sizable proportion of the population in the major "modern" cities. While this was most marked in the early decades of occupation, it remained high until the moment of independence, after which there was a precipitous decline. It is this sudden decline which accounts for the hiatus in urbanization rates just after independence—a watershed moment before indigenous impulses to urbanization more than fill the gap. The transition was particularly abrupt in Algeria because of the very large number of Frenchmen who departed, but, even in Morocco and Tunisia, cities changed virtually overnight from containing 25% or more foreign residents to currently containing about 2%.

Because of its small size, Tunisia is deviant in exhibiting the most centralized urban hierarchy (see dramatic maps in Groupe Huit, 1971, vol. 2), but this is often the case in tiny countries which might even be termed "city states." Both Algeria and Morocco have a wider population base and encompass a large geographic area with regional diversity of terrain, economic base, and provincial centers of considerable size. While each has one primate city of voer a million, the existence of fairly sizable cities outside the orbit of the primate assures a more balanced hierarchy. Tunisia is also poorer in natural resources than the other two, lacking the oil which is stimulating development in the interior of Algeria and the mineral wealth which could potentially accomplish as much for the inland mountain regions of Morocco.

THE MAJOR CITIES OF NORTH AFRICA

Virtually all the major cities in North Africa are seats of government and administration, confirming Santos's (1970) contention that administration has supplanted the old military-defense function in cities of the Third World. The exceptions, however, are notable. Where the administrative capital is also the economic capital, by virtue of its port functions (absolutely essential in the earlier colonial phase), it outdistances its nearest rivals. But in the cases of Morocco and Egypt, where the administrative capital is not a functioning port, the latter outdistances (Casablanca) or at least does not lag too far behind (Alexandria) the former. Table 2 presents information on the largest cities of the region and their growth rates in recent decades.

Several things should be noted from this table. First, whereas the problem of primate dominance may plague individual small countries, within the region taken as a whole there is a relatively balanced hierarchy of urban places. We have contended elsewhere (Abu-Lughod, 1973) that applying the rank-size formula mechanically is particularly distorting in the case of small states and that the scale of urbanization is everywhere growing larger; thus, the attrition of modest-sized cities from urban hierarchies is not in itself a pathological sign but may, indeed, be a measure of "modernization" or at least economic reorganization. (For an alternative postion, see, inter alia, Ibrahim, 1975.)

TABLE 2
MAJOR CITIES OF NORTH AFRICA: CURRENT POPULATION ESTIMATES AND GROWTH TRENDS BETWEEN 1950 and 1975

Cities by Current Descending Rank-Size (approx.)	Population c. 1950	Population c. 1960	Population c. 1970-1975	Recent % Annual Growth Rate
Over 500,000				
Cairo, Egypt	2,000,000 (1947)	3,500,000 (1960)	6,500,000 (1970) 8,300,000 (1975)	5.0
Alexandria, Egypt	919,000 (1947)	1,500,000 (1960) 1,800,000 (1966)	2,032,000 (1970)	3.5
Casablanca, Morocco	680,000 (1952)	965,000 (1960)	1,506,000 (1971)	4.5
Algiers, Algeria		885,000 (1960) 943,000 (1966)	1,100,000 (1970)	5.0
Tunis, Tunisia	588,000 (1956)	721,000 (1966)	1,000,000 (1975)	4.0
Tripoli, Libya	114,000 (1950)	210,000 (1964) 300,000 (1966)	551,000 (1973) 650,000 (1975)	9.0
Rabat-Salé, Morocco	200,000 (1952)	303,000 (1960)	523,000 (1971)	5.0
250,000 to 500,000				
Oran, Algeria		327,000 (1966)	368,000 (1970)	3.0
Marrakesh, Morocco		243,000 (1960)	332,700 (1971) 374,500 (1975)	3.0
Fez, Morocco		216,000 (1960)	325,300 (1971) 374,400 (1975)	3.6
Benghazi, Libya		160,000 (1966)	264,500 (1973) 310,000 (1975)	8.0
Constantine, Algeria		255,000 (1966)	280,000 (1970)	2.7
Mahalla el-Kubra, Egypt	116,000 (1947)	172,000 (1960)	255,800 (1970)	4.8
Tanta, Egypt	140,000 (1947)	184,000 (1960)	253,000 (1970)	3.0
Meknes, Morocco		176,000 (1960)	248,300 (1971) 279,000 (1975)	3.0
Sfax, Tunisia		70,000 (1966)	250,000 (1975)*	
Egyptian Canal Cities				
Ismailiya	68,000	276,000* (1960)	345,000 (1966)	
Suez	107,000	203,000 (1960)	315,000 (1966)	
Port Said	178,000	244,000	313,000 (1966)	

*Boundary change.

Nevertheless, it is readily apparent that allometric growth—i.e., the tendency for the larger cities to be growing at higher rates than the smaller ones—is still the dominant pattern, no matter how we may evaluate this fact. Whereas capital cities in recent years have generally grown at the rate of at least 5% per annum, the provincial urban centers have grown at rates that seldom exceed 3% per year.

Since this rate of growth is about equivalent to the rate of natural increase in the countries of the Maghreb, it is clear that these cities are just barely holding their own. The real net migration is going almost exclusively into the capital. It is suggestive that the two largest port cities which do not serve as national capitals, namely Casablanca and Alexandria, have grown at rates somewhere between those of the provincial capitals and the national capitals. Whether this reflects a reduction in the external orientation of trade which formerly flowed through these ports or some more specific and temporary conditions, we do not know.

The role of migration in urban growth is important but has, perhaps, been overstressed in the past. We would like to emphasize that the high growth rates of North African cities are largely accounted for by natural increase. The two chief exceptions are Cairo, because of its current role as a refugee reception center, and Tripoli, because of its role as a reception center for skilled manpower, which makes it more akin to Kuwait than to Bombay. In the other capitals, migration accounts for no more than half the growth, and in the slower growing regional centers it is an insignificant element. Because of the undue advantage in growth which the larger communities have by virtue of the dominant position that they inherited from the colonial period, it has been virtually impossible to achieve noticeable reversals during decolonization.[1] Migration is, of course, the only real source of population redistribution in the absence of urban-rural differentials in natural increase. Natural increase can only stabilize preexisting patterns.

Colonialism had its impact not only on the level and distribution of urbanization but, even more strikingly, on the internal ecological order of North African cities which, as a result, are now all more or less fragmented into subquarters of highly varying character, quality, and appearance, often juxtaposed against one another with little interaction or integration. Decolonization is bringing about major changes in the uses of this inherited shell and expansions to it. The chief transformations now taking place in North African cities may be characterized as the blending together of these disjunctive fragments and the gradual emergence of a genre of city more authentically adapted to and projective of Arab culture as it modernizes. It is to this issue that we now turn.

ECOLOGICAL STRUCTURE OF CITIES AND ITS TRANSFORMATION

It would be foolhardy to attempt a detailed ecological description of the major cities of North Africa within the scope of this brief article. Our goal in this section is much less ambitious. It is to touch upon some general principles of ecological organization, to illustrate these, and to demonstrate some of the major variations in them, by materials drawn from the larger cities of the region. I shall draw chiefly from the cases of Cairo, which I have already studied in great detail (Abu-Lughod, 1971), Rabat-Salé, for which my research is virtually complete and some preliminary findings have already been published (Abu-

Lughod, 1974; 1975b), and Tunis, which is currently under study (see Abu-Lughod, 1975a for a fuller comparison). Where relevant, references may be made to Algiers and Tripoli, but my knowledge of these cities is far from adequate. While these cities are, to some extent, atypical in being larger than most and in serving as capitals (which increases their attractiveness to migrants and burdens them with a heavy symbolic role), they are representative of many other older cities in North Africa and the Middle East which became "dual cities" during the colonial period and must now reintegrate during decolonization.

The association between the institution of colonialism and the development of the so-called "dual city" form throughout much of the Third World is now an accepted truism. The stereotype is that of a city grafted onto an older traditional form of city-building, whatever that might be; it is a "modern" city better adapted to a contemporary life that is a legacy left by the departed "superior" colonial power. However, when identifying the emerging problems in non-western cities with this dual structure, western observers have singled out *two* types of zones as problematic: (1) the oldest "native" core which is defined as a (perhaps picturesque but nevertheless unsavory) slum requiring selective clearance, rehabilitation, or perhaps restoration as an "historic quarter," and (2) a second type of indigenous quarter, also defined as a slum, which, despite, of course, the noblest efforts of the colonial power to prevent it, had proliferated illegally on the outskirts of the built-up area.

Thus, even the so-called dual city is acknowledged to contain at the minimum not two but three subcities: the historic precolonial core; the modern colonial appendage; and the unregulated indigenous quarters on the outskirts, known variously as bidonvilles in Morocco and Algeria, as gourbivilles in Tunisia, as shantytowns in Libya, and (although not exactly parallel in appearance) as the "cities of the dead" in Cairo. While colonial scholars writing on North Africa take pride in the modern city, they tend to absolve colonialism of responsibility for the other two substandard types whose existence is blamed on the traditionalism and "backwardness" of the underdeveloped country itself.

My argument is that this position is not only oversimplified but basically incorrect. There are not two or three "subcities" in each of these capitals but indeed six fairly distinctive and coexistent urban arrangements which make up the ecological structure of North African cities. Furthermore, while these basic forms of urbanism exist in different proportions in each of the cities and may have grown up under somewhat different circumstances, in each city the fate of every one of the types is deeply implicated with developments in the others. The evolution of these subcities cannot be understood without reference to the economic and political conditions of colonialism and without tracing the interrelated effects of each quarter upon the others. The six types which must be distinguished are:

1. *The Medina Core.* This is a relatively small nucleus dating back in origin to the 10th to 12th centuries. It was, at the height of its medieval importance, a

walled, compactly settled but rationally organized unit which contained the entire population, spatially segregated according to kinship-ethnicity, place of origin, clientage-affinity, and economic sector. Spatial segregation by economic class per se was rare, except for outcasts. The city was physically discrete from but nevertheless strongly linked to the surrounding countryside by significant economic and social exchanges. Only secondarily and more selectively (through a small elite) was it linked to the external world–in the cases of Cairo, Tunis, and Algiers at least, chiefly by sea. The urban community in North Africa was designed for foot and pack-animal traffic only. Wheeled carts were unknown, which resulted in a somewhat different physical layout than developed in European medieval cities. A common legal system, quite different from Roman law, regulated the relationships of neighbors within the smaller cells of the city, defined responsibilities with respect to public and private space, and handled the financing of common utilities and services. While the *physical* structure of this community still persists, the *social organization* has altered drastically.

2. *The Modern Appendage.* This appendage was either designed by Europeans or built on the European style, often by Europeans for their own use during the 19th century or, in the case of Morocco and Libya, early 20th. These settlements were considerably less dense, were layed out with straight and wide streets to accommodate wheeled vehicles, and were costly to construct and service. Efforts were made, consciously or unconscioulsy, to maintain a castelike segregation of the population according to the joint criteria of economic class *and* nationality. The chief linkages of this "city" were to the outside world (particularly Europe) via developed ports. Linkages by rail or road were to other ports or to resource extraction sites in the interior. The immediate hinterland was relatively unimportant except as a source of food products not being imported.

3. *Rapidly Proliferating Uncontrolled Settlements.* These quarters consist chiefly of self-built housing, primarily on the periphery but in some cases reaching deep into the heart of the modern quarters in wedge or checker patterns, depending largely upon the locations of land in royal, state, or religious-foundation ownership. (Insalubrious terrain was almost a prerequisite.) The housing is single- or at most two-story, with heavy land coverage and a creative use of salvage. Unlike the case in Latin American cities, these zones seem to have been settled originally by newly arrived rural migrants, although later their ranks were expanded by persons fleeing the overcrowding and progressively deteriorating medinas. Some occupants of these quarters may work in large industrial establishments, but most remain only marginally integrated into the class structure and economic activities of the city (Naciri, 1973). Wherever possible, occupants are spatially ordered by kinship, ethnicity, place of origin, and "degree of respectability," i.e., by the same principles which originally ordered the medinas. There appear to be marked variations by country, both in the importance of these settlements and their disposition. They never did develop in Egypt, because the tomb zones offered a functional

equivalent, whereas in Libya prosperity made them relatively short-lived. Elsewhere, publicly subsidized substitutes have been provided, but never at a rate sufficient to supplant them entirely.

4. *Peripheral Suburbs.* This is the most recent urban form to make an appearance, built for an indigenous upper-middle and upper class. In zones which were formerly beyond commuting distance to the center of the city and which are often around the nuclei of preexisting villages or resort communities, suburban quarters composed largely of either "villas" or large-scale apartment complexes are growing up, to which government officials and the modest industrial and commercial elite who came to power mostly after independence have been gravitating. Although suburban quarters were not totally unknown before (Cairo's two major suburbs began in the early 20th century), it was not until the recent rapid expansion of this indigenous middle class that these zones became the fastest growing in the city. They expand, picking their way cautiously among the bidonvilles, with which a competition for land has already begun.

5. *Rural Fringe.* This is a residual category, the frontiers of which are continuously but selectively being pushed back with each advance of the rapidly growing city. Occasionally, entire village settlements are engulfed, but, if it is the middle-class peripheral suburb that is advancing, these villages will be replaced rather than incorporated, at least in the long run. In the short run, whole wedges of agricultural land remain in the interstices between radial expansions along highways, and strange juxtapositions are likely.

6. *Transitional Working Class Zones.* These zones, which gradually develop between the original medina and the modern quarters nearby, serve first as buffers and later as connectors between the two and may be carved spatially out of either or both.

In the treatment that follows I shall not discuss the latter two types, since they do not present particularly unique problems, nor are they tied to specific spatial or social arrangements as are the other four. A further caveat is necessary. While the four basic urban types can be analytically distinguished and described serially, it is the tight interrelationship between the fates of each and every one that we wish to stress. They constitute a symbiotic (albeit often conflictual) set; changes in one induce changes in the others.

First, let us look at the relationship between the medinas and the modern "western-style" appendages and see how these have altered over time. In Cairo, the line between the medina and the new city is quite blurred and was never demarcated by a physical wall (a canal had been the barrier between the two, but even that had been filled in and converted to a traffic artery by the end of the 19th century), and the social barrier was not exclusively ethnic. While it is true that the "native" population was segregated entirely on the eastern side of the barrier (i.e., the medina), the foreign and Egyptian non-Muslim population, even at the height of their power, never constituted more than half of the population residing in the early modern city. That peak was reached by World

War I when foreigners made up only 10-15% of the population of Cairo. Only by combining forces with Egyptian ruling classes and Copt minorities could the facilities required by "modern urban living" be supported.

In Tunis the barrier was maintained longer and was established according to more castelike principles. Two sets of walls ringed the medina of Tunis before the advent of colonialism. The first or smallest ring surrounded the original medina dating back to the 9th century; the larger ring encircled the two suburban quarters to the north and south (the so-called ribat) which were medieval extensions to the city. Soon after the establishment of French colonial power, the inner walls were removed to make room for a *Ringstrasse* or circular traffic belt, along which the French erected pseudo-Ottoman ministry headquarters. At a right angle from the oval medina, like a long handle to an ax, the French established a modern appendage on the grand rectangular pattern of Europe which was intended to connect the medina with the port of La Golette, due east. Growth soon converted this main axis to a north-south one parallel to the original oval. Early removal of the walled barrier permitted a more blurred boundary to the medina, and the location of the Italian, Corsican, and Sicilian "proletariat" along the western edge of the medina prevented the emergence of full castelike segregation between the old and new cities. Nevertheless, because foreigners constituted a goodly third to a half of the total population, they were able to sustain a relatively self-contained modern quarter without soliciting assistance from the Tunisian elite.

In Moroccan cities, on the other hand, full castelike segregation was encouraged and indeed enforced. In Rabat-Salé what was hinted at in Cairo and what was somewhat more evident in Tunis was carreid to its fullest extreme. Not only were the walls surrounding the 10th-century city of Salé and the 12th-17th-century city of Rabat left intact, but indeed they were repaired and strengthened by the French, once the newly imposed protectorate was established in 1912. Initially, Lyautey, the new Résident Général, intended to establish the more important city of Fez as his capital, but a military retreat from an interior which would take a long time to "subdue" led to a hasty designation of Rabat at the new capital. The new policy adopted by Lyautey, to "avoid the unhealthy mixture of the races," which he had criticized in Tunisia and Algeria, was to separate as fully as possible the life of the colonial from the life of the Moroccans. *Cordons sanitaires,* called "green belts," were created wherever possible, and new cities for French occupancy were built, often at a substantial distance from existing medinas, e.g., Marrakesh and Fez. In the cases of Casablanca and Rabat, however, preexisting developments prevented the carrying out of this scheme. Nevertheless, the demographic predominance of foreigners helped for some time to maintain the self-sufficiency of the apartheid quarters (Abu-Lughod, 1975b). Whereas Cairo never had enough foreigners to make the modern zones viable without the assistance of the local royal family and the Copts who were so heavily utilized in the British administration, and whereas Tunis's modern quarters were internally differentiated by the melange of subgroups which made up the half of the population which was *not* Tunisian Muslim (Jews, Italians, and a variety of "Levantines"), Rabat-Salé had a large and high-caste population of French settlers, enough to sustain the intended segregation. Salé remained an exclusively Moroccan city. Foreigners were enjoined from living in the medina of Rabat. Thus, fully one-third of the total population were foreigners, mostly French, and *all* of them were concentrated either in the elaborately equipped quarter designed for their use or in the somewhat less elegant quarter of Océan, to which (chiefly Spanish) working-class foreigners gravitated.

The bidonvilles, in all cases, were not a mere accidental corollary of underdevelopment, unrelated to the exploitative colonial system of the 19th and 20th centuries, as they have so often been treated, but rather a direct effect of that system. Rather than being the result of a "population explosion" or a lemminglike and almost perversely irrational massive migration to the cities from the countryside, the vast quarters of makeshift housing which accumulated on the outskirts of each metropolitan area must be viewed as an integral part of a larger pattern of changing social stratification and power.

In both Egypt and Tunisia, which were parts of the Ottoman domain, land ownership in agricultural areas was drastically "reformed" during the second half of the 19th century. Lands which had formerly been held communally or whose use assignment was traditional, despite the actual title, were placed into "private ownership" on an unprecedented scale. These newly titled lands were first concentrated in the hands of a landed aristocracy (which indeed consolidated its power through this step). Later, when foreigners overcame restrictions and wrested the right from the Porte to hold property, more and more land was accumulated in the form of large foreign-owned estates. It was not so much population *growth* in the countryside but rather this new form of land tenure which led to a "flight" from the land—very similar in fact to that experienced in 18th-century England in relation to the enclosures movement.

In Morocco the process began somewhat later and followed the Algerian rather than the Ottoman pattern. Foreign agricultural plantations or estates took title to communal lands, and herders and peasants were displaced in unprecedented numbers. The victims of "enclosures" tied to world capitalist markets were forced into the cities where, as one might have expected, no provision had been made for them. Their cheap labor was certainly welcomed by the small construction and industrial sector capitalized from abroad and geared toward Europeans, and their newly monetized, if modest, buying power was similarly welcomed as a demand market for cheap European textiles, etc., but little thought had been given to their absorption into the physical plan of the city.

Peripheral "spontaneous" quarters differed from city to city, depending in part upon terrain and in part upon rates of growth and the absorptive capacity of the old medinas. In Cairo, bidonvilles in the proper sense of the term never did develop. The rural fringe was too valuable for intensive agriculture to be preempted, and only the cemetery zones on desert plateaus east and south of the city offered convertible shelter for a burgeoning population. In Tunis, both submarginal hills and the odoriferous shores of the saline lakes attracted (permitted) squatters who jerry-built on previously unoccupied land. The development of massive bidonvilles was most marked in Morocco, almost *because* of the more advanced city planning there. Despite the lavish attention to city planning in the French administration, the influx of Moroccans caught the French by surprise. In Casablanca they met the influx with a hastily planned "New Medina" toward the outskirts, but for the most part the influx was absorbed into squatter settlements, both forbidden yet tolerated as an inexpensive alternative to real capital investment.

The first bidonville of Rabat appeared in 1927, when a group of Zaers who had been driven off the land encamped quite noticeably at the city gates near the sultan's palace. To avoid this embarrassing sight, they were removed to land hidden from view by the sharp embankment that descended to the Bou-Regreg River. This precedent encouraged other settlements, no doubt. Originally, the bidonvilles were settlements of ethnically related migrants, but gradually this early pattern gave way; at present, many of the bidonvilles are inhabited not by new migrants but by persons who have moved out of the crowded medinas to more commodious peripheral quarters where they can "own their own homes." Thus, the changing nature of the medinas was directly related to what was happening in the remaining parts of the expanding and changing North African capitals.

We must therefore back up a bit and see what happened to the medinas under the impact of the growth of both modern quarters and peripheral bidonvilles. It will be recalled that originally the medina was organized according to principles of kinship, ethnicity, and clientage, but *not* basically according to class segregation. Instead, each subquarter of the medina was relatively self-contained and had a population linked together by bonds of occupational and religious involvement, ethnic origin, family ties, and clientage, which might be thought of as a form of "fictive" kinship. Each important family was surrounded by its clients–retainers or families tied to it through service or through marriages or alliances which might have been forged in the countryside. These social units, by definition, encompassed persons at a variety of levels of status and wealth who were held together by a network of mutual interdependence. The form of social organization, then, was the antithesis of the cash nexus class segregation of the modern urban community.

With the advent of European power in the region and with the founding of the modern quarters, this system began to break down. The first to desert the medinas were marginal groups which made alliances with the new power and took advantage of an almost pariah role to advance themselves along lines previously blocked. The Copts in Egypt, who had always concentrated in their own quarters of Cairo anyway, began to leave the older parts of town to live as near the foreign community as possible. In Tunis, the aristocratic Livournais Jews were the first to desert the Jewish *hara* within the medina for more modern quarters. And in Moroccan cities, there were the "protegés," Moroccan citizens who agreed to "front for" European commercial and land firms, as well as a clever merchant class, many of whom came from Fez but set up operations in other cities and who settled in zones near those of the Europeans.

Gradually, these few "defectors" were joined by upwardly mobile Egyptians, Tunisians, and Moroccans who hoped to join the ruling elite. The process, which has by now resulted in an almost total proletarianization of the medinas, had begun. As late as 1917, the medina of Cairo still housed a wide range of social classes and contained a vital industrial and commercial base which clearly dominated the economy, even though it was just beginning to be challenged by an evolving modern economic structure largely controlled by foreign interests under pressure from a bourgeois nationalist movement. But by now, the Cairo medina has become a lower-class slum. Many of its residents eke out marginal livings from a sadly decimated and declining economic base; and the high proportion of migrants has begun to give it a semirural cast.

The same is true, albeit to as yet a lesser extent, in Tunis and the medinas of Rabat and Salé. In both, local planners discuss the phenomenon of "taudification" (slum-creation). (See Ekert and El-Kefi, 1974.) More and more the upper classes and then the middle classes of Tunisians and Moroccans have deserted the medinas. While this trend was already under way before independence, it was considerably intensified and speeded up when independence resulted in a relatively massive repatriation of foreigners. The housing gaps left in the newer quarters by this departing population were filled by persons fleeing the medinas.

One of the reaons why people left the medinas–either for the modern quarters or the bidonvilles–was that they had become extremely overcrowded, a phenomenon not unrelated to the "enclosures" which had driven large numbers of rural persons into the cities. Most of these newcomers went at first to the medinas, filling the *funduks* and *waqalas* (dormitories for itinerant merchants and travelers in the earlier days) and then piling up in existing structures. As larger palaces and commodious dwellings were vacated by the departing upper class and the high bourgeoisie, these were subdivided to accommodate the new demand, often into rabbit warrens housing dozens of families. But as densities increased, not only the middle class but ordinary medina-dwellers–some true urbanites declassed through a decline in the traditional industries and trades, some fairly recent migrants on their way up to steady jobs–began to look to the peripheral squatter and public housing zones as an escape from the heightened densities and "ruralization" of the medinas.

In short, there has been in all these cities a tremendous reshuffling in the allocation of urban space, between and among subcities 1, 2, and 3. To these must be added subcity 4, the suburban apartment-house or villa quarters now favored by indigenous commercial and governmental bureaucrats, who have often bypassed many of the previous "stages" of population succession to settle initially in new housing built neither on the old medina pattern nor on the unmodified European model. One can anticipate that this group will come into increasing conflict with the bidonville dwellers for attractive peripheral sites.

The net result of all this sifting and sorting has been the evolution of North African cities which are ordered no longer by ethnicity or caste (and my definition here of caste is simply economically stratified ethnicity) but increasingly according to more contemporary principles of class. One can trace this transformation easily, seeing it most advanced in Cairo, somewhat less so in Tunis, barely under way in Rabat-Salé. The emerging cleavages are different in kind than those which were the dominant form in the past. Ethnically they are less sharp, and perhaps therefore less visible, but in many ways they are far more dangerous cleavages for the future of North African societies.

In Egypt, class differences have declined rather sharply over the past few decades, which has significantly mitigated the problem. One notes in today's Cairo a certain blending together of the separate "cities" and a steady widening of intermediary buffer belts of working-class but modern status. In Tunis, as we have shown, the caste differences between old and new quarters were never quite so sharply drawn as in Cairo or Rabat-Salé, and even the past ten years have succeeded in blurring the lines a bit more. However, new lines seem to be appearing in the suburbs. In Morocco, on the other hand, the contrast is still extreme and obvious; the old foreign-native cleavage has been converted to a Moroccan class cleavage which is almost as great. This is particularly apparent in the results of an analysis that I have just completed on the factorial ecology of the city, using data from the most recent census. In that analysis, the first factor (accounting for 50% of the variance in a correlation matrix of 27 variables computed for 532 census tracts) is clearly a *caste* factor, in which ethnic and socioeconomic variables are so tightly entailed that they cannot be separated analytically (Abu-Lughod, 1974).

This appears to be an initial stage of transformation, found as a society begins decolonization. However, if the castelike structure persists and is not modified by a gradual social "revolution" in which power and resources are redistributed in a more egalitarian fashion, these sharp class cleavages, which have highly visible projections in the physical form of the city, may translate into more violent change.

It is interesting to note that the event which presignaled the start of the Egyptian revolution, predating by some months the military action by Nasser's group, was a highly significant and symbolic urban incident. On this day Egyptian Cairenes surged out of the "medina," crossed the invisible but nonetheless important barrier which divided the social city into "native" and "modern," and proceeded to burn down, selectively, many symbolic buildings and institutions tainted by their past and persistent association with colonialism. This was not an act of national liberation against a foreign colonizer, for the English had ostensibly left decades before, but a reaction against what might now be termed "neocolonialism," i.e., against the fact that things change but remain the same. In Algeria, the abrupt nature of the French exodus and the rapid manner in which squatters, independent of the usual processes of "sifting and sorting," filled up those spaces without regard to class, tended to eradicate the tracemarks of the colonial era without this symbolic gesture. In Tunisia and in Morocco, however, the more "orderly" process whereby an indigenous elite gradually moved into the positions of power formerly held by Frenchmen and into the zones of the city formerly occupied by them has left ragged lines in the physical and symbolic structure of the cities which must be knitted together in the future.

SUMMING UP

The cities of North Africa are at a critical turning point in their history. Even as they continue to grow, the ideological underpinnings of their former functions are being questioned, and there is a tendency to disinvest in them, on the assumption that the most basic inequities lie not between classes within the city so much as between the haves of the city and the have-nots of the rural areas. Neocolonialism may, indeed, widen the latter gap, as adherents to the "development of underdevelopment" theory contend. There is no doubt that attention must be paid to internal decentralized development, if migration to the largest cities is to be prevented. But as we have tried to point out, this will not stabilize the cities in which class cleavages grow wider. Reversing the widening class gap does require investment in the cities—differential investment to improve the lot of those at the base of the urban class structure. This should be done not in preference to development outside the cities but in addition to it.

NOTE

1. A concerted national policy in Tunisia has resulted in the sole exception, the phenomenal development of Sfax as a secondary port.

REFERENCES

ABU-LUGHOD, J. (1965). "Urbanization in Egypt: Present state and future prospects." Economic Development and Cultural Change, 13(April):313-343.

--- (1971). Cairo: 1001 years of the city victorious. Princeton, N.J.: Princeton University Press.

--- (1973). "Problems and policy implications of Middle Eastern urbanization." Pp. 42-62 in Studies on development problems in selected countries of the Middle East, 1972. New York: United Nations.

--- (1974). "Factorial ecology of Rabat-Salé, Morocco." Unpublished paper, Northwestern University Comparative Urban Studies Program.

--- (1975a). "The legitimacy of comparisons in comparative urban studies: A theoretical position and application to North African cities." Urban Affairs Quarterly, 11(September):13-35.

--- (1975b). "Moroccan cities: The serendipity of conservation." In I. Abu-Lughod (ed.), African themes. Evanston, Ill.: Northwestern University Press.

ADAM, A. (1968). Casablanca, éssai sur la transformation de la société marocaine au contact de l'occident (2 vols.). Paris: Centre National de la Recherche Scientifique.

AMIN, S. (1970). The Maghreb in the modern world: Algeria, Tunisia, Morocco. Middlesex, Eng.: Penguin.

ATALLAH, B., and FIKRY, M. (1972). "Le phénomène urbain en Libye." Annuaire de l'Afrique du Nord, 11:79-103.

BENYOUSSEF, A. (1967). Populations du Maghreb et communauté économique à quatre: Equisse d'une théorie démographique de l'integration. Paris: Saeb.

BLAKE, G. (1974). "Urbanization in North Africa: Its nature and consequences." Pp. 67-80 in D.J. Dwyer (ed.), The city in the Third World. New York: Barnes and Noble.

BRAUDEL, F. (1972). The Mediterranean and the Mediterranean world in the age of Phillip II (2 vols.; S. Reynolds, trans.). New York: Harper and Row.

Centre de Recherches et d'Etudes Démographiques (1975). Projections de la population marocaine. Rabat: Moroccan Secretariat of State for Planning and Regional Development, Office of Statistics.

CHANDLER, T., and FOX, G. (1974). 3000 years of urban growth. New York: Academic Press.

CLARKE, J.I., and FISHER, W.B. (eds., 1972). Populations of the Middle East and North Africa: A geographical approach. New York: Africana.

DAVIS, K. (1969). World urbanization, 1950-1970. Vol. 1: Basic data for cities, countries and regions (Population monograph series, no. 4). Berkeley: University of California, Institute of International Studies.

ECKERT, E.H., and EL-KEFI, J. (1974). "L'Espace traditionnel de la ville de Tunis: La Medina et les deux rbat, faubourg or gourbiville?" Pp. 211-235 in Les influences occidentales dans les villes maghrebines à l'époque contemporaine. Aix-en-Provence: Editions de l'Université de Provence.

EL-SHAKHS, S. (1974). "Spatial development and the future system of settlements." Paper prepared for the Libyan Arab Republic Ministry of Planning. Rome: Italconsult.

Groupe Huit (1971). Les villes en Tunisie (2 vols.). Tunis: Tunisian Ministry of National Economy, Office of Territorial Planning.

IBRAHIM, S. (1975). "Over-urbanization and under-urbanism: The case of the Arab world." International Journal of Middle East Studies, 6:29-45.

ISSAWI, C. (1969). "Economic change and urbanization in the Middle East." Pp. 102-119 in I. Lapidus (ed.), Middle Eastern cities. Berkeley: University of California Press.

JOLE, M, KHATIBI, A., and MARTENSON, M. (1974). "Urbanisme, ideologie et segregation: Example de Rabat." Pp. 161-175 in Les influences occidentales dans les villes maghrebines à l'époque contemporaine. Aix-en-Provence: Editions de l'Université de Provence.

MIEGE, J-L. (1961). Le Maroc et l'Europe (1830-1894). Paris: Presses Universitaires de France.

MONTAGNE, R. (ed., 1952). Naissance du proletariat marocain. Paris: Peyronnet.

NACIRI, M. (1973). "Les Formes d'habitat sous-integrées: Essai methodologique." Unpublished paper, Rabat.

SANTOS, M. (1970). Dix essais sur les villes des pays sous-developpés. Paris: Editions Ophrys.

TIANO, A. (1968). Le développement économique du Maghreb. Paris: Presses Universitaires de France.

Tunisian Ministry of National Economy, Office of Territorial Planning (1972). Tunis 72-76: Composantes actuelles et objectifs quadriennaux d'aménagement. Tunis: Author.

9

West African Urbanization

M.L. McNULTY

☐ WEST AFRICAN URBANIZATION presents an interesting pattern of diversity and contrasts, deriving from the rich variety of indigenous cultures and the differential impact of French and British colonial policies. Yet permeating this pattern of diversity are several distinct and significant similarities which stem from the basic process responsible for the growth of West African urban centers—colonialism. The initial structure of the urban system was established with colonial objectives in mind, and subsequent growth has been more of an elaboration of this pattern than a transformation.

The remarkably rapid growth of urban populations in the decades since the Second World War is presently straining the capacity of these colonial systems and creating massive problems for the now independent governments. The urban legacy of colonialism was never meant to provide the basis for achieving the objectives of economic development and national integration set by independent governments. Little wonder, then, that the urban infrastructure is proving inadequate to the task. What we are witnessing in West Africa may not be so much the "transformation of the urbanization process" (Berry, 1973) as the logical consequence of a much more global process culminating in the "underdevelopment" of the Third World. That global process is associated with what Wallerstein (1974) has characterized as the expansion of the "modern world system," composed of a core of Western industrial nations and a periphery of Third World primary producing nations linked together in a dominance-dependence relationship.[1] The argument that emerges from the literature on dependency theory is summarized by Friedmann and Wulff (1974):

> Basically, it involves the notion that powerful corporate and national interests, representing capitalist society at its most advanced, established outposts in the principal cities of Third World countries essentially for three related purposes: to extract a sizable surplus from the dependent economy, chiefly in the form of primary products, through a process of "unequal exchange"; to expand the market for goods and services produced in the home countries of advanced monopoly capitalism; and to insure stability of an indigenous political system that will resist encroachment by ideologies and social movements that threaten to undermine the basic institutions of the capitalistic system.

Whether one fully accepts this argument or not, there is little question that urbanization in West Africa must be viewed in the international context which has so profoundly affected its basic nature and form. The purpose of this essay is to examine the pattern of urbanization in contemporary West Africa, to discuss the major forces which have contributed to its development, and to identify contemporary changes which will affect the future of West African urbanization.

THE CONTEMPORARY MAP OF WEST AFRICAN URBANIZATION

The contemporary map of West Africa is revealing in the scarcity of major urban centers. Only about 40 cities had populations greater than 100,000 during the early 1970s. Of these, more than half were located in Nigeria, where a strong urban tradition among the Yoruba and a large population (estimated at 80,000,000 in 1973) gave rise to major urban concentrations (Mabogunje, 1968). In the other 14 countries which comprise the region, few have more than one major urban center.

With the exception of Nigeria, the countries of West Africa have relatively small populations. Nigeria alone accounts for well over half of the estimated 112,000,000 people in West Africa. Population growth rates for the period 1960-1970 range from 1.4% for Sierra Leone to 3.9% for Nigeria. The percentage urban is relatively modest throughout West Africa with Ghana (34%), Liberia (30%), Ivory Coast (29%), and Senegal (27%) ranking as the most "urbanized." However, these can be considered only approximations since no standard definition of urban population is used in the reporting of figures.

Despite the relatively low percentage of urban residents and a continued reliance upon agriculture, the countries of West Africa are experiencing very rapid rates of urbanization. In most countries, urban populations are growing at rates two or three times those for total population. Much of this urban growth is occurring in the major cities of the region. The principal cities are growing at substantially higher rates than the total population and often contain as many as half of the total urban population of the countries in which they are located. The result is that several West African cities are among the fastest growing in the world.

This rapid growth of the principal cities is accounted for in large measure by the continuing influx of migrants from rural areas and smaller urban centers (Hance, 1970). Estimates suggest that the population of Abidjan increased by 129,000 between 1955 and 1963 with migrants accounting for approximately 76% of the total population increase. Similar figures for Lagos indicate that as much as 75% of the increase in population between 1952 and 1962, an estimated 393,000, was due to migration (International Bank, 1972). This migration to urban centers in West Africa involves relatively young migrants, often school-leavers, seeking employment opportunities and social environments more in keeping with expectations that have been raised by education and discussion with former migrants.

Despite the often serious inadequacy of the urban infrastructure, the environment of the city has been conducive to birth rates which are often as high as those in rural areas. Thus, the growth of West African cities is assured, even if massive rural to urban migration were to be greatly reduced—a prospect which seems highly unlikely under present conditions.

The distribution of principal cities on the contemporary map of urbanization is equally revealing. Except in the landlocked countries of Mali, Niger, and Upper Volta, the largest city in each country is located on the coast and serves as the capital city. With few exceptions, the principal cities of West Africa (those with more than 100,000 people) are located within 100 miles of the coast, reflecting the general pattern of development in the region. The map of West African development is characterized by marked regional inequalities. The level of development, as measured by standard indices, is greatest along the coast and rapidly decreases as one moves toward the interior (Riddell, 1970). Even along the coast, however, development is highly concentrated in and around the principal cities, giving rise to a number of "development islands" surrounded by relatively poorly developed areas (Green and Fair, 1961).

The proportion of the total population living in urban areas is positively related to per capita gross domestic product (GDP), although not as highly correlated as in some other world regions. The relationship between urbanization and percentage share of manufacturing in total GDP is much stronger and indicates that, despite the low levels of manufacturing employment throughout West Africa, it is a significant correlate of urban growth.

A LITANY OF URBAN ILLS

Almost any account of Third World urbanization or cities reads like a litany of seemingly intractable problems. What is more, by interchanging a few names and adjusting some figures slightly, the litany is depressingly similar throughout much of Asia, Africa, and Latin America.

> In Metropolitan Lagos, for example, chaotic traffic conditions have become endemic; demands on the water supply system have begun to

> outstrip its maximum capacity; power cuts have become chronic as industrial and domestic requirements have escalated; factories have been compelled to bore their own wells and to set up stand-by electricity plants; public transport has been inundated; port facilities have been stretched to their limits; the conditions have degenerated over extensive areas within and beyond the city's limits, in spite of slum clearance schemes; and city government has threatened to break down amidst charges of corruption, mismanagement and financial incompetence. Moreover, although employment opportunities have multiplied in industry, commerce and public administration, there is no doubt that thousands of in-migrants have been unable to find work, and the potential for civil disturbances has increased. [Green and Milone, 1972:14-15]

To understand this pattern of development, one must examine some of the forces which have given rise to urban and economic growth in the region.

THE FORCES OF URBANIZATION AND DEVELOPMENT

The urban systems of West Africa reflect the influences of several forces which in the past, as well as in the present, affect both the relationship between cities and the internal structure of urban land use. These influences derive from (1) the indigenous cultural systems, (2) the impact of colonialism, and (3) the drive to national independence.

While the particular influences of one or the other of these forces differs from one county or city to another, all are evident to some degree. The interaction of these forces have produced the contemporary map of West African urbanization and will continue to influence urban development in the future. The elements of urban structure derived from the indigenous cultural systems and colonial impact continue to influence the growth of the contemporary city, and conflicts arising from these forces are creating serious problems for future growth.

MAJOR PHASES OF URBANIZATION

In order to understand the impact of each of these forces, it may be useful to describe briefly the major phases of West African urbanization and the cities associated with each. Rayfield (1974) distinguishes three phases of urbanization in West Africa. Each phase was associated with trade and gave rise to cities which functioned as administrative and commercial centers. Each phase blended into succeeding phases and the growth of newer cities often displaced existing centers or rendered them obsolete.

Phase One–Cities of the Western Sudan. Some of the earliest cities in West Africa were associated with the rise of powerful kingdoms in the western Sudan where control over the southern termini of the trans-Saharan trade in gold, slaves, salt, and forest products provided an important impetus to urban and political development. Important urban centers such as Kano, Gao, Kukawa, and the legendary Timbuktu grew as the major entrepots at the southern edge of the trans-Saharan trade routes (Bovill, 1968; Hopkins, 1973). These towns represented a core area whose periphery extended into the forest region to the south connected through a series of smaller markets and trade routes.

Phase Two–Cities of the Guinea Coast. The opening of sea routes to the Guinea Coast by Portuguese navigators and the growing political instability along the trans-Saharan routes set in motion a major reorganization of the spatial structure of West Africa. Trading posts and castles proliferated along the coast as competing European interests began to stake out trade areas, and settlements such as Axim, Elmina, Cape Coast, Saltpond, Accra, and Keta grew around the ports. With the decline of the slave trade in the 19th century, many of the smaller settlements became defunct. However, some of the more favorably located centers began to prosper as the search for new products gave rise to a diversification of the coastal economies.

Phase Three–Rise of the Colonial Cities. The increasing demand for raw materials and the opening of new markets led to an intensification of trade between West Africa and Europe. During the 19th century, European colonialism was formalized, and the rival powers were able to set about the exploitation of West Africa in earnest. The need for improved transport and a colonial administrative structure favored the growth of those centers such as Abidjan, Sekondi, Accra, and Lagos which served as major ports, administrative headquarters, or both. This phase of urbanization was characterized by a growing economic dependence on Europe and the creation of a system of cities to serve colonial ends. Those towns, even important indigenous centers such as Oyo in Nigeria, which did not fit into the colonial spatial scheme suffered serious economic and political decline.

Rayfield (1974) sees the contemporary city as a continuation of the colonial city. While this is in large measure true, it may be that West African urbanization is entering a fourth phase, characterized by the attempts of African governments and planners to create urban systems which more adequately reflect and serve national economic and social objectives. The success of these efforts would have a profound effect upon the future pattern of development in West Africa.

Throughout the major phases of urbanization, interregional and international trade has been the principal impetus to the growth of urban centers. West Africa has a long history of contact with areas outside the region, and these contacts were greatly intensified by colonial penetration, which created dependent economies and established cities to facilitate colonial administration and economic exploitation.

THE COLONIAL IMPACT

In terms of the contemporary urban system, the colonial experience has had the most profound effect. The old cities and capitals of the western Sudanese empires have either disappeared or remain as mere shadows of their original importance. Once the principal trade routes were refocused toward the coast, these northern cities withered and were rapidly overshadowed by new competition along the coast. The reorientation of the space economy of West Africa, occasioned by the opening of sea routes by the Portuguese and hastened by political turmoil along the Saharan trade routes, resulted in a complete reversal in the roles of the respective areas—the core become the periphery.

The West African cities which form the nuclei for this new core area along the coast reflect the particular economic-political-cultural environment in which they developed and the functions that they were expected to perform. Thus, despite the existence of some precolonial cities in West Africa, most contemporary cities reflect their origin as part of a colonial economic and administrative structure. Where they existed, traditional urban patterns have been overlain by a colonial pattern (Mabogunje, 1968).

This impact of colonial rule is apparent in the urban systems of West Africa as well as in the internal spatial structure of individual cities. The effect at the national level will be briefly outlined before addressing the nature of land use within colonial cities.

COLONIAL CITY SYSTEMS

The criteria under which the West African urban system began to develop were primarily established to serve colonial ends. In large measure, the present city systems of West Africa are composed of centers which were intended to effect efficient organization of an export-oriented colonial economic system. This role is evident in a number of system characteristics, including (1) coastal orientation, (2) peripheral locations, (3) simplistic structure, and (4) polarization.[2]

Coastal Orientation. The growth of the urban systems and transport networks of West Africa reflects their heavy dependence on foreign markets and sources of imported goods. The major terminal points for these systems are located on the coast, and transport links push directly inland to sources of exportable products. The origin of many of these railway lines, which played such important roles in affecting the spatial structures of the colonial economies, lay in the need to tap areas of economic wealth (for shipment to Europe) and to establish "effective control" as designated by the Berlin agreements. The result was a transport system which afforded relatively rapid movements to the coastal towns and ports but which severely limited lateral movement. Commenting on this colonial structure of railway lines, Green and Seidman (1968) note:

> Where raw material exports were not of high value, railways tended to be attenuated, many of them narrow gauge, with low carrying capacities and speeds due to inadequate road beds and rails. West Africa was carved up by competing colonial powers for whom inter-territorial railway systems had no economic significance. No railways cross monetary zone boundaries. Few even cross national frontiers.

In some instances, this coastal orientation pattern has a potentially beneficial, albeit probably unintentional, influence. In West Africa, at least, ecological zones are arranged generally in bands that vary from the south to the north, so that the colonial transport systems traverse diverse agricultural regions and provide the basis for interregional trade. This may, as in the case of Nigeria, stimulate the growth of interregional trade and serve to foster a greater degree of economic integration (see Hay and Smith, 1970), although this does not ensure the development of equally strong political and social ties, as the case of the Nigerian civil war tragically illustrates.

In other areas:

> Export-oriented planning and imperially constituted frontiers have led to transport systems . . . which are weirdly irrelevant to population patterns, potential market areas, and natural sites for the development of industrial-agricultural complexes. [Green and Seidman, 1968]

Whereas the primary task of the leaders of independent African nations is in the *integration* of regions and peoples, the colonial administrations often functioned on a principle which had the directly opposed intent of maintaining regional separateness—a principle which served both their economic and their political interests. Is it surprising, therefore, that the colonial urban systems are in many cases inadequate or inappropriate to the task of fostering the growth of strongly integrated regional economic systems?

Peripheral Locations. Even a cursory examination of a map of West African urban systems reveals the tendency for the major urban centers (and often the national capitals) to be located on the periphery of the national territory. Again, there were generally good reasons for this pattern in a colonial situation and the resulting urban systems reflect the political, economic, and technological milieus under which the systems developed. Major administrative and trade centers developed in a manner so as to minimize the distance between the colonial economy and the metropolitan countries; thus, locations near the coast with access to sea routes to Europe were favored. They also derived from the first impact of new technologies which entered the countries through coastal entrepots and diffused inland. The result was a system of cities organized and located so as to maximize contact with the foreign power rather than to maximize access to the national territory.

Needless to say, this pattern of development has greatly benefited the inhabitants of these coastal areas, and now many of them are reluctant to forego this privileged position which a reorientation of the national space economy might require. However, it is clear that the colonial period has left large areas of the continent with very little in the way of economic structure, social services, and the like. This has resulted in the rather marked patterns of regional inequities which characterize most West African nations.

The effect on national consciousness and pride of a capital city being located on the national periphery has been a matter of concern in a number of countries.

Such considerations have led to discussion of alternative sites for new capital cities, located in more accessible positions from the perspective of the national territory. *West Africa* (1975) recently reported on Nigerian arguments concerning this issue.

> The Federal Government has been warned against allowing itself to be "stampeded" into accepting Lagos as permanent capital. The editor of *New Nigerian,* Malam Turi Mohammadu, launching a book *Dialogue on a New Capital–A Political Analysis* by Dr. Nnamdi Azikwe at Kaduna, said that "as Nigeria's capital, Lagos makes a poor comparison with other capitals of the world." Apart from the environment, Lagos lacked the space for expansion. The difficulty of living in Lagos had discouraged many from accepting Federal jobs. This had affected establishment of a truly Federal Civil Service.
>
> Dr. Azikwe suggests that the federal capital should be moved to Kafanchan; Malam Turi Mohammadu proposed Kaduna.
>
> The President of Ahmadu Bello University (ABU) Students Union, Malam Adamu Waziri, said the agitation for a capital was the oldest "political riddle." Removal of the federal capital from Lagos would be a final break with colonial experience.
>
> The Nigerian army should decide on a more suitable capital for Nigeria before handing power over to civilians, according to Alhaji Abubakar Abutu, Vice-President of the United Labour Congress–one of the four central labour organizations. Alhaji Abubakar said Lagos was vulnerable to external aggression and lacked space and facilities to expand. The capital should be in a more central point in the interest of security and national development.

Simplistic Structure. The structure of the urban systems of West African countries (here considered as being composed of urban vertexes and transport edges) reflects the generally low level of development of most of these systems. In this regard, the urban and transportation systems of many West African nations exhibit most of the characteristics of poorly developed areas examined by Kansky (1963). The relationship between network characteristics and level of development is not unidirectional but rather reciprocal–low levels of development restricting the growth of the network and these, in turn, adversely affecting the level of development. Thus, poorly developed countries have inadequate transportation and urban systems which tend to retard the progress of development.

In general, as the level of development of a nation increases, the urban and transport systems will exhibit higher degrees of *connectivity*. Clearly, the African patterns of transport illustrate generally low levels of connectivity, with large areas unserved by transport facilities and with movement between even some of the major towns being possible only by circuitous routes. Interconnections between many centers, particularly along the coasts, are difficult, if possible at all, thereby impeding interurban movements of goods and people. Although this problem is severe within countries, it is even more pronounced with respect to movement between urban areas located in different countries, thus curtailing the opportunities for inter-African trade. This is of immediate and grave concern for those industries whose success will depend upon ready access to the largest possible markets. From a colonial perspective, of course, the system was well connected. That is, the transport routes and urban centers served as an effective funnel for the transshipment of raw materials to the coast. However, national development will depend upon the growth of regional specialization and the integration of centers of production, requiring that the national city systems be well articulated internally. In this regard, the present systems may well act as impediments to such development. In order to effect such changes within the systems, vast resources will be required to reorient the existing transport routes.

The characteristics discussed above provide the basis for examining what might be termed the overall *coherence* of the urban and transport systems in West Africa. Coherence may be considered as a general measure of the degree of interdependency among the nodes in the system and reflects the extent of integration between the elements of the system. For the most part, the coherence of the Africa urban systems is quite low. While only sketchy data are available to assess the degree of interdependence within African urban systems, those for which there is data tend to reflect a pattern of flow which demonstrates the role of West African centers as part of a collection and distribution system which is still largely focused upon the ports and dependent upon foreign markets. Internal circulation is generally at a low level and reflects the poor degree of coherence. Adams (1972) examined the external linkages of West African nations by analyzing data describing the international flow of information, people, and trade. The results indicate the continued dependence upon former metropolitan countries in Europe and the meager linkages which exist between West African countries. He concluded that:

> West Africa, at the present time, has two primate cities, London and Paris. National jealousies would appear to preclude the development of a higher-order centre within West Africa itself.

In many of the West African city systems, there is an absence of the characteristics of reciprocity of growth evident in the growth of towns in the more developed countries. Owing to this lack of coherence in the urban system, growth generated in one part of the system may not be effectively transmitted to other parts of the system. What happens in Port Harcourt may be of less significance to Lagos than what happens in Liverpool. This means that the generative effects of urban growth, the "spread effects" of economic development, may not have as widespread an impact as might be hoped.

Polarization. Finally, the marked polarization of development around one or

TABLE 1
PERCENTAGE SHARE OF MANUFACTURING IN WEST AFRICAN CAPITAL CITIES

City	Percentage
Dakar (Senegal)	87
Bathurst (Gambia)	100
Conakry (Guinea)	50
Freetown (Sierra Leone)	75
Monrovia (Liberia)	100
Abidjan (Ivory Coast)	63
Accra (Ghana)	30
Cotonou (Dahomey)	17
Lagos (Nigeria)	35

SOURCE: A.L. Mabogunje (1973).

a few centers has led to the growth of West African cities exhibiting a high level of primacy. Most countries exhibit one or perhaps two centers in which industry, commerce, capital, talent, and the various amenities of urban living have become concentrated. Below this level in the system, urban centers, if indeed they can be so classified, are practically devoid of employment opportunities, services, and other urban facilities. Very often the centers of such concentration are the coastal cities. Given the structure of the transport gradients in poorly developed areas, there are strong economic reasons why such concentrations along the coast have proceeded as they have. These few large centers often emerged from the numerous coastal towns established during the early periods of European trading activities after a period of port concentration (Gould, 1959; Dickson, 1971).

The concentration of colonial administrative functions in the capital city, and the fact that these were often the principal ports for their respective countries, contributed to their growth as the principal commercial and industrial centers as well. Throughout West Africa, the capital cities have become the major foci of commercial and industrial activity. This importance is reflected in the percentage share of manufacturing accounted for by the capital cities (see Table 1). This continuing polarization within the principal cities has become a matter of concern, and several governments are currently seeking ways of encouraging subsequent growth in secondary urban centers.

INTERNAL STRUCTURE OF COLONIAL CITIES

In addition to the effect upon the urban system characteristics discussed above, colonialism has also left an important mark upon the internal spatial structure of many West African cities. This is an often cited characteristic of colonial cities throughout the Third World. As Friedman and Wulff (1974:53-54) point out:

> Constructing broad generalizations about third world cities is always a hazardous business. Yet, if a single fact stands out, it is that cross cultural

studies of urban land use have consistently reported the existence of a "dual city." These colonial cities, as McGee (1967) has called them, owe their dualistic structure to the intrusions of Western capitalism into forms of traditional culture. Existing side by side, and only weakly interrelated, both "modern" and traditional cities display their own morphological patterns and residential behavior.

Brand (1972) has provided an assessment of the colonial impact upon Accra, Ghana, and suggests that it has had a profound effect upon the spatial organization of land use:

> Contemporary Accra is the product of a century of contact and interaction between indigenous and colonial economic systems. However, the urban landscape reflects more evidence of the persistence of colonial forms of organization than it does the accommodation or parallel development of the two systems. Traditional life styles and preindustrial modes of occupance are still apparent in parts of central Accra and in the vast "stranger" communities on the periphery, but, on the whole, the evidence confirms the crystallization of an ecological structure and pattern dominated by exogenous norms and institutions. The spatial impress of European imperialism is a salient feature of the townscape.

Similar descriptions have been given of other West African cities. Mabogunje (1968) commented on this interaction between indigenous and European colonial forces of urbanization with respect to Ibadan, Nigeria:

> Today, and in spite of recent development, Ibadan remains a city with a dual personality. Its pre-European foundation constitutes a significant proportion of the city. Although this has been outstripped by the newer development in terms of area, it still commands attention because of its almost unbelievable density of buildings, their spectacular deterioration and virtual absence of adequate sanitation. Moreover, its inhabitants live their lives apart from the modern immigrants. The differences in their wealth, education, acquired skills, social customs and attitudes emphasize the social distance between the two sections of the city.

The continued existence of these two forms of urban development creates difficult problems for African planners. To what extent should urban concepts derived from European experience be allowed to dictate the future development of West African cities? What elements of traditional urbanism should be preserved? Are the two patterns basically inimical to one another, or is it possible to combine elements from each to create modern, yet distinctively African urban centers. These issues at present are rarely addressed. Contemporary developments seem to suggest that elements of traditional urbanism must be uprooted and replaced with what currently passes for "modern" development. This is perhaps most evident in the capitals of former

French colonies, which have adopted European standards and styles much more readily than is the case in former British colonies. The contrast between the two is illustrated in the following comment by Tamar Golan (1975):

> One needs to fly directly from Abidjan, capital of a former French colony, to Lagos, a capital of a former British colony, to realize that the differences between the two are, perhaps, bigger than those which exist between Paris and London. . . .
>
> Traffic in Lagos has rendered life there intolerable. Abidjan also has its share of rush-hour traffic; but the Houphouet-Boigny Bridge on the lagoon was built just before the area was choked—and not afterwards. As you drive into the main commercial zone of the Plateau, by the newly constructed "rocade," you cannot help admiring its beauty and modern planning. Driving into Lagos centre, on the other hand, be it from the Mainland, Ikoyi or Victoria Islands, is an experience not to be repeated, and only the ordeal of coming into the city from Ikeja Airport can be compared with it. . . .
>
> Arriving in Abidjan, you drive past newly built houses, all modern and lacking in personality. They could have been built anywhere in the world, and have little to do with their particular environment. Driving into Lagos, you have ample time to look around, as your car is stuck in innumerable traffic jams, your driver gets into long arguments with others who are trying to get one inch ahead of him. What you see around you is, however, unmistakably authentic, colourful, unique. . . .
>
> In Abidjan, you see the names of reputable French firms, and may think for a moment that you are driving through a pretty little town in the south of France. Not so in Lagos, where Olatunji "The Greatest Hairdresser in the World," invites you to choose your hairstyle from a selection painted on the door of his "boutique." In a way, flying from Abidjan to Lagos is not much different from flying from Paris to Lagos—in both cases you feel like a tourist first coming into contact with "native Africa."

Is it inevitable that West African cities must give way to buildings that are "all modern and lacking in personality"? Is it necessary that African cities come to remind one of "pretty little towns" in France or Britain? Presently there are indications that where the traditional and European forms of urbanization come together, the former is gradually being replaced by the latter. Urbanization as it affects West Africa is thus tending to produce "convergent paths," creating urban institutions and in some cases urban spatial structure, which are becoming more, not less, similar (McNulty and Horton, 1975).

This issue is well illustrated by the controversy which has centered on the Ghanaian government's attempts to "redeem" Nima, a low-income migrant community in Accra. Nima was one of the areas identified in Brand's study (1972) as representing the strong indigenous sections of Accra's residential structure. Paul Jenkins (1975) recently discussed the planned redevelopment of the area:

> The key facts about Nima are easily reviewed. In an area of half a square mile the 1970 Census found 33,500 people living. It is not a "shanty town"; though there are hardly any multi-story buildings most of the people live in single story, swish, compound houses. Nima has been primarily a "Zongo," crowded with immigrants from the grassland areas of Ghana and the neighborhing states, most of whom were poorly paid or unemployed.
>
> The other key fact about Nima is that for decades it has suffered from government neglect. This can be seen most clearly in drainage and sanitation. Until recently the area had no properly constructed concrete drains, and even now only two drains have been built. Throughout the rest of the area rainfall runs off through gulleys eroded in the laterite, or lies on the surface in stagnant pools. Furthermore, for Nima's population there are only 10 12-hole public latrines (with tanks which need frequent pumping out).
>
> Col. Acheampong's speech at the opening of the NRC's Nima project in late April called these "deplorable conditions . . . a blot on the prestige of Accra as the capital city," and he promised "redevelopment . . . which will materialise in the form of houses including blocks of flats, for hire purchase, outright purchase, or rental." Construction was to be undertaken by "Government as well as private individuals and groups" and the NRC chairman invited "private Ghanaian entrepreneurs to form housing finance and estate management syndicates to supplement the Government's efforts." . . .
>
> Behind what seems to be a miscalculation on the government's part concerning the impact of rebuilding Nima apparently lies a decision not to attempt the main practical alternative. This is, to enlist the involvement of the existing people of Nima in an evolutionary improvement of their own area.

Nima is only one recent example of the process of "rationalizing" urban land use in West African cities. The "practical alternative" to the removal of low income populations living in older, often uncontrolled settlements, has been gaining some respectability in planning circles, but it flies in the face of an important locational consideration. The settlements often are located on sites in or near the central parts of the city—prime development land which powerful private and corporate interests are anxious to acquire. More often than not, these interests have won out, resulting in the forced removal of residents and the development of so-called "low and moderate income" housing which is in fact beyond the means of all but a handful of former residents.

At the heart of such controversy is the question of how urban land will be allocated to particular uses. The idea of a land market is alien to West Africa, yet it is generally becoming an important feature of urban areas. Indeed, many of the foreign consultants, planning firms, representatives of international financial institutions, or representatives of aid agencies point to the lack of such a land market as one of the primary obstacles to improving urban conditions in West Africa.

The author of a recent United Nations publication (1973) makes the following point in his introduction:

> Although most African countries do not face the problem of high population densities as in Western Europe and parts of Southeast Asia, several reasons make it imperative for African countries to adopt urban land policies. First, with cities growing at the fastest rates ever recorded in the present century, the scramble for urban land among various competing land uses has intensified. This scramble is giving rise to poor land management, misuse and abuse of land. Indeed, the scramble would not be necessary if there were an urban land market to regulate land distribution. The structure of land market, however, is burdened with customary land holding practices which impede land transactions. In several instances, urban land prices have been so inflated that over fifty percent of urban families cannot afford to buy plots, let alone shelters. No doubt, there is no easy way to regulate the urban land market and to control land use without an urban land policy.

In a later section of this UN report the author writes:

> In summary, the clouded and unmarketable land titles, the juxtaposition of "imported" land law with customary land law, the relatively undeveloped urban land market and clogged channels of real estate financing are among the major bottle-necks hindering the systematic development of urban land. These problems do not defy solutions, provided that land tenure systems are modernized to facilitate transactions and legal registration of titles and governments take risks to establish real estate financial institutions.

Such arguments for the "rationalization" of the land market have led governments to attempt registration of land, formalization of sales, and establishment of real estate institutions. The operation of such competitive land markets, yielding the "highest and best use" (at least theoretically), is a major force in creating convergent paths of urban form.

Convergence in urban form is being accomplished, at least to some degree, by the adoption of several specific institutional frameworks and strategies of urban planning. These include:

(1) Planning methodologies and instruments carried over from former colonial administrations.

(2) Current attempts to "rationalize" the urban land market, principally along lines established in North America or Europe.

(3) Continued reliance upon foreign planners and consultant firms.

PLANNING WEST AFRICA'S URBAN FUTURE

Urbanization, despite the relatively modest levels attained to date, is likely to become a major feature of West African development. Rapid rates of urban growth, particularly in the largest cities, assure that problems of urbanization will not be easily avoided in the future. The central question is whether African governments will be able to come to grips with the problems presented by rapid urbanization. This partly depends on the way in which this increasing urbanization is viewed and on the extent to which sound planning strategies are formulated.

In some countries it is popular to discuss the need for an "agricultural" revolution and a need to revitalize small towns and rural areas. When 80% or more of the population is presently rural and agricultural, this makes sense. But an unwillingness or inability to prepare for the increasing levels of urbanization could well stymie efforts at future development including agriculture. Herbert (1975) has recently argued that:

> The hope that "the urban problem" will be solved by agricultural development and the development of small towns is a false hope. The arithmetic of population growth, long term agricultural export potential, the relative income elasticities of demand for urban and agricultural output and the prospects for increases in agricultural productivity suggest, to the contrary, that the future of the LDC's is likely to be predominantly an urban future.
>
> One of the great dangers of continuing to harbor the illusion that urbanization will somehow slow down or disappear is that governments in both the LDC's and the MDC's will be slow in moving to develop constructive urban strategies. It is already obvious that a very dangerously low status is given to urban affairs and urban management in most of the LDC's at present.

The low status given to urban affairs is evident in the lack of explicit mention of urban planning in many of the development plans of West African nations. In commenting on this aspect of Nigeria's First National Development Plan (1962-1968), Green and Milone (1972) noted that, although the plan "allocated a total capital expenditure of 42 million Nigerian pounds to towns and country planning (including housing), it contained no national strategy for urban and regional development." Nor did it attempt to assess the likely impact of the sectoral investments which were advocated. The Second National Development Plan could be similarly criticized since "the planning objectives of the industrial sector for 1970-74 make no reference to either urbanization or migration, and the location of government-sponsored industries is to be decided 'purely on economic considerations.' "

Recently, the Nigerian government has taken steps to address some of the criticisms leveled against previous plans. The most recent plan shows considerably more understanding of and need for urban and regional planning. Moreover, the government has recently established a Ministry of Housing, Urban Development, and Environment, thus dramatically elevating the level of concern to the national stage. This will at least begin to establish the administrative machinery necessary for coping with urban growth at the national level. Like Ghana and other former British colonies, Nigeria had very little in the way of planning legislation, depending almost solely on Town and Country Planning ordinances (enacted 30 years ago) which rarely dealt with problems beyond the local level and were addressed more to the needs of physical planning than comprehensive urban and regional planning.

The recently published guidelines for the Five-Year Development Plan, 1975-1980, in Ghana evidence a similar interest in addressing the problems of urbanization more directly. The guidelines identify the main problems in the physical planning aspects of housing to be (1) obsolete planning legislation and defective urban planning machinery and (2) an ineffective land management system. Thus, during the plan period, efforts will be made to (1) promulgate a comprehensive physical planning act to replace the Town Planning Ordinance of 1945 and (2) revise building regulations, define town boundaries, set simple rules and adequate standards, coordinate groups concerned with planning, and accelerate the rate of plan preparation.

The guidelines also articulate a regional planning strategy based on the creation of "growth foci." The adoption of a growth pole strategy is outlined, with a four-tier hierarchy envisaged. The guidelines identify three "national level" growth poles: Accra-Tema, Sekondi-Takoradi, and Kumasi. It further suggests that regional level centers will be selected and services upgraded beginning with Tamale, the principal center designated at this level to be developed during the plan period. The towns to be designated as district and local service centers have not been identified, but apparently a comprehensive study of town functions and service requirements is currently being completed along the lines similar to that employed by Grove and Huszar (1964) is an earlier attempt at classifying Ghana's service centers.

This approach to urban and regional planning is being coupled with efforts to decentralize administrative and planning functions to lower-order centers. The proposals for decentralization are impressive and ambitious and, if successful, could revolutionize the planning process in Ghana. Since the enactment of the Town Planning Ordinance of 1945, there has been slow progress toward expanding planning capacity and extending planning functions to cities outside the capital.

During the period 1945-1959, only Accra, Kumasi, and Sekondi-Takoradi had planning establishments located within them. Between 1960 and 1966, planning offices were established in Tamale and Koforidua. In 1967 efforts at decentralization resulted in the addition of offices at Cape Coast, Sunyani, Bolgatanga, and Ho. Present Plans call for the provision of planning offices in all 60 district capitals as well. The principal difficulty with implementing these ambitious plans for decentralization would appear to be a severe shortage of properly trained planning officers.

The Ghanaians have also established a Regional Resource Planning department within the Ministry of Economic Planning with responsibility for urban and regional planning. Unfortunately, the common separation of physical planning and regional economic planning is maintained, with the Town and Country Planning Department attached to the Ministry of Works and Housing.

Perhaps the most encouraging aspect of these improvements in planning is the fact that they signal an awareness on the part of the government that urban problems are worthy of, and demand, high priority considerations. Moreover, they represent attempts to decolonize the planning process as a prelude to forging a new national system of cities consistent with African needs and priorities. Quite obviously, the goals, values, priorities, and plans of African leaders often differ radically from those of the colonial regimes which they have superseded (although this had not been true in all countries). Consequently, one must ask whether the urban systems which developed under one set of goals and criteria can effectively be used to implement those of another. There is no general agreement on this question. The objectives of a strong, integrated, and developing economy are shared by most of those concerned with such issues; however, there is considerable disagreement as to the most appropriate strategies to be used in pursuit of those objectives. But whatever strategy is adopted, explicit recognition of the spatial implications of the strategy must be made. There is ample evidence that the impact of spatial structure upon the success of development plans has not been fully realized.

Part of the problem of planning for change within an ex-colonial country involves the degree to which former institutions and structures are to be allowed to dictate future ones. Are the colonial structures to be dismantled *in toto* through a general process of decolonization, or should the existing structures be built upon and extended? The former position has been presented by Genoud (1969). Although his comments were directed primarily to considerations of political and economic structures, they are not inappropriate to our discussions of spatial structure:

> Decolonization is not just growth. To decolonize does not mean to improve upon the former colonial administration in the running of the economy. To decolonize, at the very least, is to remove obstacles preventing the autonomous development of the country. And the obstacles are not only economic ones. . . . The first stage . . . is to prepare the bases for the development of a *new* country. The infrastructural investments should not be compared to actual production at the present time but to the new economy-to-be. In a way, the bankruptcy of the old colonial economy is a pre-requisite for the successful implementation of a strategy of decolonization.

Others have argued that the existing facilities, whatever their shortcomings, represent a great deal of sunk capital which cannot be totally dismissed. Whatever the faults of the colonial situation, it has resulted in the existence of a number of growth areas. Logan (1970) has argued that:

> It is essential to keep the growth areas functioning, and, on that basis alone, it is possible to argue that investment in services should first be allocated to these growth areas. In other words, instead of trying to *change* a spatial form that has evolved over a long time, we are arguing that we should extend it and make it more productive. This is not to deny, of course, that welfare investments should be made in other areas, but even here, decisions should be made on the basis of an area's relation to the growth area.

Both approaches lead to the development of systems different than those presently existing; however, they differ in the manner in which this is to be achieved. Realistically, even the most radical departures in planning will not lead to a complete reorientation of the existing systems. As Genoud (1969) points out in reference to the Ghanaian experience,

> the choice of the Accra-Tema-Akosombo triangle as the main pole of industrialization seems to indicate that, even when the development decisions are taken by national planners, the very same elements of comparative costs and benefits and external economies play a prominent role.

It is possible, and in fact highly likely, that the developmental policies of the national planners may incorporate elements of the colonial pattern, but their existence can be rationalized on the basis of national interests. This by no means is meant to imply that planning decisions will be either easier or wiser because they are done with noncolonial objectives and criteria. In fact, it is likely that the decision process will be complicated by the fact that a greater number of interests have to be served under the independent government than under the colonial regime. Regional interests tend to be stronger and more important in the new nations than in the colonial territories. Strong pressures develop to encourage a dispersal strategy. This is particularly true of those areas which were relatively disadvantaged during the colonial period. One hears a great deal of talk about an "equitable" distribution of national resources and numerous references to "sharing the national cake." Meeting these demands while at the same time attempting to maximize the rate of economic development is likely to prove a difficult task for African planners.

One fact which emerges from our consideration of West Africa is that the city has played, and will continue to play, an important role in shaping the map of development. Because of the central role played by urban centers, it may be that, as Herbert (1975) has argued, "The essential tasks have to do not with the slowing down of urban growth, but, on the contrary, with (1) the husbanding of existing resources to upgrade the quality of urban life and (2) aggressive and imaginative use of urbanization to increase the total resource pool."

NOTES

1. The literature on dependency has grown rapidly in recent years. Initially presented by Latin American social scientists (see Frank, 1967; Girvan, 1973), dependency theory has more recently been applied in other contexts (Amin, 1973).

2. Some of the material in this section and the final section is drawn from an earlier paper by the author (McNulty, 1972).

REFERENCES

ADAMS, J.G.U. (1972). "External linkages of national economies in West Africa." African Urban Notes, 6(summer).

AMIN, S. (1973). "Underdevelopment and dependence in black Africa." Social and Economic Studies, 22(March):177-196.

BERRY, B.J.L. (1973). The human consequences of urbanization. New York: St. Martin's Press.

BOVILL, E.W. (1968). The golden trade of the moors. London.

BRAND, R. (1972). "The spatial organization of residential areas in Accra, Ghana, with particular reference to aspects of modernization." Economic Geography, 48(July): 284-298.

DAVIS, K. (1969). World urbanization, 1950-1970, Vol. 1: Basic data for cities, countries and regions. Berkeley: University of California Press.

DICKSON, K.B. (1971). A historical geography of Ghana. Cambridge: University Press.

FRANK, A.G. (1967). "Sociology of development and underdevelopment of sociology." Catalyst, 3:20-73.

FRIEDMANN, J., and WULFF, R. (1974). The urban transition: Comparative studies of newly industrializing societies. Los Angeles: UCLA Comparative Urbanization Studies.

GENOUD, R. (1969). Nationalism and economic development in Ghana. New York: Praeger.

GIRVAN, N. (1973). "Dependence and underdevelopment in the New World and the Old." Social and Economic Studies (special number 22).

GOLAN, T. (1975). "From Abidjan to Lagos." West Africa, (May 5):505-507.

GOULD, P. (1959). The development of the transportation pattern in Ghana. Evanston, Ill.: Northwestern University Press.

GREEN, L., and FAIR, D. (1961). Development in Africa. Johannesburg: Witwatersrand University Press.

GREEN, L., and MILONE, V. (1972). Urbanization in Nigeria: A planning commentary. New York: Ford Foundation.

GREEN, R., and SEIDMAN, A. (1968). Unity or poverty: The economics of Pan-Africanism. Baltimore: Penguin African Library.

GROVE, D., and HUSZAR, L. (1964). The towns of Ghana. Accra: Ghana Universities Press.

HANCE, W. (1970). Population, migration and urbanization in Africa. New York: Columbia University Press.

HAY, A., and SMITH, R.H.T. (1970). Interregional trade and money flows in Nigeria, 1964. Ibadan: University of Ibadan Press.

HERBERT, J.D. (1975). "The urban avalanche." Paper presented at the 141st annual meeting of the American Association for the Advancement of Science, New York, January 30.

HOPKINS, A.G. (1973). An economic history of West Africa. London: Longman.

International Bank for Reconstruction and Development (1972). Urbanization sector working paper. Washington, D.C.: Author.

JENKINS, P. (1975). "Urban redevelopment: The right plan?" West Africa, (June 23):709.

KANSKY, K. (1963). The structure of transportation networks. Chicago: University of Chicago Press.

LOGAN, M.I. (1970). "The process of regional development and its implication for planning." Nigerian Geographical Journal, 13:109-120.

MABOGUNJE, A. (1968). Urbanization in Nigeria. New York: Africana.

--- (1973). "Manufacturing and the geography of development in tropical Africa." Economic Geography, 49.

McGEE, T.G. (1967). The Southeast Asian city. London: Bell.

McNULTY, M.L. (1972). "African urban systems, transportation networks and regional inequalities." African Urban Notes, 6(summer):56-66.

McNULTY, M.L., and HORTON, F.E. (1975). "West African urbanization: Patterns of convergence or divergence?" Technical report no. 44, Institute of Urban and Regional Research, Iowa City.

RAYFIELD, J.R. (1974). "Theories of urbanization and the colonial city in West Africa." Africa, 44(April):163-185.

RIDDELL, B. (1970). The spatial dynamics of modernization in Sierra Leone. Evanston, Ill.: Northwestern University Press.

ROSSER, C. (1972). Urbanization in tropical Africa: A demographic introduction. New York: Ford Foundation.

United Nations (1973). Urban land policies and land-use control measures. Vol. 1: Africa. New York: Author.

WALLERSTEIN, I. (1974). The modern world system. New York: Academic Press.

West Africa (1975). "Calls for new capital." (July 7):793.

10

Urbanization and Underdevelopment in East Africa

EDWARD W. SOJA
CLYDE E. WEAVER

☐ URBANIZATION IN EAST AFRICA is almost entirely a 20th century phenomenon and quintessentially the product of European colonialism and economic exploitation. East Africa today is a region of more than 35 million people, only a tiny proportion of whom are non-African. It has experienced over a thousand years of agricultural history and has been the locus of significant indigenous state formation beginning more than four centuries ago. Some of its rural areas are among the most densely populated in the world, and the ports along its ocean fringe have been in contact with the Mediterranean, the Middle East, India, and even China almost since the time of Christ. Yet perhaps nowhere else does the city appear so artificially and uncomfortably intrusive on the indigenous cultural landscape as it does over most of East Africa.

In the following pages, we will focus upon the patterns of urbanization which have been unfolding in East Africa since the early 1960s, when the three constituent territories of Kenya, Tanzania,[1] and Uganda achieved their independence from Great Britain.

The entire study, however, is embedded in a much larger historical and spatial perspective. Contemporary urbanization in East Africa cannot be understood apart from the more pervasive process of *underdevelopment* initiated during the period of European colonial contact and continuing today within the context of international economic relations. The nature, causes, and directions of urban growth in East Africa are so tightly entwined with the underdevelopment process that to attempt a more conventional descriptive analysis of population growth rates, employment figures, city-size distributions, and intraurban land-use changes divorced from this interpretive perspective would seem inappropriate and tunnel-visioned.[2]

Urbanization in East Africa will therefore be examined from three related points of view: first, *as an expression of the set of societal values,* which most forcefully influences locational decision-making at a given point in time; second, *as a primary component in the system of spatial organization,* which enmeshes the population of East Africa into particular patterns of social and spatial interaction; and third *as an empirical framework,* within which to assess, on a normative basis, the causes and consequences of contemporary problems of economic, social, and political development and the efforts made by the independent states of Kenya, Tanzania, and Uganda to deal with them.

EAST AFRICAN URBANIZATION: SOME GLOBAL COMPARISONS

Several distinctive features make East Africa an unusually attractive case study in comparative Third World urbanization and an exceptionally clear illustration of the imprint of colonial underdevelopment on emerging urban patterns. East Africa, for example, forms a central part of the least urbanized major region in the world. The eastern flank of Africa, from the Sudan and Ethiopia in the north to Malawi and Mozambique in the south, contains over 110 million people, at least 90% of whom were classified as rural in 1970. On virtually every statistical measure of urbanization used by Kingsley Davis in his survey of world urbanization, "Eastern Africa"[3] stands at the bottom of the list along with the special case of Oceania. On a country-by-country basis, Kenya, Tanzania, and Uganda fall fairly consistently below the regional average, and, of all countries with equal or larger populations, only three (Nepal, Afghanistan, and Ethiopia) are more rural.[4]

Given this small percentage base, the *rates* of urban growth in East Africa are very high, as is true for most African countries. But East Africa as a whole is more like parts of south-central and southeastern Asia than the rest of Africa in the degree to which rural population growth has kept pace with the growth of the urban population. From 1950 to 1970, for example, the reduction in the percentage rural for East Africa was 4.3%, a figure less than half that of any other African region and lower than anywhere else except south-central Asia and Oceania (Davis, 1969). Thus, despite high urban growth rates, the rural-urban ratio has been declining comparatively slowly.

The persistence of the rural sector would be even more pronounced statistically in East Africa were it not for the phenomenal growth of a few large cities. Table 1 contains an intriguing list of cities over 100,000 in 1960 which have grown more rapidly in the subsequent decade. Along with Danang, Brasilia, Tijuana, and Mexicali are Kampala and Dar es Salaam, each more than doubling its size in the 10-year period. Not too far behind were Nairobi and Mombasa in Kenya, with annual growth rates of 6.2% and 5.8% respectively. These four cities

TABLE 1

ANNUAL PERCENTAGE GROWTH RATES (1960-1970) FOR SELECTED CITIES GREATER THAN 100,000 IN 1960

City	% Growth/Annum 1960-1970	Population 1960	Population 1970
1. Koriyama, Japan	16.7	103,000	484,000
2. Danang, South Vietnam	14.3	105,000	400,000
3. Brasilia, Brazil	13.5	142,000	503,000
4. Oita, Japan	12.6	125,000	410,000
5. Tijuana, Mexico	11.9	152,000	467,000
6. Tucson, U.S.A.	11.6	227,000	682,000
7. Goiania, Brazil	11.5	154,000	456,000
8. Ghulna, Pakistan	11.4	119,000	350,000
9. Suita, Japan	11.0	117,000	331,000
10. Funabashi, Japan	10.7	135,000	372,000
11. Sagamihara, Japan	9.9	102,000	262,000
12. Hai-k'ou, China	9.6	200,000	500,000
13. Mexicali, Mexico	9.6	174,000	436,000
14. Narayanganj, Pakistan	9.5	148,000	367,000
15. KAMPALA, UGANDA	9.2	135,000	325,000
16. Rabat, Morocco	9.2	228,000	550,000
17. DAR ES SALAAM, TANZANIA	9.0	148,000	350,000
18. Shimuzu, Japan	8.9	143,000	334,000
19. Lyalpur, Pakistan	8.8	402,000	931,000
20. Sapporo, Japan	8.7	524,000	1,206,000

SOURCE: Kingsley Davis, *World Urbanization, 1950-1970,* Vol. 1.

alone accounted for nearly 60% of all urban growth in East Africa during this period.

The overwhelmingly rural character of East Africa is given an added distinctiveness when viewed from a deeper historical and cultural perspective. There are perhaps no other comparable areas in the world in which large, indigenous populations have traditionally shunned so persistently any form of nucleated settlement. Aside from the ancient trading centers along the Indian Ocean coast, East Africa was virtually townless until the last half of the 19th century. Occasionally a chief's compound or the center of one of the larger states in southern Uganda and northwestern Tanzania might cluster together a few thousand people, but this was rare and usually not long-lasting. Furthermore, the predominant settlement pattern in precolonial East Africa was extraordinarily dispersed, commonly consisting simply of separate, relatively autonomous homesteads. Even among those groups described as having some village settlement, homesteads were typically widely scattered over the landscape and showed little outward expression of functional nucleation.

Thus not only were there no cities or towns in the interior of East Africa much over a hundred years ago, there were precious few settlements which could clearly be classified as substantial rural villages. Although significant changes have taken place in the past century, the traditional ethos of settlement dispersal and nonnucleation remains a powerful factor on the contemporary scene, frequently observed but rarely examined explicitly. At the same time, the historical legacy of colonially generated systems of cities emplanted on one of the most nonurban traditional cultural landscapes in the world also continues to be a vital force pervading almost every aspect of urban life.

THE HISTORY OF EAST AFRICAN URBANIZATION

PRECOLONIAL URBANIZATION

Long-distance trade was primarily responsible for the growth of towns in East Africa prior to the colonial period. Trading centers were established along the Indian Ocean coast of what is today Kenya and Tanzania as early as the 1st and 2nd century A.D. Coastal urbanization, however, had very little impact on the interior until the mid-19th century, when Arab and Swahili traders settled along the caravan routes to create such towns as Tabora, Mpwapwa, and Ujiji.

Prior to the mid-19th century, the only significant settlement concentrations in the interior were the royal capitals of such large, centralized states as Buganda and Ankole. Some of these capitals may have contained several thousand inhabitants but were little more than temporary encampments appearing and then disappearing with the vicissitudes of individual political power. By far the majority of the people in East Africa lived in family compounds scattered over the countryside.[5]

A crude regionalization of settlement patterns in precolonial East Africa would thus consist of four major regional types: a coastal belt of trading centers and compact rural villages; areas of ephemeral urban nucleation in the zone of large centralized states in the fertile crescent north and west of Lake Victoria; small pockets of nuclear villages, emerging primarily in response to the threats of disease and external attack, located mainly in Tanzania; and a vast sea of almost totally dispersed homesteads covering virtually all of Kenya and most of Tanzania and Uganda as well. The extent to which the dispersed settlement pattern continued to predominate in Tanzania up to its recent efforts at creating nucleated *ujamaa* villages is indicated in one study (Georgulas, 1967) which estimated that, at about the time of independence, around 175,000 people lived in "gathered villages" in the interior, averaging about 350 per settlement; 280,000 occupied compact settlements along the coast; and 8.8 million lived in dispersed homesteads.

THE EARLY COLONIAL PERIOD

The location, size, and distribution of urban centers in contemporary East Africa is almost entirely the product of British and German decision-making during the first few decades of colonial rule. From the 1880s to the First World

War was a period of radical spatial change in East Africa, during which effective administrative control was established (often over mighty indigenous resistance) and the basic infrastructure of colonial dominance was emplanted. Through the siting of administrative headquarters, the routing of transport lines, and the identification of areas of strategic and economic importance to colonial interests, a whole new system of locational advantage and productive potential was superimposed over East Africa. By the end of the period, the broad outlines of the present pattern of urbanization in East Africa had already become solidified.

The key urban foci of this imposed spatial system were the major ports and administrative headquarters, and their location directly reflected the central strategic and exploitative objectives of the colonial powers. The major seaports of East Africa emerged as the basing points for colonial penetration and as the centers for the export of commercial products from the interior: Mombasa at the head of the railway line through the Kenya Highlands to Uganda; Tanga on the line through the Usambaras to the fertile Kilimanjaro region; and Dar es Salaam on the ocean end of the long Central Railway Line to Lake Tanganyika and the Congo (Zaire) basin beyond. With the possible but only temporary exception of Lindi, which served the extreme southern port of Tanganyika for several more decades, the growth of these ports immediately induced a lasting period of stagnation along most of the rest of the coast. Once bustling centers such as Bagamoyo, Malindi, and Lamu fell into relative decline in the shadow of their growing neighbors, settling once and for all the thousand-year rivalry among the Indian Ocean ports.[6]

The location of the colonial capital was an even more important influence on patterns of urban development. In 1891, the Germans shifted their administrative headquarters from Bagamoyo, a major caravan port, to the tiny, declining settlement of Dar es Salaam, largely because of Dar's more attractive harbor and weaker ties to the Sultanate of Zanzibar. By the time that the Central Railway Line was completed in 1914, Dar was already the largest town in colonial Tanganyika. The growth of Nairobi was even more dramatic. Nonexistent before 1895, Nairobi grew from a small railway construction depot to the capital of the East African Protectorate and the major center for European settlement in the Kenya Highlands within a little more than a decade. Although Mombasa lost its administrative function to Nairobi like Bagomoyo did to Dar, it continued to grow as the major port for both Kenya and Uganda as well as large section of northern Tanzania. Mombasa remained the largest urban center in the region until surpassed by Nairobi in the 1940s and then by both Dar and Kampala since independence. (See Figure 1.)

The situation in Uganda was a little more complicated. Before the turn of the century, the key strategic interest of the European colonial powers in East Africa was the Kingdom of Buganda, dominating the Upper Nile. A major British trading station and fort was established at Kampala, close by to the Baganda royal encampment at Mengo, and the combined Baganda-British colonial

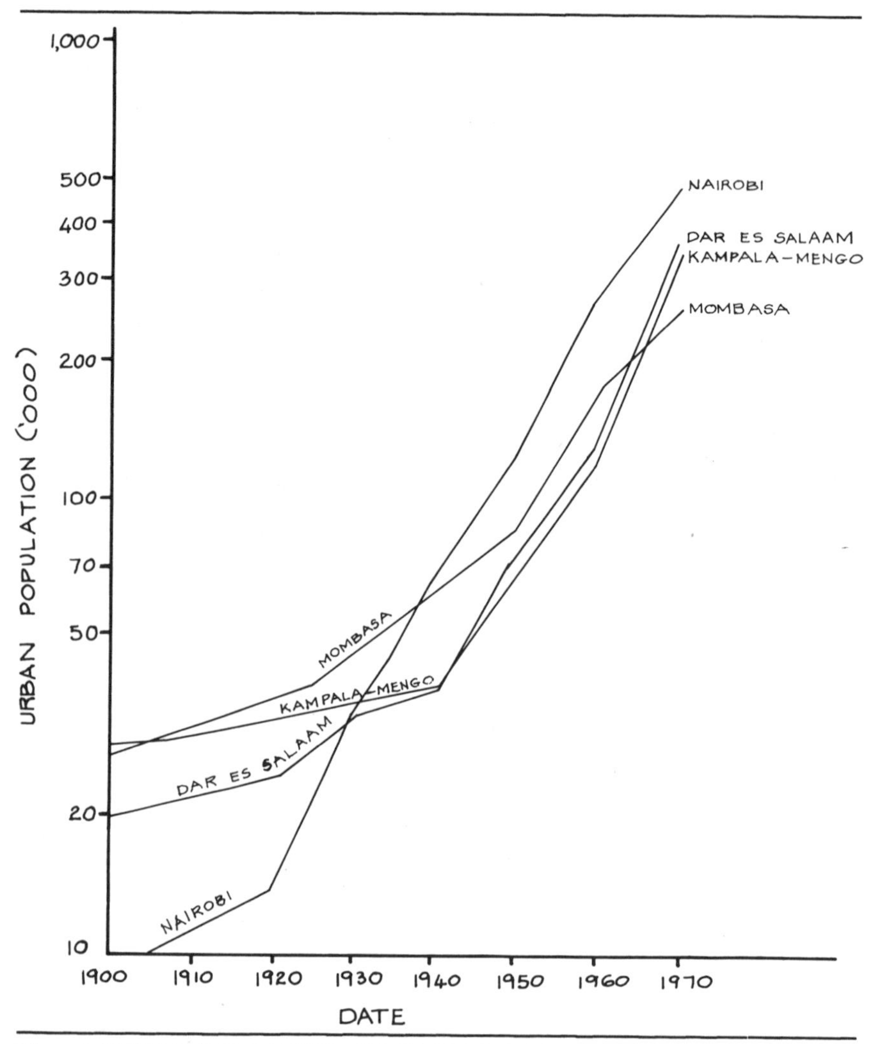

Figure 1: POPULATION GROWTH FOR MAJOR EAST AFRICAN CITIES, 1900-1970

settlement developed rapidly as the major Ugandan town, despite the location of the *de jure* administrative center at Entebbe, some 20 miles to the south. Frozen by the *Pax Brittanica* and colonial rule, the traditionally itinerant royal capital of Buganda remained on Mengo hill and continued to exert an influence on the urban evolution of the Ugandan primate city. In this sense, Kampala resembled the "dual" cities of West Africa (e.g., Kano, Ibadan) more than it did Nairobi or Dar.

Nevertheless, the urban structure of the Kampala region, with its multiple nuclei revolving around the commercial center in Kampala, a traditional African focus at Mengo, several separate church and mission settlements, the harbor at

Port Bell, and the administrative station at Entebbe reflected a general pattern of urbanization identifiable throughout East Africa, particularly in Kenya (Kimani and Taylor, 1973). Like traditional homesteads, town-building functions in East Africa outside the largest centers tended not to cluster together but to be scattered over the landscape. There was first a layer of trading centers distributed in such a way as to tap accessible resource producing areas and to help link the East African interior with international markets. Occasionally these coincided with a preexisting system of traditional marketing points, but more often they were located in accordance with the emerging modern transport network, especially the railways.

Another layer was created through the development of a network of administrative centers, probably the most powerful generator of town growth in East Africa. Strategic and health factors tended to dominate the location of administrative centers, the first leading toward accessible points near the center of major ethnic groups, the second to areas of high ground and good drainage, most commonly some distance away from both local African markets and the Asian *dukas* (shops) of the trading centers.

Mission schools and hospitals created still another layer of nucleation at the local level, but also tended toward separate sites removed from other town-building forces. The result of this remarkable dispersion of urban functions and services was a pattern of small, polynucleated towns, consisting of a few thousand people spread out around several nodal centers, and a rural landscape dotted with isolated services, a school here, a market center there, an administrative station somewhere else. Whether this was a product of the traditional culture of dispersed settlement, a deliberate colonial policy, or simply something which just "happened" is difficult to judge. It is not surprising to discover, however, that at independence the only towns in East Africa larger than 40,000 were the three colonial core cities of Nairobi, Dar es Salaam, and Kampala, plus the large port of Mombasa, and that urban and rural development policy in the three independent states has become particularly concerned with encouraging greater village and town nucleation and promoting the growth of small and medium-sized urban centers.

The period up to the end of World War I, when all of East Africa came under British control, was thus an era of colonial penetration, locational selection, and infrastructural development (Soja and Tobin, 1974). The result was a new network of settlements capped by the colonial core cities and structured by a hierarchy of administrative centers enforcing colonial control and representing imperial interests. Administrative functions were the primary generative seeds of town growth throughout East Africa, particularly when they were locationally associated with the main cash-crop-producing areas. Virtually every city over 20,000 in East Africa today was an established administrative station before 1910. The only partial exceptions are Nakuru and Thika, centers of early European settlement in Kenya, the former already by 1910 considered the "capital of the White Highlands."[7]

URBANIZATION AND UNDERDEVELOPMENT: COLONIAL CONSOLIDATION BEFORE INDEPENDENCE

During the next half century, the system of social and spatial relations emplanted earlier was reinforced and elaborated to incorporate most of the population of East Africa into the colonial political economy and thus more directly into the international economic system. This process of underdevelopment brought with it many beneficial side effects, such as improved health and nutrition, modern educational facilities, better farming techniques, and easier transport and communications. But the *distribution* of these benefits served primarily the interests of British and other colonial powers rather than indigenously expressed demands and local development potential (Leys, 1974; Brett, 1973; Luttrell, 1972). As a result, external dependency was profoundly increased, and great social and regional inequalities than had ever before existed were generated. More important, the underdevelopment of East Africa worked to consolidate a system of unequal exchange—between areas and between different population groups—which continues to dominate the social and spatial fabric of contemporary Kenya, Tanzania, and Uganda.

Urbanization in East Africa, as noted earlier, can thus be best understood as an expression of the more comprehensive and pervasive process of underdevelopment rather than as a phenomenon which can be extracted from its larger societal context and examined in isolation, as is often the case in conventional comparative urban analysis. Several excellent studies of underdevelopment in East Africa have recently appeared, and we will not attempt to repeat the rich historical detail now available in these works. Nor do we feel it necessary here to recapitulate the arguments which attempt to justify the validity and appropriateness of what might broadly be called "underdevelopment theory" to the East African context—the empirical evidence is too strong in our view to choose any alternative perspective. Our main objective in this section is to illustrate the imprint of underdevelopment on the urban systems of Kenya, Tanzania, and Uganda prior to independence and thereby to describe in greater detail the powerful historical legacy so central to an understanding of present-day urban patterns.

East Africa was never a major source of capital accumulation for Britain, and it would be foolish to speak of enormous flows of resources being drained from the three territories during the colonial period. But whether the drainage was large or small is not as important as the fact that the social and spatial structure which evolved were designed to distribute scarce resources in such a way as to promote an exploitative emphasis in economic relations, to facilitate control and domination by a nonindigenous colonial elite, and to solidify a condition of dependency within the international market system. Whatever the benefits of underdevelopment might have been, it has engrained a framework of social and spatial relations and an associated pattern of urbanization which is inherently antithetical to an autonomously controlled and socially just process of development.

As Brett (1973) has noted, there were four main "layers" to the colonial political economy: primary producers almost entirely located in agriculture; processors and traders handling their products; metropolitan export industries established within the colonial market; and the colonial administrative and political apparatus responsible for managing the entire system. The ways in which these layers or subsystems interrelated and affected the allocation of human and capital resources largely determined the patterns of East African urbanization. Each layer, for example, was hierarchically structured and geographically focused around the colonial primate cities of Nairobi, Dar es Salaam, and Kampala-Mengo-Entebbe. These centers functioned during the colonial period to concentrate the wealth produced in their increasingly expanding hinterlands–largely through decisions involving infrastructural location and institutional regulations–and to channel this accumulated wealth for the primary benefit of the colonial elite, both within and outside East Africa. By independence, these late 19th-century creations, plus the Nairobi "outport" of Mombasa, contained almost half the entire urban population of East Africa and proportionately much more of its industry, employment, and urban social services.

Aside from these four cities, plus Tanga (a kind of miniature Mombasa serving the rich, plantation areas of northeastern Tanzania) and one or two industrial centers such a Jinja (Uganda), virtually all other urban settlements were either administrative centers, serving to maintain political control and efficient economic management by the colonial government, or trading centers, functioning essentially as geographical "middlemen" between agricultural producers and metropolitan export industries represented in the colonial primate cities. There was neither need nor reason why any of these centers should grow to substantial size, and not one, even today, contains much more than 50,000 people.

The colonial urban systems thus each contained a single "core" city, a specialized secondary center (Mombasa, Tanga, Jinja), and an array of small administrative and commercial centers acting as control and transmission points within the highly centralized colonial space economy. More than anything else, urban size reflected administrative function modified somewhat by location with respect to the major export-producing enclaves. Transport and communications lines emanated spokelike from the primate cities to tap controllable commercial agricultural areas and a few mineral sites and to connect this skeletal network to the major ports. The entire pattern of spatial organization became geared to an efficient funneling outward of primary resources.

In this context, urbanization and infrastructural development are little more than instruments for the effective management and maintenance of the colonial export economy and the unequal distribution of its derived wealth. Africans were discouraged from settling permanently in urban centers since their labor was needed primarily on the settler estates, on the plantations, and where necessary, on their own peasant farms. Infrastructural services were notably absent from those African agricultural areas which could compete with European plantations and settler farms and thus both divert needed labor and increase wages.

These patterns were most pronounced in Kenya, where the exclusion of an independent African peasantry went furthest due to the necessity of capital accumulation in the settler sector and the resultant net transfer of capital and labor to it from the peasantry.[8] Here was a classical example of underdevelopment and unequal exchange which produced an Africa-subsidized island of European controlled production amidst a series of densely populated, land-starved, subsistence farming "reserves" faced with few alternatives other than to supply a dependable source of cheap labor. This was done in a variety of ways: through the direct expropriation of land and the creation of restrictive "native reserves"; through mass taxation to encourage wage employment, and the diversion of such revenues to construct settler-oriented infrastructural services; through the prohibition of certain cash crops in the key Kikuyu and Luo labor reservoirs; through the imposition of special customs and tariff regulations and legal restrictions which, for example, protected European-growth maize from African competition; through the destruction of indigenous crafts industries with cheap imports; through the maintenance of low wages regardless of labor supply; and often through coercion. The most urbanized subregion in East Africa still remains the former "White Highlands," while most of the adjacent erstwhile African reserves, despite their extremely high population densities and great agricultural productivity today, contain no towns with more than 10,000 people. (See Figure 2.)

In Uganda, precapitalist modes of production remained much more important than in Kenya, and the colonial political economy came more to resemble West Africa, where, in the absence of significant European settlement, the economy revolved around peasant agricultural production dominated by the monopolistic control of colonial import-export houses. Also like West Africa, Uganda saw considerable use made of traditional African organizational systems, particularly among the Baganda, to facilitate and legitimize the new colonial pattern. This resulted in a somewhat less servile form of underdevelopment than in the labor reserves of Kenya, but one which nonetheless created and accentuated social and regional inequalities, protected large European firms engaged in marketing and processing from African and to some extent Asian competition, and prevented the accumulation of sufficient capital and entrepreneurial skills on the part of the African population to sustain the structural changes necessary for autonomous and equitable economic development (Brett, 1973). And the urban system which emerged was not much different from that in Kenya. Kampala was subtantially smaller than Nairobi, which served a much larger settler population, and nothing comparable to the Baganda town of Mengo existed in Kenya. But Ugandan urbanization was equally meager, and infrastructural services were overwhelmingly focused on the Kampala core area.

Tanzania mixed both peasant agriculture and a limited settler economy but

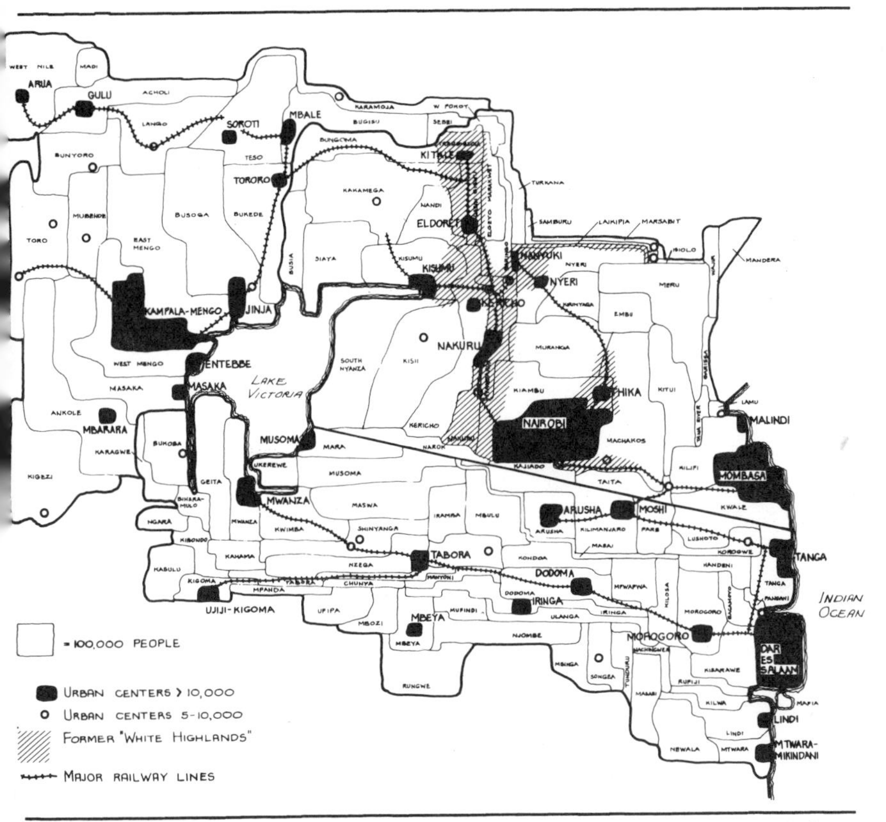

Figure 2: URBAN CENTERS IN EAST AFRICA MAPPED IN "POPULATION SPACE"

was predominantly affected by the establishment of European agricultural plantations. Although plantation agriculture did not require so substantial a net transfer of resources from the peasant sector as did settler farming, it did depend upon the availability of cheap labor and thus upon the inability of certain areas to produce directly for the world market. As in Kenya, infrastructural services were withheld from the major labor reservoirs in the southern highlands, the Kigoma region, and several other areas near the railway lines, setting in motion a process of selective rural decay and neglect akin to what occurred along the coast with the growth of the large colonial port cities (Luttrell, 1972). At independence, therefore, Tanzania represented somewhat of a mixture of both the Kenya and Uganda models.

Industrialization was not totally discouraged in East Africa, but it became primarily oriented to the protection of British capital from lower-cost "foreign" (mainly Japanese) competition and did not help to stimulate African industrial leadership. As Brett notes (1973:299):

> African enterpreneurship was stifled by the effects of marketing and processing policies; this failure to provide them with opportunities to

> accumulate capital, combined with the commitment to British firms, meant that the industrialization process would have to occur through the importation of capital-intensive technology managed by expatriates and controlled from abroad. This then laid the foundation for future dependence on the multi-national corporation for the capital and skills required for further industrialization; it also produced the massive structural imbalances now visible throughout the East African economies which have resulted from the low capacity of this technology to provide non-agricultural employment.

This special form of dependent industrialization, as much subject to oligopolistic control as commercial agriculture, also worked to accentuate the primacy of the core cities, especially Nairobi, where the majority of the industries were located. More important, it generated another source of wage employment for the vast labor reservoirs which, when independence removed many of the colonial restrictions on spatial mobility, poured out thousands of job-seekers who contributed significantly toward making Kampala, Dar es Salaam, and Nairobi among the fastest growing cities in the world.

It can be argued, then, that the underdevelopment of East Africa—with its pervasive nonurban tradition—produced a concomitant pattern of dependent urbanization, filled with paradoxes and major problems for its legatees. Specifically, urban policy during the colonial period was aimed almost entirely at creating healthy oases for Europeans and limiting competition from Asian traders and merchants, who were able to carve an economic niche for themselves in the first place because Africans were considered incapable of providing similar services or, if they could, would divert needed labor from agricultural production. African urbanization was virtually ignored throughout most of the colonial period, while attention was paid to keeping Asians in their "place" —literally and figuratively. Asians were confined to special residential areas in the larger towns and to "gazetted townships" elsewhere. Major efforts were made to prevent large-scale Asian agricultural settlement anywhere in East Africa. Although a certain preferential role was given to the Asian community as middlemen, it was tightly circumscribed by the colonial government to prevent undesired economic and political competition.

Meanwhile, Africans outnumbered all others in virtually every city and town in East Africa from the beginning of the colonial period! Long recognized as "target laborers" who would venture to the alien city for short periods until enough money was obtained to purchase a particular item, Africans also adjusted to their "impermanent" urban roles in other ways. Sprawling peri-urban settlements emerged in the areas around the larger cities, synthesizing one form of rural-urban bond, while a characteristic pattern of circular migration evolved to create another (Weisner, 1972). The latter is perhaps more widespread in East Africa than in any other area in the world. Thousands of African families today maintain essentially two permanent residences, one rural and one urban, and circulate between them on a regular basis, effectively defying an easy classification as either rural or urban.

URBANIZATION SINCE INDEPENDENCE

Urbanization in East Africa since independence has been tightly constrained by the colonial legacy of underdevelopment and the dependent spatial structures it imposed. Despite the efforts of the independent African governments to transform their colonial inheritance, urban development since 1960 has been little more than an intensification of already existing patterns and relationships and the more open manifestation of problems inherent to the political economy of colonialism. This has been especially evident in Kenya, the most deeply underdeveloped of the three territories during the colonial period. The recent and influential ILO Mission on Employment, Incomes, and Equality, for example, asserted straightforwardly that:

> Since independence, economic growth has largely continued on the lines set by the earlier colonial structure. Kenyanisation has radically changed the racial composition of the group of people in the centre of power and many of its policies, but has had only a limited effect on the mechanisms which maintained its dominance—the pattern of government expenditure, the freedom of foreign firms to locate their offices and plants in Nairobi, and the narrow stratum of expenditure by a high-income elite superimposed on a base of limited mass-consumption. [International Labour Organization, 1972:11]

Whereas Kenya has clearly followed a preeminently capitalist path, permitting the entry of some Africans into the dominant elite but changing little else, Tanzania and to some extent Uganda have attempted radically different courses and to a greater degree have managed to weaken the structures of dependency upon foreign and domestic capital. But it has nevertheless been extremely difficult to reorder the powerfully engrained spatial infrastructure of underdevelopment, which persists despite substantial ideological, attitudinal and institutional divergence.

Table 2 shows the population of all East African cities with over 10,000 people at about the time of independence and according to the most recent census data. It is clear from these figures that the degree of population primacy of the three colonial core cities has not only maintained itself but actually increased since independence, both in terms of size with respect to the second largest urban center and in proportion of the total urban population. On most other measures of economic dominance, a similar pattern can be seen, suggesting very little if any major decentralization of industry, employment, urban services, and income, at least by 1970.[9] Economic expansion, however, has not kept pace with population growth, thus swelling the capital cities with thousands of unemployed laborers and creating huge squatter settlements and enormous strain on the provision of adequate urban services.

The mushrooming growth of East Africa's largest cities should surprise no one. The colonial political economy was specifically designed to create a large

TABLE 2

MAJOR TOWNS (10,000+) IN EAST AFRICA

A. KENYA

	1969	1962	Average Annual % Change
1. Nairobi	509,300	343,500*	6.9
2. Mombasa	247,000	179,600	5.4
3. Nakuru	47,200	38,200	3.4
4. Kisumu	32,400	23,500	5.4
5. Thika	18,400	13,600	5.0
6. Eldoret	18,200	19,600	–0.1
7. Nanyuki	11,600	10,400	1.6
8. Kitale	11,600	9,300	3.5
9. Malindi	10,800	5,800	12.3
10. Kericho	10,100	7,700	4.5
11. Nyeri	10,000	7,900	3.8

NAIROBI as % of total town population: 1969 = 55; 1962 = 46
Nairobi-Mombasa ratio : 1969 = 2.06; 1962 = 1.91

B. TANZANIA

	1967	1957	Average Annual % Change
1. Dar es Salaam	272,800	128,700	11.2
2. Tanga	61,100	38,000	6.1
3. Mwanza	34,900	19,900	7.5
4. Arusha	32,500	10,000	22.5
5. Moshi	26,900	13,700	9.6
6. Morogoro	25,300	14,500	7.4
7. Dodoma	23,600	13,700	7.6
8. Iringa	21,700	9,700	12.4
9. Ujiji-Kigoma	21,400	16,300	3.1
10. Tabora	21,000	15,400	3.6
11. Mtwara-Mikindani	20,400	15,300	3.3
12. Musoma	15,400	7,200	11.4
13. Lindi	13,400	10,300	3.0
14. Mbeya	12,500	6,900	8.1

DAR as % of total town population: 1967 = 45; 1957 = 40
Dar-Tanga ratio : 1967 = 4.46; 1957 = 3.38

C. UGANDA

	1969	1959	Average Annual % Change
1. Kampala-Mengo	330,700	122,700*	16.9
2. Jinja	55,500	29,700	8.7
3. Mbale	23,500	13,600	8.0
4. Entebbe	21,100	10,900	9.4
5. Gulu	18,200	4,800	27.9
6. Mbarara	16,100	3,800	32.4
7. Tororo	16,000	6,400	15.0
8. Masaka	13,000	4,800	17.1
9. Soroti	12,400	6,600	8.8
10. Arua	10,800	4,600	13.5

KAMPALA as % of total town population: 1969 = 64; 1959 = 59
Kampala-Junja ratio : 1969 = 5.95; 1959 = 4.13

*These figures have been adjusted as far as possible to reflect intercensal boundary changes and thus to make them comparable with more recent population data. Although the boundaries for Dar es Salaam did not change substantially between the two census dates, there have been some changes in the boundaries of the smaller East African centers, especially in Uganda. It has not been possible to adjust for these changes, and thus some of the annual urban growth rates may actually be artifically inflated.

reservoir of cheap wage laborers primarily for the agricultural export economy and secondarily for the small modern industrial sector, which by its nature was almost entirely concentrated in the largest urban centers. Except as needed for industry and related commercial and administrative services, African urbanization was discouraged and controlled. Independence, however, led to many changes. Most restrictions on African urban settlement were removed, and the Asian monopoly of wholesale and retail trade was weakened by government regulation, eventually inducing a large-scale Asian exodus. Continued rapid population growth combined with substantially increased educational services created a ballooning group of school-leavers and graduates unable or unwilling to find rural wage employment in an agricultural sector already weakened by the departure of many European settlers and landowners. Even in the few pockets of African peasant agriculture which were able to expand export production–in Kikuyuland, Buganda, Sukumaland, and the Chagga area around Kilimanjaro–the numbers of landless were increasing. Under these conditions, it is no wonder that the expanding pool of laborers so effectively generated by colonial policy flooded the big cities searching for work.[10]

A substantial–if not predominant–proportion of urban growth in East Africa since 1960 can therefore be attributed to the rapid expansion of a class of African urban poor which was assiduously restricted from urban settlement during the colonial period. This group has been the source of most of the major changes which have shaped East African urbanization since independence, including the rapid population growth of the capital cities, the emergence of sprawling squatter settlements, and public policy formation with regard to housing and infrastructural development. Particularly in Nairobi, the expansion of the "protoproletariat" (McGee, 1974) can also be linked to the increasing role of multinational corporations and foreign capital investments in general. By 1967, 57% of all manufacturing industry in Kenya was foreign-owned and accounted for 73% of all profits–figures which have almost surely increased since in the view of the ILO Mission on Employment, Incomes, and Equality. To a significant extent, the growing supply of cheap labor inherent in what has come to be called the "informal sector," under constant pressure from the similarly growing "reserve army" of job seekers, has created since independence the urban counterpart to the rural labor reserves which permitted the underdevelopment of East African agriculture.

Beyond the growth of the urban poor, changes in the hierarchy of urban centers have been relatively minor and due primarily to modifications in the distribution of political and administrative powers, paralleling a similar pattern which prevailed before independence. This in part explains why Kampala and Dar, economically and politically subordinate to Nairobi during the colonial period, have grown at a more rapid rate than the Kenyan capital. The phenomenal growth of Arusha–from a little over 10,000 in 1957 to 32,400 in 1967–also reflects an administrative readjustment, with the town being designated the headquarters of the East African Economic Community.

Administrative reorganization has been largely responsible for the rapid growth of Iringa, Musoma, Singida, and many smaller Tanzanian towns, which have been elevated into regional capitals. And all the major administrative centers in Uganda except Fort Portal and Kabale, in the southwestern corner of the country, have significantly increased in size, reflecting their added importance in the postindependence political structure. The only cities in East Africa which have grown substantially without the primary stimulus of expanded administrative responsibilities and powers are the industrial towns of Thika and Tororo and the tourist center of Malindi.

Another striking feature in the population growth figures is the relatively slow growth of medium-sized cities in Kenya. The nine centers between 10,000 and 50,000 in 1969 had an average population of 18,900, up from 15,000 in 1962. The 12 comparable centers in Tanzania, however, averaged 22,400 in 1967 and 12,700 in 1957. The eight centers in Uganda had an average population of only 16,400 in 1969, but this marked an increase from just 7,000 ten years earlier.

Part of the explanation for this pattern lies in the fact that all the medium-sized Kenyan towns, except Malindi, were primarily service centers for the White Highlands and were thus located outside the densely populated African areas. As a result, they were more "artificial" and unlikely to grow as rapidly as the less-settler-focused centers in Tanzania and Uganda. To view the other side of the coin, there can be few areas in the world today which compare with the former African reserves of Kenya adjacent to the now Africanized "White Highlands." The old Kikuku Reserve in particular has become one of the most modern and productive African agricultural areas on the continent, contains over one million people and densities of over 1,500 to the square mile, and yet does not have a single urban center larger than 6,000 inhabitants as of the last census. (See Figure 2.)

At the level of the urban system, therefore, there has been relatively little change in postindependence East Africa beyond the addition of an expanded sector of African urban poor. The colonial primate cities have become even more dominant within their territories, and Nairobi in particular has tightened its ties to and dependency upon foreign capital and multinational companies. The specialized secondary centers of Mombasa, Tanga, and Jinja have continued to grow, but less rapidly than the national capitals. Significantly, the smaller towns have grown fastest in Uganda and slowest in Kenya, paralleling differences in the degree and nature of colonial underdevelopment.

Moreover, the nonelite urban population of East Africa remains highly itinerant (Parkins, 1969; Jacobson, 1973; Elkan, 1967; Weisner, 1972) and probably more rural in character and life-style than in most other areas of the world. To be sure, there are now many more "permanent" African urban dwellers than during the colonial period, and several recent studies have indicated increasingly longer average periods of urban residence (Ross, 1973; Gugler, 1972). But the pattern of "one family-two homes" is still deeply engrained in Kenya and elsewhere in East Africa, as is the pattern of dense peri-urban settlement.[11] In essence, then, the broad systemic framework of urbanization in East Africa today represents more an elaboration than a transformation of the colonial legacy of underdevelopment.[12]

SPATIAL PLANNING AND THE SETTLEMENT SYSTEM

Given the severity of current urban problems and the still tenacious hold of colonially generated social and spatial structures, it is not surprising that the East African states, particularly Tanzania and Kenya, have devoted major policy attention to the planned restructuring of their settlement systems. Each of the capital cities has been the subject of a major internationally funded planning analysis (Nairobi Urban Study Group, 1973; Project Planning Associates, 1968; Uganda, 1968), and both Kenya and Tanzania have incorporated innovative spatial planning programs[13] into their national development plans.

The most radical spatial policies have been introduced in Tanzania, where a series of weakly integrated programs have been established to (a) inhibit the growth of Dar es Salaam, through migration controls if necessary, (b) deconcentrate industry toward nine other "growth poles" (Tanga, Mwanza, Arusha, Moshi, Morogoro, Dodoma, Tabora, Mbeya, and Mtwara), (c) decentralize administrative responsibilities to all the regional capitals, which have been increased significantly in number since independence, and (d) transform the dispersed rural settlement pattern through the creation of *ujamaa* villages.

Of these policies, only ujamaa has led to substantial changes in the settlement system, with over 4 million people in nearly 6,000 villages by the end of 1974.[14] But even ujamaa has been severely constrained by the colonial legacy, being least well received just where its ideological component is most seriously needed: in the major export enclaves of the colonial period (Luttrell, 1972). Whereas from 20% to 60% of the population in such peripheral regions as Mtwara, Lindi, Mara, and Kigoma lived in ujamaa villages by March 1973, the figure for Tanga was 8.8%, Mwanza 2.6%, Shinyanga 1.5%, Morogoro 3.2%, Arusha 2.7%, and Kilimanjaro 0.6%.[15]

Despite government policy and a pronounced anti-urban national ideology, Dar es Salaam has continued to grow[16] to the point that a decision has been made to adopt the most drastic form of spatial "surgery"—the shift of the national capital. Planning and building are currently under way to move the capital city of Tanzania from Dar to Dodoma in an attempt to weaken the Dar-focused spatial and economic structures inherited from the colonial period. In combination with the new Tanzania-Zambia railway line through the southeastern regions, this may cause a significant reorientation of growth toward previously neglected areas of Tanzania. But an adequate evaluation of the impact of these changes must await future studies.

The most comprehensive and coordinated spatial planning apparatus in East

Africa has been developed in Kenya. Whereas Tanzania has suffered from the disjointedness of its spatial policies, with little recognition of the interdependencies between the primate city, the medium-sized towns, and the ujamaa villages, Kenya has created an integrated "growth center" program with specific and related objectives for all levels of the settlement hierarchy. Indeed, the 1970-1974 *Development Plan* gave major priority to this development strategy, with an elaborate chapter containing a detailed map, a list of over 1,000 designated centers, and a requirement that "ministries, Local Authorities and other development agencies . . . consult the Town Planning Department at the initial stage in the consideration of their schemes." Elaborate provincial physical development plans have recently been produced with detailed statistical information on infrastructural services and the distribution of urban functions, and efforts are now being made to prepare an integrated physical development plan for the entire country.

But, although major emphasis is given to a more equitable spatial distribution of urbanization and development, little attention has been paid to actually controlling the growth of Nairobi and Mombasa. Without effective regulations to assure compliance by such influential decision-makers as the Ministry of Education and the political elite and without effective regulations to control migration and industrial location, there appears to have been little impact on the economic primacy of Nairobi and on the pattern of regional inequalities in income and services. It is perhaps too early to gauge the success or failure of the growth center program, but there are indications that it is being deemphasized in more recent economic planning. Whereas the chapter outlining the growth center strategy was the first substantive and specialized chapter in the 1970-1974 development plan, for example, the new plan for 1974-1978 puts it third, behind chapters on "Development of Rural Areas" and "Employment, Population, and Manpower."

Spatial planning in Kenya and Tanzania has contained a number of innovative strategies which have been praised and built upon elsewhere in the world. The overall effect of these policies on the system of settlements, however, has been disappointingly meager thus far, except perhaps for the lowest level of the settlement hierarchy in Tanzania. Despite the powerful national commitment of Tanzania and the comprehensive and coordinated planning strategy of Kenya, the hierarchy of settlements that each inherited from the colonial period has remained stubbonly resistant to planned reorganization.

THE INTERNAL SPATIAL ORGANIZATION OF CITIES

In addition to having special problems of their own, the large East African cities have mirrored internally the same patterns of evolution and change which have characterized the larger space economy of East Africa both before and after independence. As noted earlier, Nairobi, Dar es Salaam, and Kampala-Mengo grew as the key taproots of colonial exploitation and served as the focal points for the economic and political reorganization of East African society. Just as the organization of space at the macrolevel became rigidly compartmentalized and functionally differentiated to support the objectives of a dominant colonial elite, so too did the internal spatial organization of the colonial capitals come to reflect the space-forming imperatives of alien rule.

The most prominent feature in the social ecology of all three cities, for example, was a pronounced racial compartmentalization associated with equally striking differences in residential density, infrastructural services, and income levels. These racial compartments functionally reflected the social and spatial stratification of the larger colonial political economy. The dense pockets of African settlement were, particularly in Nairobi, the urban equivalents of the rural labor reserves, supplying unskilled workers for the European enclave. Asians occupied their own separate niche as traders and middlemen, but were tightly circumscribed in their residential choice and degree of direct competition with European interests. Except for the largest European enterprises (and many smaller ones in Nairobi, where the city core was more dominantly European), Asians dominated most of the central business district retail trade and commerce and similarly penetrated into the African areas, effectively blocking the growth of African enterpreneurial skill and experience.

The European residential areas occupied the choicest land available: in the high ground north and west of the central business district in Nairobi; along the cool Indian Ocean coast in Dar es Salaam; atop the attractive unoccupied hills of Kampala. Even within walking distance of the central business district, densities remained low and the flavor suburban. Here were concentrated the best urban services outside the central business district: paved roads, hospitals, schools, sewers, piped water, electricity, parks, etc.

There have been some notable changes in the colonial capitals of East Africa since independence. Increasing numbers of elite Africans have been able to penetrate the high-income areas as overt racial restrictions were removed and wealth emerged more clearly as the most important determinant of residential choice. Monuments to independence, large tourist hotels, huge new administrative buildings, and foreign consular facilities have been constructed. The Asian presence has been significantly reduced, and the rise of an African commercial class has been felt throughout the urban area. At the same time, squatter settlements have burgeoned as never before within the inner city and around the expanding industrial estates; housing shortages and unemployment within the nonelite African population have magnified to unparalleled levels; and, despite some modification in racial mix, the contrasts in density, income, and public service between the three residential sectors have remained marked. In Nairobi, at least, they have probably increased.

As was argued earlier for the entire system of cities, the internal spatial organization of the three East African captials continues to be shaped by the colonial legacy of underdevelopment. For the most part, the patterns of change which have been unfolding since independence represents an intensification and elaboration of the colonial legacy rather than a significant transformation. The degree to which there has been some transformation, however, is primarily related to the divergent political and ideological paths adopted by the independent governments of Kenya, Tanzania, and Uganda, and this necessitates a separate and more specific analysis of their respective capital cities.

Nairobi. At independence, Nairobi was a classic colonial city. Residential patterns were segregated by race, with Europeans overwhelmingly dominant in Upper Nairobi, Asians in Parklands, and Africans in Eastlands and Eastleigh.[17] Most of the land was owned by Asians, Europeans, and various business organizations, even within the primarily African areas. The modern central business district almost exclusively served the expatriate population, while Africans were forced to depend upon Asian-owned shopping areas to the east of the central business district and scattered through the rest of the city. Large African housing estates had been constructed but were nowhere near sufficient to meet the growing demand of African urban migrants. Squatter villages emerged periodically during the colonial period, but almost all were systematically demolished by the government, often without any attempt to provide alternative accommodation. Only during the closing years of colonial rule was there any serious attempt to provide housing for Africans and their families. Most of the older housing estates (Pumwani, Kariokor, Shauri Moyo, Bahati, Ofafa I) were little more than bachelor quarters made up of single rooms shared by several men.

The major changes in this pattern which have occurred since independence are (1) the penetration of formerly European residential areas and Asian commercial areas by Africans and (2) the rapid growth of "uncontrolled" squatter settlements within the old city boundaries and in the newly incorporated areas on the urban fringe. With regard to the first of these changes, there has probably been much less modification of land ownership patterns than is often presumed. A study by Kimani (1972b), for example, showed that in 1971 individual Asians and Europeans still owned nearly 45% of the city's land, with primarily Asian-owned businesses accounting for another 20%. Moreover, most of the Asian owned land was highly valuable commercial and residential property representing over half the total assessed value of urban land surveyed. Africans, now making up 80% of the total population, owned less than 4% of the city's area, rated at 2.7% of the assessed value.[18]

In another interesting study, Kimani demonstrated that land values in Nairobi were closely correlated with distance from the city center and secondarily with urban density, mirroring the characteristic tentlike structure found in most Western cities (Kimani, 1972a). At the same time, however, analysis of residuals indicated that land in the European sector was consistently undervalued given its location and density, while land in the African sector was just as consistently overvalued despite its much lower infrastructural and amenity standards. This pattern almost surely evolved early in the colonial period and worked effectively over time to subsidize European and to some extent Asian urban settlement in much the same way that the White Highlands were subsidized through the underdevelopment of the African rural economy. Furthermore, given the soaring property prices and rentals which have occurred in the European residential sector since independence due to increased demand from expanding foreign companies, consular offices, and the new elite of wealthy Africans, a situation has been created in which European and Asian sellers have been reaping major financial windfalls while continuing landlords gain disproportional rental profits.

As indicated in Table 3, there has been a substantial decrease in the Asian and European population of Nairobi since independence and a major increase in the numbers of Africans. The Kikuyu in particular have jumped from less than a quarter of the population in 1962 to 37.5% in 1969, reflecting in part the incorporation of primarily Kikuyu peri-urban settlements into the city after independence but also due to other factors which have enabled the Kikuyu to dominate the economic and political life of independent Kenya. As mentioned earlier, a major proportion of African urban growth has been accommodated within illegal squatter settlements. Before independence, there may have been about 20,000 to 30,000 illegal squatters in Nairobi, but by the early 1970s most reliable estimates indicated that fully one-third of the city's population lived in "uncontrolled" settlements, with at least 53,000 in the Mathare Valley area alone (Etherton, 1971).

Over the years, the Nairobi City Council has evolved a policy of selective containment of urban squatters, periodically demolishing most squatter villages while simultaneously trying to upgrade the more persistent and politically influential. In 1971, for example, Mathare Valley was "improved" with major urban services after a cholera outbreak in some sections of the city; and similar improvements, often combined with government-sponsored tenant purchase and site and service schemes, have occurred more recently in the Mathare-Kariobangi squatter area. But like earlier experiments with self-help housing, there are indications that many of the benefits are flowing into the hands of middle-income entrepreneurs and landlords who repurchase and subdivide the original allotments and then charge high rents to the low-income residents (Stren, 1975).

Aside from Mathare-Kariobangi, the other major area of substandard housing which has survived City Council demolition programs is Dagoretti, a peri-urban area west of the old city which was formerly part of the old Kikuyu Reserve. Dagoretti has been expanding largely through privately built rental housing on

TABLE 3
POPULATION CHANGE IN NAIROBI, 1962-1969

	1962 Population	%	1969 Population	%
TOTAL	343,500	100.	509,300	100.
African	196,900	57.3	407,700	80.0
–Kikuyu	(81,700)	(23.7)	(191,400)	(37.5)
Asian	86,900	25.3	67,200	13.1
European	28,100	8.1	19,200	3.7

individually owned land. Powerful political representation in Parliament and the prevalence of private individual land ownership protected Dagoretti from City Council influence and permitted its rapid expansion from 24,000 inhabitants in 1962 to almost 42,000 in 1969. The recently completed Nairobi Urban Study Project points to even more accelerated urban growth, with Dagoretti expected to reach a population of 300,000 in 1980 (Nairobi Urban Study Group, 1973).

In recent years, therefore, squatter settlement has been geographically contained and, for the first time, officially acknowledged to the point of providing some vitally needed social overhead facilities in certain selected areas. But the rapid urban influx continues; new squatter villages are still being demolished; and the low-income urban areas of Nairobi are becoming increasingly overcroweded and unhealthy. In their analysis of the "informal" urban sector, the ILO study claimed that the Kenya government appears to maintain three key beliefs with regard to the urban squatters: they are essentially temporary urban residents or occasional migrants who can be induced to return to rural areas; they are unemployed or sporadically employed and thus contribute little to the urban economy; and any significant attempt to improve their living conditions would only induce additional migration (ILO, 1972:226).

All available survey information, however, suggests a very different picture. The majority of squatters surveyed in Nairobi in 1971 had been there since independence, and two-thirds of the adults had come from other parts of Nairobi and not from the countryside (40% having lived in other, demolished squatter areas).

Less than a third owned land outside Nairobi, and the land that was owned was already directly supporting over 3.5 persons per hectare. Furthermore, only a tiny fraction (5%) described themselves as unemployed, and almost all said that they would stay in Nairobi even if their homes were demolished again. Short of a major transformation of the entire Kenyan economy, all evidence suggests that the squatter settlements will almost surely remain, shifting over the geographical space of the city in response to evolving government actions.

The overall result of government policies in Nairobi, despite a public commitment to low-income housing, has favored middle- and high-income groups in the city. Urban economic inequalities have become intensified through (a) increasing land values—and hence rents—on uncleared land, (b) the co-optation of the tenant purchase system by a middle class of property owners and landlords, and (c) the provision of land with amenities at less than market prices to potential investors instead of to the occupiers through the site and service schemes. An economically depressed and rapidly expanding low-income population has become entrenched in inadequately serviced high density areas while simultaneously helping to sustain through high rental costs the growth of a propertied African bourgeoisie.

Another related aspect of urbanization in Kenya as a whole involves the disproportionate share taken by Nairobi of total urban investment, especially with regard to prestige development. From 1967 to 1971, Nairobi increased its proportion of total private construction in urban areas from 78.3% to 88.7%, and a similar pattern can be presumed for its share of public building, which in any case represents only 15% of all construction (Stren, 1975). On nearly all measures of investment, trade, employment, and earnings, Nairobi's share in comparison to the other major "growth centers" in Kenya has been increasing, accentuating its primacy within Kenya and sustaining its emergence as an international center of business and tourism.[19]

Dar es Salaam. Up to the late 1960s, the patterns of urban development in Dar es Salaam did not differ significantly from those of Nairobi. Urban growth was extremely rapid (see Table 4), racial barriers fell, a small African propertied class began to emerge, and Dar's national primacy was intensified. Government urban policy became publicly committed to low-income housing but nevertheless tended to favor high- and middle-income groups in its implementation. Huge squatter villages mushroomed along the edges of the older African residential sector and adjacent to the growing industrial areas and were periodically demolished in "slum clearance" programs. And Asians continued to monopolize the commercial sector, while the major European residential areas were increasingly penetrated by wealthy African businessmen and politicians. Dar was a smaller-scale, less prosperous version of Nairobi.

What has happened since in Dar es Salaam is a major alteration in the legal and institutional basis of land ownership, housing policy, and urban growth in general, aimed at shifting the social distribution of the benefits of urbanization toward the lower income groups. Due to the absence of sufficient financial resources, a much weaker administrative bureaucracy than exists in Kenya, and a pervasive antiurban bias in the emerging national ideology, the innovative programs which have been introduced have not yet been successfully implemented. But the foundations have been laid for a radical transformation of the social and spatial structures inherited from the colonial period and thus for an increasing divergence in the social ecology of Dar from the Nairobi model.

The Arusha Declaration of 1967, for example, introduced a leadership code whereby political party leaders and higher government officials are prohibited from owning shares or directorships in any private company, receiving more than a single salary, or owning houses for rent. This effectively separated the political leadership from major urban-based enterprise and protected the implementation of urban policy against the private business interests of the politicians and

TABLE 4
POPULATION CHANGE IN DAR ES SALAAM, 1957-1967

	1957 Population	%	1967 Population	%
TOTAL	128,700	100.	272,800	100.
African	93,400	72.5	228,900	83.9
–Zaramo	(46,900)	(36.4)	(62,700)	(23.0)
Asian	27,400	21.3	28,200	10.7
European	4,500	3.5	3,400	1.3

high-ranking civil servants who have been so influential in the urban growth of Nairobi.

In 1971, the central government announced the takeover of all rental buildings with a value of £5,000 and over–roughly 2,900 buildings, in all worth £32.5 million (Stren, 1975). An Office of the Registrar of Buildings was created to administer this program, and arrangements were made to compensate the former owners. The end result was the virtual elimination of wealthy urban landlords, the vast majority of whom were Asians, and a collapse of the Asian-controlled private urban construction industry, from which Tanzania is currently attempting to recover.

Other major innovative programs include the formation of a Sites and Services Directorate particularly concerned with the improvement of squatter settlements, the institution of a new civil service rental policy based on income, and the establishment of a new Tanzania Housing Bank to assist in the construction of low-cost housing in urban areas and ujamaa villages through low-interest loans. Associated with these programs has been a more open acceptance of squatter settlements as permanent components of the urban landscape and of the obligations that this recognition imposes on urban development policy. This has resulted in further expansion of the city's squatter areas, which in 1972 contained 44% of the total population of greater Dar es Salaam (Stren, 1975:282).

The Tanzania government, symbolized most dramatically by its decision to move the capital to Dodoma, has attempted as far as it can to turn its back to Dar es Salaam. This has not yet successfully curbed Dar's increasing primacy or substantially improved the living standards of its low-income population. Tanzania simply does not have the financial resources, the experienced bureaucracy, and the elaborate infrastructure that Kenya evolved largely due to the latter's historical domination of the wider East African economy. As Stren (1975:293) has concluded:

> The tendency in Kenya has been to consolidate the pattern begun during the later colonial period, although relatively successful performance has only covered up problems of equity. Policy in Tanzania has moved progressively away from the past, but performance has bogged down because of inadequate resources.

Nevertheless, the severely curtailed role of wealthy landlords and self-serving politicians, the pronounced shift in the balance of investment and attention toward the low-income population, and the expanded public control over urban development has begun to produce a very different Dar es Salaam from the city of 10 years ago–less comfortably modern to its wealthier inhabitants but more socially equitable for the majority of its population.

Kampala-Mengo. Unstable political conditions since independence, major differences both with regard to the city of Kampala-Mengo and the political economy of Uganda as a whole, and the relative weakness of available statistical data make an assessment of the Uganda capital comparable to that of Nairobi and Dar es Salaam extremely difficult. As Southall (1967:326) notes:

> Kampala-Mengo is interesting because it contains within itself most of the major factors, combined at different strengths, which are found in African cities of quite varied type, such as the older, more traditional West African cities and the newer, European-dominated cities of East and Central Africa. It combines both segregation and the political dominance of a particular African tribe; it includes both European and African controlled land, traditional and modern roles, local African residents of long standing and high status, as well as thousands of temporary migrant labourers of many ethnic backgrounds.

In no other city outside West Africa has a traditional African settlement had such a pervasive and long-lasting influence on the growth of a large, modern urban center. The Baganda royal capital at Mengo blunted and modified the full expression of colonial structures of underdevelopment, while the strength and influence of the Baganda injected a new component into the racially stratified social and spatial structure prevalent throughout the rest of East Africa.

Reflecting this complexity, the urban morphology of Kampala-Mengo is multicellular and functionally dispersed, with much less of the rigid triracial compartmentalization of Dar and Nairobi. Land use, economic status, residential density, and the distribution of various ethnic and racial groups came to be patterned around a series of flat-topped hills and intervening valleys distributed within the dual structure of traditional Mengo and colonial Kampala (Southall, 1967). The center of the Baganda traditional city is on Mengo Hill, where the palace of the former Kabaka (king) is located. To the east and west are the four "ecclesiastical" hills: Rubaga (the center of Baganda Roman Catholic society and culture), Nsambya (a secondary Roman Catholic community now encroached upon by additional activities from the nearby central business district), Namirembe (center of the Anglican Church of Uganda), and Kibuli (capped by the central mosque of the African Muslim community). Around each center have settled the members of the affiliated religious group, with economic status fairly clearly graded according to distance and elevation. Further north, but still partly within the *Kibuga* (the general name for the Baganda royal capital area), are Mulago Hill, site of the main hospital and medical school and occupied predominantly by doctors, nurses, hospital staff, and servants; and Makerere Hill, inhabited almost entirely by the students, teachers, and auxiliary staff of the oldest university in East Africa.

Near Namirembe, but within the former city of Kampala, is Kampala Hill, site of the original British fort (now a museum) and known today as Old Kampala. In sequence eastwards are Nakasero (the slopes of which form the central business district); Kololo (the high-income residential area for all races, which grew after the top of Nakasero was filled with bungalows and gardens), and the lower hills of Nakawa and Naguru, where government housing estates were built

TABLE 5

POPULATION CHANGE IN KAMPALA-MENGO, 1959-1969

	1959 Population	%	1969 Population	%
TOTAL	122,700	100.	330,700	100.
African	91,800	74.8	293,300	88.6
–Baganda	(43,200)	(35.7)	(159,500)	(48.2)
Asian	26,700	21.7	31,500	9.5
European	3,600	2.9	4,300	1.2

for Africans beginning around 1950. On each hill, status tends to decline with lower elevation, thus creating a series of low-income, primarily African residential and commercial districts in the valleys between. The most prominent of these are Katwe, strategically located between Mengo, Nakasero, and Nsambya along the major road link to Entebbe; and Wandegeya, between Mulago, Makerere, and Nakasero. In the southeast, toward Port Bell, is a major area of non-Ganda Africans, particularly the Kenya Luo.

This remarkable urban ecology has probably not changed substantially since independence except for the modifications related to rapid population growth and the recent expulsion of Asians. Table 5 illustrates the marked increase in the African population, especially the Baganda, but is not indicative of the mass Asian exodus which has taken place since 1971.

Up to about 1970, Kampala-Mengo was the only one of the East African capitals which experienced an increase in the number of both European and Asian inhabitants. Since the accession of President Idi Amin, however, a series of government decisions has been introduced aimed at transforming the underdeveloped Uganda economy by fiat more than by the slower, evolutionary path chosen in Tanzania. This has created enormous economic and political turmoil, as well as extensive violence, and prevents any effective assessment of urban change in Kampala-Mengo during the 1970s. What is clear, however, is that the expulsion of noncitizen Asians in 1972, the takeover of British-owned firms in 1973, subsequent efforts to assist and enforce an African takeover of Asian business, and the drastic reduction of Baganda political and economic influence have combined to radically modify the distribution of power in Uganda. With regard to the social ecology of contemporary Kampala-Mengo, little else can be said.

CONCLUSION

Urbanization is not an isolatable process which can be hacked out of its larger societal context and examined through narrow disciplinary filters. The location, size, distribution, and internal social and spatial organization of cities are at once an empirical record of past decisions and a contextual environment which channels and constrains contemporary social action and behavior. To understand the dynamics of urbanization therefore requires a perspective which can encompass the continuous and holistic interaction between social process and spatial form as it unfolds over time.

To survey the patterns of urbanization in East Africa since independence from such a comprehensive perspective is clearly an enormous task and one for which the present analysis can only be considered a beginning. At one level, East Africa represents a very special case in comparative urban studies. Its extremely low level of urbanization, the exceptionally rapid growth of its capital cities, the relative recency of major urban growth everywhere except along the coast, and the remarkable degree of settlement dispersion which has existed throughout its history combine to produce a distinctive urban character which demands specific attention. At the same time, however, the region shares with many other areas of the world a common heritage of colonialism and underdevelopment which has shaped—and has continued to shape—its urban geography. To embed an emphasis on this persisting legacy in an analysis of contemporary urbanization is not simply a matter of ideological orientation or liberal castigation. The process of underdevelopment provides the most comprehensive and persuasive framework for explaining both the present pattern of urbanization in East Africa and the dynamics of urban change.

The pronounced and increasing primacy of the former colonial capitals, the weak development of middle-sized towns, the distribution of urban centers along the main railway lines nearest to the major centers of European influence, the areal concentration of infrastructural services in certain regions and urban residential sectors, the explosive growth of squatter settlements and unemployment, the pervasive pattern of circular migration between rural and urban homes, and many other features of East African urbanization are primarily the products of the political economy of colonialism imposed over a particular local context. Furthermore, these inherited structures of colonial urbanization generate significant costs and constraints for the independent states of East Africa. As a consequence, the degree to which they can and should be transformed has become one of the central developmental issues throughout the Third World and the major source of political and ideological divergence in comparative development strategies.

NOTES

1. Zanzibar, which forms part of the United Republic of Tanzania, will not be discussed in this paper. The name "Tanzania" will therefore refer essentially to the mainland portion of the republic and will be used also as a substitute for Tanganyika when discussing the pre-union period.

2. There are many conceptual and ideological controversies associated with the term underdevelopment. Basically, it will be used here to refer to a process of societal change associated with the creation of social, economic, political, and spatial structures which inherently lead to a dependency upon foreign interests and powerful external influence over local decision-making. Furthermore, these structures—which find one mode of expression in

the patterns of urbanization–revolve around a system of unequal exchange between population groups or classes and between different locations in space. This tends to promote the continuation of substantial poverty, social and spatial inequality, and subordination to the interests of Western capital. Except for the particular emphasis on the spatial structure of underdevelopment, the view adopted here closely matches the approach taken in two recent books on East Africa: *Colonialism and Underdevelopment in East Africa* (Brett, 1973) and *Underdevelopment in Kenya* (Leys, 1974).

3. Eastern Africa includes–in addition to Kenya, Tanzania, and Uganda–Burundi, the Comoro Islands, Ethiopia, the Territory of the Afars and Issas (formerly French Somaliland), Madagascar, Malawi, Mauritius, Mozambique, Reunion, Rwanda, the Seychelles, Somalia, Zimbabwe (Southern Rhodesia), and Zambia. Zanzibar is also included separately from mainland Tanzania.

4. It is also noteworthy that with the recent growth of Lagos (Nigeria) and Kinshasa (Zaire), East Africa and Oceania remain the only regions without a single city of more than a million people.

5. Although dispersed, these family compounds were not socially isolated. An intricate web of social relationships based on clans, lineages, age-sets, and other organizational forms clustered families into larger units, many of which, for want of a better term, have been called "villages." In a few spots, these social units resembled compact rural settlements, but by far the predominant patterns were one of pronounced dispersal.

6. All major urban centers are mapped in Figure 2.

7. It is also interesting to note that none of the long string of ancient coastal trading centers (except Mombasa, Dar, and Tanga) have a population greater than 20,000 today.

8. Although Africans produced 70% of the exports of Kenya up to the First World War, this percentage dropped to less than 20% by 1928, and for many years African productivity declined in absolute value as well. This pattern persisted until the Second World War, when the economy of Kenya began to expand much more rapidly (Leys, 1974:31).

9. Although Nairobi's share of total wage employment in Kenya has been falling, its share of total earnings has increased. Further discussion of this pattern of increasing concentration will be presented in later sections.

10. The apparent "anomaly" of continued rural-urban migration in the face of increasing urban unemployment and the presumed existence of a positive marginal product in agriculture has recently led to a major reevaluation of traditional economic models of labor migration in developing countries. Among the most important of the new approaches is that of Harris and Todaro, who worked primarily in Nairobi. The Harris-Todaro models focus upon urban-rural differences in "expected earnings" and tend to suggest that minimum urban wage laws, increased salaries and fringe benefits, and other improvements in urban wage unemployment, such as have been occurring in Kenya since the 1950s have accentuated the rural-urban contrast and thus accelerated both migration and unemployment.

It is unfortunate, however, that this more "rational" economic explanation of migration and unemployment can too easily be interpreted so as to lose sight of historical reality–to imply, for example, that urban salaries are too high rather than rural wages artificially depressed; to neglect the powerfully engrained cultural dependency upon wage employment, which continues to exist in East Africa and has become redirected from agriculture to the industrial needs of the city by the same political and economic forces that created it in the first place; to let the belief that there does not exist a labor surplus in African agriculture obscure the implications of landlessness, the emerging rural class structure, and the impact of more widespread Western education on access to adequate rural employment.

It may be comforting to find that economic indicators such as wage differentials, expected earnings, and urban unemployment rates tend to covary statistically with migration patterns. But to assume that these variables explain why thousands of Africans continue to flood into Nairobi despite an insufficient number of jobs to employ them is fundamentally shortsighted. This phenomenon is the expected outgrowth of an underdevelopment process which has historically depressed rural wages, emphasized the maintenance–by force if necessary–of adequate labor supplies for export producers, become increasingly dependent upon multinational corporations in the local economy, and effectively concentrated enormous economic and political powers in the primate cities. The most innovative change in Kenya and elsewhere in the Third World has been the recognition that increased wages in the formal sector will work, in combination with government protection of foreign and domestic capital investment, to guarantee sufficient labor supplies without the more extensive and coercive regulations of the colonial period.

11. Recent boundary changes in several Kenyan towns–designed in part to make the peri-urban dwellers pay for urban services–have incorporated dense but essentially rural areas and resulted in extraordinary population increases. There are indications that such towns as Kisumu and Machakos now claim over 100,000 inhabitants, more than quadrupling their size with the stroke of a pen. A similar boundary shift at about the time of independence accounted for a jump in Nairobi's population from 266,800 to 343,500.

12. One is tempted to describe what has happened since independence as "involution," a term already used by Geertz (1963) for agriculture and McGee (1971) for urbanization in underdeveloped countries. As noted by Geertz, the term was originally used by Alexander Goldenwieser to refer to "those culture patterns which, after having reached what would seem to be a definitive form, nonetheless fail either to stabilize or transform themselves to a new pattern but rather continue to develop by becoming internally more complicated.

13. Spatial planning, broadly defined, refers to a process aimed at the reorganization of sociospatial systems (including both individual cities and the settlement system as a whole) to meet particular policy objectives. In the present context, it is applied specifically to the system of settlements and to policies designed to redress interurban and interregional inequalities in income, infrastructural services, and general welfare.

14. The figures were reported by the Economist Intelligence Unit, *Quarterly Economic Review* (Tanzania-Mauritius), 1975, no. 2.

15. These percentages were reported in *The Daily News,* September 26, 1973, p. 7.

16. Of the 30 industries established in the first two years of the growth pole plan, 20 went to Dar, 7 to Arusha and Moshi, and only 3 to the remaining centers. (See Stren, 1975:283).

17. Residential segregation ceased to be official British colonial policy in the 1920s, but was maintained locally through the use of restrictive covenants inserted into all property transactions, which prohibited multiracial occupation and ownership in certain areas.

18. Kimani's study covered only the old city area and not the new territory added in the boundary expansion of 1963. Even using the old city population figures, however, he estimates that there was one acre of African-owned land for every 800 Africans, versus ratios of 1 : 32 for Asians and 1 : 7 for Europeans.

19. Westcott (1974) has noted that the proportion of development funds for services and infrastructure allocated to Nairobi and Mombasa increased between 1972-1973 and 1973-1974 (to 60%), and projected public investments in "main infrastructure" in the 1974-1978 Development Plan is even higher (71%).

REFERENCES

BAKWESEGHA, C.J. (1974). "Patterns, causes, and consequences of polarized development in Uganda." Pp. 47-66 in S. El-Shakhs and R. Obudho (eds.), Urbanization, national development, and regional planning in Africa. New York: Praeger.

BIENEFELD, M.A. (1970). A long-term housing policy for Tanzania (E.R.B. Paper 70.9, revised). Dar es Salaam: Economic Research Bureau, University of Dar es Salaam.

BIENEFELD, M.A., and BINHAMMER, H.H. (1969). Tanzanian housing finance and housing policy (E.R.B. Paper 69.19). Dar es Salaam: Economic Research Bureau,

University of Dar es Salaam. Reprinted in pp. 177-199 in J. Hutton (ed.), Urban challenge in East Africa. Nairobi: East African Publishing House, 1972.

BIENEN, H. (1974). Kenya: The politics of participation and control. Princeton, N.J.: Princeton University Press.

BIGSTEN, A. (1975). "The spatial economic structure of Kenya: An empirical report on the spatial distribution of population production, employment and public services." Memorandum no. 44, Department of Economics, University of Gothenburg.

BLOOMBERG, L., and ABRAMS, C. (1965). "Report of the United Nations Mission to Kenya on housing." Nairobi.

BRETT, E.A. (1973). Colonialism and underdevelopment in East Africa: The politics of economic change, 1919-1939. New York: NOK Pub.

CARVALHO, M.E.F.C. (1969). "Regional physical planning in Kenya: A case study." Ekistics, 27:232-237.

CHANA, T., and MORRISON, H. (1973). "Housing systems in the low income sector of Nairobi, Kenya." Ekistics, 34:214-222.

DAVIS, K. (1969). World urbanization, 1950-1970. Vol. 1: Basic data for cities, countries and regions (Population Monograph Series, no. 4). Berkeley: University of California.

--- (1972). World urbanization, 1950-1970. Vol. 2: Analysis of trends, relationships, and development (Population Monograph Series, no. 9). Berkeley: University of California.

De BLIJ, H.J. (1963). Dar es Salaam: A study in urban geography. Evanston, Ill.: Northwestern University Press.

--- (1968). Mombasa: An African city. Evanston, Ill.: Northwestern University Press.

De SOUZA, A.R., and PORTER, P.W. (1974). The underdevelopment and modernization of the Third World (Commission on College Geography, Resource Paper no. 28). Washington, D.C.: Association of American Geographers.

ELKAN, W. (1967). "Circular migration and the growth of towns in East Africa." International Labour Review, 96:581-589.

El-SHAKHS, S., and OBUDHO, R. (eds., 1974). Urbanization, national development, and regional planning in Africa. New York: Praeger.

ETHERTON, D. (1971). "Mathare Valley: A case study of uncontrolled housing in Nairobi." Housing Research and Development Unit, University of Nairobi.

GEERTZ, C. (1963). Agricultural involution: The processes of ecological change in Indonesia. Berkeley: University of California Press.

GEORGULAS, N. (1967). "Settlement patterns and rural development in Tanganyika." Ekistics, 24:180-192.

GILLMAN, C. (1945). "Dar es Salaam, 1860 to 1940: A story of growth and change." Tanganyika Notes and Records, (20):1-23.

GROHS, G. (1972). "Slum clearance in Dar es Salaam." Pp. 157-176 in J. Hutton (ed.), Urban challenge in East Africa. Nairobi: East African Publishing House.

GUGLER, J. (1972). "Urbanization in East Africa." Pp. 1-26 in J. Hutton (ed.), Urban challenge in East Africa. Nairobi: East African Publishing House.

GUTKIND, P.C.W. (1963). The royal capital of Buganda: The case of internal conflict and external ambiguity. The Hague: Mouton.

--- (1972). "The socio-political and economic foundations of social problems in African urban areas: An exploratory conceptual overview." Civilisations, 22(1):18-34.

HALLIMAN, D.M., and MORGAN, W.T.W. (1967). "The city of Nairobi." In W.T.W. Morgan (ed.), Nairobi: City and region. Nairobi: Oxford University Press.

HANCE, W.A. (1970). Population, migration, and urbanization in Africa. New York: Columbia University Press.

HARRIS, J.R. (1972). "A housing policy for Nairobi." Pp. 39-56 in J. Hutton (ed.), Urban challenge in East Africa. Nairobi: East African Publishing House.

HARRIS, J.R., and TODARO, M.P. (1969). "Wages, industrial employment and labour productivity in a developing ecomony: The Kenyan experience." East African Economic Review, 1.

--- (1970). "Migration, unemployment and development: A two-sector analysis." American Economic Review, 60(1):126-142.

HUTTON, J. (ed., 1972). Urban challenge in East Africa. Nairobi: East African Publishing House.

International Labour Organization (1972). Employment, incomes and equality: A strategy for increasing productive employment in Kenya. Geneva: Author.

JACOBSON, D. (1973). Itinerant townsmen: Friendship and social order in urban Uganda. Menlo Park, Calif.: Cummings.

JOHNSON, G.E., and WHITELAW, W.E. (1974). "Urban-rural income transfers in Kenya: An estimated remittances function." Economic Development and Cultural Change, 22:473-479.

KANYEIHAMBA, G.W. (1973). Urban planning law in East Africa: With special reference to Uganda (Progress in planning, Vol. 2, Part 1). New York: Pergamon.

KIMANI, S.M. (1972a). "Spatial structure of land values in Nairobi, Kenya." Tijdschrift voor Economische en Sociale Geografie, 63(2):105-114.

--- (1972b). "The structure of land ownership in Nairobi." Journal of Eastern African Research and Development, 2(2):101-124.

KIMANI, S.M., and TAYLOR, D.R.F. (1973). Growth centers and rural development in Kenya. Thika, Kenya: Maxim Printer.

LARIMORE, A.E. (1969). "The Africanization of colonial cities in East Africa." East Lakes Geographer, 5:50-68.

LAURENTI, L., and GERHART, J. (1972). "Urbanization in Kenya." Working Papers, International Urbanization Survey. New York: Ford Foundation.

LEYS, C. (1974). Underdevelopment in Kenya: The political economy of neo-colonialism, 1964-1971. Berkeley: University of California Press.

LOCK, S.C. (1972). "Urban planning in Kenya." Development Digest, 10(3):120-124.

LUTTRELL, W.L. (1972). "Location planning and regional development in Tanzania." Pp. 119-148 in J.F. Rweyemamu et al. (eds.), Towards socialist planning (Tanzanian Studies no 1). Dar es Salaam: Tanzania Publishing House. Also in Development and Change (1972-1973), 4:17-38.

MARRIS, P., and SOMERSET, A. (1972). The African entrepreneur: A study of entrepreneurship and development in Kenya. New York: Africana.

MASCARENHAS, A.C. (1966). "Urban development in Dar es Salaam." Unpublished M.A. thesis, University of California, Los Angeles.

--- (1967). "The impact of nationhood on Dar es Salaam." East African Geographical Review, 5:39-46.

--- (1970). "The port of Dar es Salaam." Tanzania Notes and Records, 71:85-118.

--- (1971a). "The growth and function of Dar es Salaam." Pp. 134-135 in L. Berry (ed.), Tanzania in maps. London: University of London Press.

--- (1971b). "Land use in Dar es Salaam." Pp. 136-137 in L. Berry (ed.), Tanzania in maps. London: University of London Press.

--- (1971c). "The port of Dar es Salaam." Pp. 138-139 in L. Berry (ed.), Tanzania in maps. London: University of London Press.

MASCARENHAS, A.C., and CLAESON, C.-F. (1972). "Factors influencing Tanzania's urban policy." African Urban Notes, 6(3):24-42.

MBOGUA, J.P. (1965). "Pumwani estate social survey." Nairobi City Council, Department of Social Services and Housing.

McGEE, T.G. (1971). The urbanization process in the Third World: Explorations in search of a theory. London: G. Bell.

--- (1974). "The persistence of the proto-proletariat: Occupational structures and planning for the future of Third World cities." Comparative Urban Studies, School of Architecture and Urban Planning, University of California, Los Angeles.

McKEE, S.I.D. (1973). "Planning and urbanization in Kenya." Journal of the Royal Town Planning Institute, 59(9):337-344.

McVICAR, K. (1968). "Twilight of an East African slum: Pumwani." Unpublished doctoral thesis, University of California, Berkeley.

MORGAN, W.T.W. (1967). Nairobi: City and region. Nairobi: Oxford University Press.

--- (1969). "Urbanization in Kenya: Origins and trends." Transactions I.B.G., 46:167-178.

MORRISON, H. (1972). "Mathare Valley report: A case study in low income housing." Nairobi City Council, Urban Study Group.

Nairobi Urban Study Group (1973). Nairobi–Metropolitan growth strategy. Vol. 1: Main report. Nairobi: Author.

OBUDHO, R.A. (1974). "Urbanization and regional planning in western Kenya." Pp. 161-176 in S. El-Shakhs and R. Obudho (eds.), Urbanization, national development, and regional planning in Africa. New York: Praeger.

OCITTI, J.P. (1973). "The urban geography of Gulu." Occasional paper no. 49. Department of Geography, Makerere University.

PARKIN, D.J. (1969). Neighbours and nationals in an African city. London: Routledge and Kegan Paul.

PIORO, Z. (1972). "Growth pole and growth centres theory as applied to settlement development in Tanzania." Pp. 169-194 in A. Kuklinski (ed.), Growth poles and growth centres in regional planning. The Hague: Mouton, for UN Research Institute for Social Development, Geneva.

Project Planning Associates (International) Ltd. (1968). National Capital Master Plan. Dar es Salaam: Author.

RODNEY, W. (1972). How Europe underdeveloped Africa. Dar es Salaam: Tanzania Publishing House.

ROSS, M.H. (1973). The political integration of urban squatters (African Urban Studies). Evanston, Ill.: Northwestern University Press.

ROTHSCHILD, D. (1969). "Ethnic inequalities in Kenya." Journal of Modern African Studies, 7(4):689-711.

RWEYEMAMU, J.F., (1971a). The historical and institutional setting of Tanzanian industry (E.R.B. paper 71.6). Dar es Salaam: Economic Research Bureau, University of Dar es Salaam.

--- (1971b). The structure of Tanzanian industry (E.R.B. paper 71.2). Dar es Salaam: Economic Research Bureau, University of Dar es Salaam.

--- (1972). "Planning, socialism and industrialization: The economic challenge." Pp. 30-49 in J.F. Rweyemamu et al. (eds.), Towards socialist planning (Tanzanian studies no. 1). Dar es Salaam: Tanzania Publishing House.

RWEYEMAMU, J.F. et al. (eds., 1972). Towards socialist planning (Tanzanian studies no. 1). Dar es Salaam: Tanzania Publishing House.

SAFIER, M. (ed., 1970). The role of urban and regional planning in national development of East Africa. Kampala: Milton Obote Foundation.

SAFIER, M., and LANGLANDS, B.W. (eds., 1969). "Perspectives on urban planning for Uganda." Occasional paper no. 10. Department of Geography, Makerere University College.

SCAFF, A.H. (1964). "The re-development of Kisenyi (Kampala)." Pp. 1-9 in Proceedings of the EAISR Conference, December 1963. Kampala: University of East Africa, East African Institute of Social Research, Makerere College.

--- (1967). "Urbanization and development in Uganda: Growth, structure, and change." Sociological Quarterly, 8(1):111-121.

SOJA, E., and TOBIN, R. (1974). "The geography of modernization: Paths, patterns, and processes of spatial change in developing countries." In R. Brunner and G. Brewer (eds.), A policy approach to the study of political development and change. New York: Free Press.

SOUTHALL, A.W. (1966). "The growth of urban society." Pp. 462-493 in S. Diamond and F.G. Burke (eds.), The transformation of East Africa: Studies in political anthropology. New York: Basic Books.
--- (1967). "Kampala-Mengo." Pp. 297-332 in H. Miner (ed.), The city in modern Africa. New York: Praeger.
SOUTHALL, A.W., and GUTKIND, P.C.W. (1957). Townsmen in the making: Kampala and its suburbs. Kampala: East African Institute of Social Research.
STREN, R. (1972a). "The evolution of housing policy in Kenya." Pp 57-96 in J. Hutton (ed.), Urban challenge in East Africa. Nairobi: East African Publishing House.
--- (1972b). "Urban policy in Africa: A political analysis." African Studies Review, 15:489-516.
--- (1975). "Urban policy and performance in Kenya and Tanzania." Journal of Modern African Studies, 13(2):267-294.
SUTTON, J.E.G. (1970a). "Dar es Salaam: A sketch of a hundred years." Tanzania Notes and Records, (71):1-20.
--- (ed., 1970b). Dar es Salaam: City, port and region. Tanzania Notes and Records, 71 (special issue).
TAYLOR, D.R.F. (1974). "The role of the smaller urban place in development: The case of Kenya." Pp. 143-160 in S. El-Shakhs and R. Obudho (eds.), Urbanization, national development, and regional planning in Africa. New York: Praeger.
TEMPLE, F.T. (1973). "Politics, planning and housing policy in Nairobi." Unpublished Ph.D. dissertation, Massachusetts Institute of Technology.
TIWARI, R.C. (1972). "Some aspects of the social geography of Nairobi, Kenya." African Urban Notes, 7(1):36-61.
TODARO, M.P. (1969). "A model of labour migration and urban unemployment in less-developed countries." American Economic Review, 69(1):138-148.
--- (1971). "Income expectations, rural-urban migration, and employment in Africa." International Labor Review, pp. 387-413.
TRIBE, M.A. (1970). "The economics of urbanisation." Pp. 149-158 in M. Safier (ed.), The role of urban and regional planning in national development for East Africa. Kampala: Milton Obote Foundation.
--- (1972). "Patterns of urban housing demand in Uganda." Pp. 138-156 in J. Hutton (ed.), Urban challenge in East Africa. Nairobi: East African Publishing House.
TSCHUDI, A.B. (1972). "Ujamaa villages and rural development." Norsk Geografisk Tidsskrift, 26(1-2):27-36.
Uganda, Republic of, Ministry of Regional Administrations, Department of Town and Regional Planning (1968). Kampala-Mengo regional planning studies. Kampala: Author.
University of Nairobi, Housing Research and Development Unit (1971). "Site and service schemes: Analysis and report." Nairobi: Author.
WEISNER, T.S. (1972). "One family, two households: Rural-urban ties in Kenya." Unpublished Ph.D. thesis, Harvard University.
--- (1973a). "Kariobangi: The case history of a squatter resettlement scheme in Kenya." Symposium paper presented to the ninth International Congress of Anthropological and Ethnological Sciences, Chicago.
--- (1973b). "Studying rural-urban ties: A matched network sample from Kenya." Pp. 122-134 in W. O'Barr, D. Spain, and M. Tessler (eds.), Survey research in Africa. Evanston, Ill.: Northwestern University Press.
WERLIN, H.H. (1966). "The Nairobi city council: A study in comparative local government." Comparative Studies in Society and History, 8:181-198.
WESCOTT, C.G. (1974). "Migration, primacy, growth centers, and politics in Kenya." Paper presented at the 17th annual meeting of the African Studies Association, Chicago.
WHITELAW, W. (1974). "Rural-urban relations, low-income settlements, intraurban

mobility and the spatial distribution of housing in Nairobi, Kenya." Working papers in economics no. 6, University of Oregon.

WITTHUHN, B.O. (1974). "An imposed urban structure: A case study from Uganda." Pp. 67-74 in S. El-Shakhs and R. Obudho (eds.), Urbanization, national development, and regional planning in Africa. New York: Praeger.

11

Dynamics of Urbanization in the Central American Common Market

E.C. CONKLING
JAMES E. McCONNELL

□ ALTHOUGH URBAN GROWTH is occurring throughout the less developed world, it has progressed more rapidly in Latin America than in Africa or Asia. If current trends continue, at least three-fourths of the people south of the Rio Grande will be living in urban places by the end of this century. Even in Central America, where urban growth had lagged behind the rest of Latin America, the pace of urbanization has accelerated recently, accompanied by changes in urban form and function similar to those found in neighborhing Latin lands.

The forces contributing to urban growth and change in Central America are much the same as those operating in other developing countries except for certain elements unique to the isthmian region. In addition to their distinctive blends of Iberian and native American cultures, the countries of Central America bear the imprint of numerous accidents of history and physical geography. Significant among these special characteristics is their tiny size. The smallest Central American nation, El Salvador, is only the size of Massachusetts. Taken together, all these countries occupy an area no larger than California. During the past two decades the most dynamic element in regional affairs has been the emergence of the Central American Common Market, which includes Costa Rica, El Salvador, Guatemala, and Nicaragua, with Honduras in a state of self-imposed suspended membership (two other Central American countries, Panama and Belize, do not belong). Although it has affected nearly every facet of the region's economy, this experiment in regional integration has produced an important impact upon the principal cities of its member nations.

This chapter will focus on what has been happening to these cities since formation of the CACM in 1960. As in other less developed countries, each of

Central America's leading cities tends to monopolize the nation's human and natural resources. At the same time, however, the principal city is the point of convergence for many national problems: the accumulated effects of colonialism, dualism, and ethnic conflicts and, more recently, the manifold economic, social, and political results from overly rapid population growth. In addition to these familiar problems of the Third World, the cities of Central America have had to share the national economic burdens of small countries in a world of giants. Because the CACM was established precisely to overcome this handicap through formation of a common market for Central American products, the impact of this organization will be central to the discussion which follows. An attempt will also be made to foresee the future prospects for the region's cities and to consider some possible solutions for the problems that they can expect to encounter.

THE CHANGING CENTRAL AMERICAN CITY

The questions of urban growth and change in Central America can be considered at various levels of observation. When comparing the progress of urbanization in the CACM with that in other parts of the world, it is likewise essential to note the many differences among the cities of member countries and among the cities of a single country. Urbanization in Central America is a complex process that differs in several respects from that of less developed countries in Asia and Africa, in its demographic, cultural, and economic components. To a greater extent than is usually recognized, the changes that integration has introduced into national economies have affected the size, functions, and internal spatial structure of Central American cities.

PATTERNS OF GROWTH

Just how urban are the CACM's cities by comparison with those in other parts of the underdeveloped world? An exact answer to this question is complicated by the lack of uniformity in the definition of "urban" from one country to another and by the effects of differing country size. Taking urban to mean all persons living in places of 2,500 population or more, then one-third of the citizens of Costa Rica, Guatemala, and Honduras, and about 40% of those in El Salvador and Nicaragua can be regarded as urban. This compares with an average figure of 22% each for Asia (excluding Japan) and Africa. Another measure less dependent upon country size is the average population of the place in which a nation's population resides. Expressed in this way, the average Central American lives in a population center only one-tenth as large as does the average Canadian and only one-fifteenth as large as the average Mexican (Arriaga, 1969).

Although the CACM countries are less urbanized than most of the rest of the Western Hemisphere, Central American cities are now growing very rapidly. In recent years urban populations of the region have been rising at an average rate of 4.3% per annum, whereas the rural populations have increased by 2.6% each year. The most rapid urban growth in the CACM has occurred in Guatemala, where the annual rate of increase is 5.3%; the slowest urban growth is in Honduras, with a rate of only 2.8% per year.

The greatest differences, however, are to be found within individual countries. Like other less developed nations, each of CACM's member countries has an urban population that is heavily concentrated in the capital city. Every one of the capitals is a primate city in the full sense of the term. Established originally as an administrative center for the Spanish invaders, it remains the modern seat of political and commercial control. The focus of all national life, the primate city reaches out to dominate the entire country through a transport network that ties it to all parts of the tributary area. It is the seat of an urban-based power elite composed of the same families who own much of the rural land; it is the center of intellectual life, the seat of ecclesiastical control, and the center for the dissemination of new ideas.

The population of each CACM primate city dwarfs that of all other urban places in the country. Guatemala City, for example, is more than ten times as large as second-place Quezaltenango. San José, the capital of Costa Rica, is more than seven times as large as its nearest competitor, Alajuela. Other urban centers in each country are not only much smaller but also perform far fewer urban functions than do the primate cities. Except for the port cities, all of which are very small, these lesser centers perform only central place functions or, at most, play a minor administrative role as regional capitals in a highly centralized governmental system.

With the lone exception of Nicaragua, the CACM's population is concentrated mainly in intermontaine basins and valleys and other broader upland areas at elevations ranging between 2,000 and 7,000 feet. Four of the national capitals are in highland basins, each city surrounded by a tight cluster of satellite communities. The metropolitan regions of San José and San Salvador are typical of the concentrations that have formed in the highlands, with their delightfully springlike climates throughout the year. The CACM capitals are thus located very differently from most Asian or African primate cities, which are usually major ports.

ELEMENTS OF CHANGE

Not only do Central America's capitals dominate the central place hierarchy in size and functions, but they have also attracted an overwhelming share of the new growth of the past two decades. Although some of the processes contributing to this growth and its unequal effects are new, many have operated for centuries. The center-periphery forces so evident in Central America today were initially set in motion by the Spanish invaders. Each of the five CACM capitals was established by the occupying forces early in the 16th century as a place from which to control the surrounding countryside.

As the major point of contact with the outside world, the capital city has acquired a comparatively modern aspect that contrasts markedly with the tradition-bound rural hinterland. The city-based commercial economy remains under the control of a small group descended from the conquistadores, together with some persons of mixed blood; whereas the more isolated regions are occupied by a largely Indian population of subsistence farmers, especially in Guatemala. Another form of dualism prevalent in the more favored rural districts results from enclaves of North American plantations and other commercial enterprises amidst the subsistence farming population.

Superimposed upon these arrangements dating from a much earlier age are a number of newer elements contributing to city growth and change. One of these stems from powerful demographic forces released mainly since World War II. The other results from modernization processes, especially those introduced by the Central American Common Market.

As a whole, the populations of these countries are experiencing some of the highest rates of natural increase in the world. Because the rapidly falling death rates of recent decades have not been accompanied by significant declines in birth rates, Central America has entered the most explosive phase of the demographic transition. Through a combination of circumstances, the greatest impact of this increase has fallen upon the cities, whose populations have risen more than two-thirds again as rapidly as in the countryside. This is especially true of Guatemala, where the urban population has grown twice as fast as the rural. Only in Honduras has urban growth failed to exceed that of the farm population.

The urban-rural growth differential is likely to be even greater than these figures suggest. Although any place of 2,500 people or more is officially defined as urban, a large number of the smaller communities are merely agricultural villages. It is not these small places but rather the major metropolises that are experiencing most of the growth. Indeed, the CACM's primate cities have rates of population increase ranging between 7% and 9% per annum. Although the migration of rural folk to the city accounts for a substantial part of this rural-urban growth differential, high rates of natural increase among the urban population contribute a surprisingly large proportion.

Just as in other less developed lands, rural-urban migration in Central America results from both a push and a pull. Whereas the Asian and African migrations are caused mainly by a severe overcrowding of the rural environment, Central America's stock of arable land is not yet exhausted (except possibly in El Salvador). Within the CACM, therefore, the cityward movement results less from the push of rural conditions and more from the relative attractiveness of the large city. Among these urban lures are the promise of better employment and educational opportunities, more favorable living conditions, and encouragement from city-dwelling relatives and friends.

Rural-urban migration in Central America has a number of distinctive characteristics that tell much about the society of the region and at the same time suggest the kinds of impacts that this transfer of populations is having upon both the sending and the receiving areas. Unfortunately for their home communities, the migrants are usually the most fit and energetic young adults. Least likely of all to move are the pureblood Indians bound by close ties of tradition and kinship to their communal settlements. The migration tends to occur in stages: those persons leaving the farm usually move first to a nearby provincial center, then, after a time, set out for the metropolis.

This behavioral pattern was tested in a study of migrants from each of Guatemala's minor civil subdivisions (municipios) to the capital, Guatemala City (Thomas, 1968). The analysis showed that the number of migrants supplied from any one municipio to Guatemala City could be predicted to a substantial degree by the population size of the municipio and its distance from Guatemala City–in other words, by the standard gravity model formulation. Adding two other variables to the model, however, considerably increased the degree of explanation. One of these was the proportion of Indians in each municipio's population–evidence of a cultural barrier to migration. The other variable that proved important was proximity of the municipio to a cabacera, or provincial capital, to which migrants might move before continuing on to Guatemala City.

Despite the importance of rural-urban migration in the recent growth of CACM cities, it seems probable that high rates of natural increase among the city dwellers themselves may be contributing at least as much. Although the data for the CACM as a whole are not sufficiently consistent to test this hypothesis, studies in other Latin American countries show that as much as three-fourths of the growth of larger cities can be attributed to natural increase of the city population alone (Arriaga, 1968).

Regardless of which of the two components contributes more, it is clear that urban growth in Central America is but a part of the whole explosive population growth problem of the region. Birth rates remain very high among both rural and urban dwellers, although there is some indication that a slight decrease in birth rates may be imminent in certain of the largest cities. Meanwhile, however, death rates are decidedly lower in the primate cities, where the medical services and public health facilities are more readily available and of better quality than in the hinterland. The spread between birth and death rates is thus increasing in the big urban centers.

The prospect of industrial employment is one of the main attractions of the cities to rural in-migrants. The accelerating pace of manufacturing growth has been an important element in urban growth since 1960. In turn, the growth of manufacturing, as well as its structure and location, reflects the policy decisions of development planners, particularly those of the Central American Common Market. Nevertheless, the final results have not always been in accordance with official plans. Planning authorities have discovered that some locations are much more attractive than others to private investors, who have consistently avoided some locations despite special inducements offered by public bodies.

Formation of the CACM has had a decided impact on manufacturing growth within the region. Whereas manufacturing had been a lagging sector prior to integration, between 1960 and 1971 industrial output grew at a rate much higher than the increase in gross domestic product. While GDP was rising at an average annual rate of 5.7% (itself a decided improvement over previous years), value added by manufacturing was increasing by 8.3% each year. In 1960 manufacturing was increasing by 8.3% each year. In 1960 manufacturing had contributed 13% to GDP in the five nations; by 1971 it was supplying 18%. This surge in manufacturing output in large part reflected the deliberate investment policies of the CACM, which allocated 39% of all investment funds to industrial development projects (Banco Centroamericano, 1971).

The heightened industrial activity unexpectedly affected even the traditional industries: the processing of domestic raw materials and foodstuffs for local consumption as well as other market-oriented manufacturing such as the manufacture of textiles, clothing, wood products, and furniture. Most of all, however, the drive for industrialization added new kinds of manufacturing. Among these were firms making intermediates, such as wires and cables, reinforcing rods, steel shapes, electrical machinery, appliances, refined petroleum and by-products, chemicals, pulp and paper, and rubber and rubber products.

Although the CACM has maintained an official policy of balanced growth, intended to ensure an equitable distribution of industrial output among the five nations, this has been impossible to enforce. In locating their new industrial establishments, investors have favored those countries where industrial development had already proceeded furthest: Guatemala and El Salvador at one end of the group and Costa Rica and Nicaragua at the other. In between, Honduras was avoided because her backward economy lacked the agglomeration advantages so important to modern industry. Thus, despite the CACM's balanced growth policy, Honduran industry actually lost ground relatively.

Within each CACM member nation it was the capital city which managed to capture virtually all the country's share of new manufacturing. At the end of the CACM's first decade, Guatemala City and San Salvador still retained an undiminished 60% of the region's total value added by manufacture. San José and Managua, whose manufacturing had actually grown even faster than that of the other capital cities, accounted for most of the rest. A direct result of this industrialization has been the influx of job seekers into the four capitals.

In these same growth centers tertiary activities have likewise expanded to provide the proliferating needs of the new industries and of an enlarged population with rising incomes. In the CACM as a whole, the value of commerical activity grew at an average annual rate of 6.8% between 1960 and 1966, and it has increased at a rate of 5% to 6% each year since. Banking and insurance grew 10.4% annually during the earlier period and between 7% and 8% in succeeding years. This growth was very largely confined to the capital cities, where the rising number of tertiary workers is much in evidence. Even more visibly apparent in the central business districts of these cities are innumerable new office buildings, hotels, department stores and shops, and business firms providing a variety of consumer goods and services.

An essential element in the growth of the CACM's cities has been transport development. It has also been one of the reasons for the unequal sharing of that growth. Typical of former colonies of European powers, each Central American capital is the principal node in the transport system which ties together the domestic economy. The system, as originally designed by the Spaniards, fans outward from the metropolis to form a branching network that taps all of the principal raw material supply areas and links the capital with lesser places in the urban hierarchy.

When the CACM was inaugurated, each national transport system was essentially independent of the other national systems, with few interconnections between them. To overcome the isolation that this imposed upon the Central American states, the CACM's investment policy has favored transport development, with particular emphasis on integrating the various national systems. A large part of the foreign aid supplied to the CACM, especially that from the United States, has gone into highway construction. Supplementing the CACM program, each of the national governments has worked to modernize and extend its road system throughout the countryside to bring a larger part of the national territory into the modern commercial economy.

The net effect of these highway construction programs has been to leave the national dominance of primate cities undiminished and, if anything, enhanced. In each instance, the transport improvements have eased travel to the capital city, thereby contributing to the flow of in-migrants from the hinterland. These developments have also extended the range of the capital's commercial establishments, at the expense of smaller centers. The highway program has likewise done nothing to reduce the metropolitan monopoly of manufacturing. Meanwhile, the CACM's program to integrate the Central American highway network has only enlarged the national role of each capital city and to enhance the advantage of the CACM's leading growth centers over those which lagged behind.

THE EVOLVING PATTERN OF LAND USE WITHIN THE CITY

In most respects the internal spatial structure of Central America's capital cities is typical of major metropolises in the less developed world. This pattern reflects certain distinctive features of the urban economic and social milieu: a large proportion of the labor force in the services; a plurality of racial, ethnic, cultural, and income groups; and a dualistic local economy that places modern and traditional forms in juxtaposition. Like other rapidly growing urban centers in developing nations, each Central American metropolis is expanding outward and taking in much new territory along its perimeter. Here is where much of the change in urban land use is occurring.

In the CACM the large work force in the tertiary and quaternary occupations includes increasing numbers of government workers–clerks, administrators, and support personnel. Retailing involves not only merchants but also swelling hordes of street-corner vendors peddling trinkets, clothing, foods, and lottery tickets. In addition, there is an expanding group of transport workers, warehousemen, and office workers.

The commercial dualism prevalent in large Third World metropolises is much in evidence in Central America. The North American-style central business district is expanding upward and outward in response to the rising demand from the growing middle-class population. Modern air-conditioned department stores and shops and a variety of service establishments are proliferating. New high-rise buildings are continually appearing in the increasingly dense urban center, where they are needed to accommodate the headquarters of domestic firms and the subsidiaries of foreign enterprises, to house shopping malls, or to provide space for government agencies.

Nearby, and equally crowded, is a very different type of business district, the *mercado central.* Often this is a huge, cavernous, barnlike structure covering a city block and crammed with endless numbers of stalls and booths manned by small merchants catering to the daily needs of the indigenous population for agricultural produce, locally made handicrafts, housewares, and cheap clothing. Crowding the outside of the building are innumerable hawkers and street sellers. In some Central American cities–San Salvador, for example–the market is held in the open air, occupying several city streets on the margins of the modern central business district. In such fast-growing metropolises as Guatemala City, the mercado central can no longer accommodate all of the population and other mercados have therefore sprung up in outlying neighborhoods.

The changing land use pattern in the CACM's primate cities includes large new concentrations of manufacturing, which are confined mainly to the city margins. Modern industrial parks have located adjacent to major transport arteries, and, in some instances, a scattering of manufacturing plants continues along these routes for some distance into the country. The highway leading to the airport has been a favored location for such industrial strip development, especially in San José and San Salvador. Nearly all these firms are housed in very modern, attractive buildings that give visible evidence of the recency of industrial growth. Meanwhile, industrial dualism persists even in this changing environment, as shown by the scattering throughout the city of innumerable cottage industries producing leather goods, handicrafts, and handwoven textiles.

Even in Guatemala City, the largest and most rapidly growing of the CACM's capitals, much of the colonial residential pattern remains intact. Many very wealthy families continue to maintain sumptuous homes in separate quarters near the urban core, and low-income housing spreads outward to the city margins. Nevertheless, the rising incomes of the expanding middle class and the flood of rural in-migrants have brought important changes to the residential pattern. Unlike North American cities, central population densities remain high in the Central American metropolis. Those families recently added to the urban population must therefore form new communities along the periphery.

Two kinds of new residential concentrations are prominent on the city margins. One of these is the North American-style suburb built to accommodate the emerging middle-income group, who can afford to commute to work by private motorcar. Examples of such middle-class suburbs are Guatemala City's newly constructed Kaminal Juyu housing developments, which straddle Highway CA-1 as it enters the western end of the metropolitan area. Among the residents of the Kaminal Juyu subdivisions are the medical personnel who staff the nearby San Vicente and Roosevelt Hospitals.

Much more attention has focused, however, on another new type of residential concentration in Central American cities. This is the squatter settlement, into which newly arrived migrants from outlying villages and towns are pouring. Crowding into the metropolis in search of employment in the new industries and commercial establishments, they are confronted by the critical shortage of housing. The only recourse for these impoverished villagers is to build shacks and lean-tos from sheet metal, boxes, and other discarded materials on the edge of the built-up area.

Because the intermontaine basin occupied by Guatemala City has already become crowded, the only unused land available to the poorer in-migrants are the barrancas—steep-sided, very deep ravines carved by the several streams that have cut back into the basin floor well into the heart of the city. Each barranca is now lined with incredibly dense concentrations of squatter shacks. Bearing such names as La Limonada, El Esfuerso, Quinze de Agosto, La Trinidad, and La Verbena, these communities provide a certain minimal shelter in this mild tropical highland climate. Here also the inhabitants manage to preserve much of the social organization of the villages from which they came.

URBAN PROBLEMS AND PROSPECTS

As this description of changing economic and social conditions in Central American cities implies, recent developments have generated a number of urban problems. These problems of the present, however, could well be dwarfed by those yet to come if current trends cannot be halted or at least brought under control. Agencies of both the CACM and the individual national governments have undertaken or porposed solutions for some of these difficulties, but the need for additional measures seems indicated.

CURRENT ISSUES

From the day of its inauguration, the CACM has struggled with the inequalities between member nations, which find direct expression in the fortunes of their primate cities. As yet no way has been found to provide certain of the Common Market's cities with the necessary conditions to attract new manufacturing and commerce. This has been particularly true of Tegucigalpa and other Honduran population centers. On the other hand, Managua, the capital of Nicaragua, was doing very well until virtually destroyed by the severe earthquake of 1972. Unfortunately, in this part of the world, Managua's tragic experience cannot be regarded as a rare aberration; the entire region lies within a zone of volcanic activity and violent earth movements. Each of the CACM's major cities experiences frequent tremors, and two of the cities lie directly in the shadows of active volcanoes.

For both kinds of special problems—in the one case the inability to reach the threshold of economic activity required by modern industry, and in the other vulnerability to devastating acts of nature—the answer would seem to lie in cooperation among the Central American nations. The CACM's doctrine of balanced growth appears to hold the obvious solution to Honduras' inability to acquire industry. Yet Honduras has effectively removed herself from such assistance by her de facto withdrawal from the CACM in the aftermath of her "Soccer War" with El Salvador. Other members of the group have also undermined the effectiveness of the organization with unilateral acts motivated by shortsighted considerations of a nationalistic and political nature.

Although, to some extent, sovereign nations like Honduras have the ability to defend themselves against what they perceive to be unequal competition from foreign sources, the lower order centers in the central place hierarchy of an individual country have no effective way of meeting the competition from primate cities in their own land. The growth of the latter often occurs at the expense of the former. The provincial towns and villages lose to the capital their most productive workers, the savings of their wealthy landowners, and many of the material resources they produce. Very little do they receive in return. They are unable to attract capital, and they cannot provide the conditions required for establishing most types of manufacturing. The provincial towns are thus relegated to the role of service centers to the surrounding rural territories, which in most instances are populated only by impoverished subsistence farmers. To date, about the only attention many of them have received from national and CACM authorities has come from the road construction program, the net effect of which has been to speed the drain of their resources to the national center.

Political pressures for the equalization of benefits have been growing recently, as a result of which governmental planners in some of the CACM countries have begun to enact programs designed to ease the problems of these disadvantaged hinterland areas. Regional development efforts have been aimed principally at modernizing agriculture, which it they should succeed, would reflect beneficially on the local service centers as well. In addition, the national governments have begun to help certain of the provincial towns with infusions of aid for developing their infrastructure.

But what of the much favored primate cities of the CACM? They, too, have grave problems but a very different sort. Where the provincial town contends with economic stagnation, the capital struggles with the problems of the opposite kind. Indeed, the question is usually raised as to whether the Central American nations have become overurbanized. The answer seems to be that, at least for the present, they are not. According to the usual measures, such as the amount of arable land available for the support of the urban population, the CACM's cities have not yet become too large by comparison with those of other less developed nations (Kamerschen, 1968). Instead, the real difficulty with Central America's major centers is that they are growing much to rapidly.

As the previous figures have shown, the CACM's primate cities have caught most of the brunt of an exceedingly high rate of population growth. As a result they have found it almost impossible to assimilate the additions to their numbers. The pressures of urban population growth assume several forms, one of the most basic being the necessity to provide employment for all the newcomers. The rural-urban population movement for which these cities are the targets is in fact mainly a migration of labor. However, although the principal attraction of the capital city is thus the opportunity for work, the great majority of rural in-migrants are unskilled and illiterate. The paradox of this is that the migrants are usually the best persons that their home communities have had to send. Most are hard-working, energetic, and ambitious; they are just not trained for urban types of employment.

Despite the CACM's rapid industrial growth of the 1960s, which was much greater than that of most less developed countries during that period, the rate of job formation was not nearly fast enough to accommodate the massive additions to the urban labor force contributed both by rural migrants and by the natural increase of the city population. Perhaps the job deficit would have been less had it not been for the CACM's emphasis on developing import-competing industries of a kind that have low capacities for labor absorption but high skill requirements. Compounding the problem at this time, the rate of industrialization has begun to slacken because the CACM has essentially exhausted its opportunities for "easy import substitution." Those types of production most readily attracted to the Central American region are now almost fully represented.

As a result of these circumstances, the principal cities of the CACM are filled with large numbers of disappointed job seekers. The true dimensions of the unemployment problem are concealed under the cloak of underemployment: countless individuals scratching out a meager existence in unproductive ways. The only hint of this in the official urban statistics is the abnormally large numbers of service employees, whose ranks include legions of street vendors and shoeshine men and boys. The job shortage also tends to depress wage levels, which are kept down by competitive pressures from the excessive number of unemployed. Low family incomes are further strained by the very high dependency rates resulting from the abnormally large proportion of the population under 15 years of age, which in turn is a function of high birth rates.

Another challenge for the CACM's urban planners is the need to improve living conditions for newcomers. Population growth in the principal cities has greatly outpaced the relatively meager investment in new residential construction. Although housing is generally very bad throughout Central America, perhaps 50% of the houses in metropolitan communities are below minimal standards. Much of this substandard housing consists of makeshift affairs constructed from scraps of wood and metal. lacking toilets or plumbing, and having no access to safe drinking water. These conditions are typical of Guatemala City's barrancas.

Public services throughout the CACM countries are grossly inadequate, largely because governments at all levels have insufficient taxing authority to generate the necessary capital. Most of the existing hospitals, clinics, and other health and medical services are concentrated in the cities. Urban dwellers are therefore much better off in this respect than the country folks; yet even the cities are unable to add such services fast enough to accommodate their multiplying populations. Because city sewerage systems have become strained far beyond their capacities, raw sewerage is being dumped directly into local streams that are the sources of water supply for bathing and household use. The threat of severe public health problems therefore constantly hangs over most of the large cities.

Population growth in Central America has outstripped the ability of these countries to build and staff schools. Only Costa Rica has managed to keep abreast of the expanding needs of education; elsewhere the levels of literacy are slipping backward. Conditions are worst of all in Guatemala. Nevertheless, the problem of education is not as bad in the cities as in the country. Each of the CACM capitals provides facilities for education at all levels, including universities, even though most of these centers lack the capacity to accommodate the entire school-age population.

One of the explosive political issues in Central America at this time of writing is the soaring cost of living, a grave problem for a population living so close to the margin of survival. Because a large proportion of the rural population maintains a subsistence way of life outside the money economy, the weight of inflation falls mainly on city dwellers. Inflation reached new highs during 1974, ranging up to 20% and 30% per annum in CACM countries (and even higher in Guatemala). A number of things combined in that year to drive up the cost of living: the escalating prices of imported manufactured goods from the developed lands and oil from the OPEC countries, severe droughts in several main crop areas, and an excessive increase in the money supply by central governments.

Finding solutions to this formidable list of problems confronting Central American cities is made the more difficult by several obstacles to effective urban planning. In addition to the chronic shortage of public financial resources, there is a general lack of planning experience, inadequate data, poor coordination among agencies, severe pressures from politically and economically powerful political interests, and political instability.

PROBLEMS OF THE FUTURE

Today's urban issues in Central America bear the seeds of potentially graver problems in the years immediately ahead. Most of the national problems now facing the CACM countries bear most heavily on their primate cities. Because these cities are growing far more rapidly than the populations of the nations of which they are a part, this focus can be expected to sharpen rather than diminish. Indeed, the first issue with which the CACM's cities must contend in the near future is the question of growth itself.

Central America's population as a whole is doubling every 18 to 20 years, and to date there has been little if any success in curbing this growth. Efforts to introduce family planning face formidable obstacles in these countries. Even though national ministries of health have formally sponsored programs of this type, there is in fact little real official support for them. One reason for this lack of governmental initiative is the determined opposition of organized religion, which, in turn, tends to reinforce the general lack of popular interest in the question. Few Central Americans are aware of the dimensions of the problem or the future consequences of current trends. Customary attitudes toward marriage and the family are deeply imbedded in popular thought and will not give way quickly or easily.

Even if this were not the case, if all barriers disappeared and family planning were introduced widely and at once, there are far too many persons in the reproductive age group at this time for any important downward trend in population growth to appear. It is therefore safe to predict that current growth rates will persist for some time to come; indeed, it is likely that they will actually continue to rise in the near future. Of the five countries, only Costa Rica has passed the maximum growth period of the demographic transition; the other four are only now entering it.

For reasons cited earlier, it appears certain that the CACM's capital cities will continue to bear the principal impact of this population growth. Although rising incomes may eventually reduce family sizes among middle-income city dwellers, little change can be expected in the attitudes of the great masses with low incomes. There is little likelihood that rural-urban migration will abate; more probably it will increase. Although further industrialization, even at the present reduced rate, will remain the chief lure to migrants, new incentives for migration will probably appear. Government programs to modernize agriculture will encounter serious barriers, if only because of the difficulty of making such wholesale changes in the work practices of farmers, who comprise two-thirds of the CACM's labor force. Too little time remains to effect this agricultural revolution before rural population growth exhausts the supply of unused land. When that point of saturation is reached, rural underemployment will begin to worsen, and the surplus population of the hinterland will probably migrate to the cities in even greater numbers than at present. Thus, although the element of "pull" is greater than that of "push" in today's rural-urban migration, the "push" factor will gain importance in Central America as it already has in most of Asia and Africa.

If it is assumed that the saturation of rural land is a few years off, it can be predicted that by 1980 the total population of the CACM will have nearly doubled since 1960, the year in which the integration treaty was signed. Within that same 20-year period, however, the urban population of the five nations will have grown by nearly one and one-half times.

The greatest proportionate increase in urbanization will have occurred in Guatemala, the CACM's most populous nation and the one with the largest city. Various estimates show the projected 1980 population of that country to be 83% to 93% greater than it was at the time of integration; the urban component of that population will meanwhile have risen by possibly 156%. Whereas only 32% of Guatemala's people lived in urban places in 1960, 45% will be city dwellers by 1980. With an estimated million and a third people in 1975, Guatemala City is now about the same size as the Standard Metropolitan Statistical Area of Buffalo, New York. The national capital accounted for about 41% of the country's urban population in 1965; by 1980 it will have nearly 62% of the total. During the 15 years the city will have grown from an estimated 618,000 people to almost two million, thus more than tripling in size.

Costa Rica's urban population is increasing by a smaller percentage than the other CACM countries. City dwellers were 35% of the nation's total population in 1965 and will have risen to only 36% by 1980. Nevertheless, Costa Rica's captial city, San José, is the CACM's second largest city. It holds 24% of the country's total population, and the other principal urban places are near at hand in the volcano-rimmed basin known as the Meseta Central.

It should be emphasized that these projections are based on current growth rates. When the agricultural land of Central America at last becomes saturated, urban growth rates could accelerate further. This would complicate the CACM's urban problems even more, compounding the difficulties of assimilating additional populations into city economies.

The policy-makers of the CACM and its member countries are continually striving for answers to these problems and have developed specific programs to deal with some of them. The question of squatter settlements is one issue that receives much attention from national authorities. These neighborhoods of deep poverty are festering places for political unrest. But if the residents of these shack communities are unhappy, so are the legal owners of the land they have occupied. The combination produces an explosive situation that prudent politicians cannot long ignore.

Guatemala City, being the CACM's largest city, has the greatest squatter problem. A number of proposals for improving conditions in these communities have reached the Guatemalan congress, and legislative action on at least some of them is expected soon. Among the measures being considered are proposals for government surveys of the disputed lands, securing legal titles for occupants of squatter properties, a census of social and demographic information, and the provision of social overhead such as streets, utilities, and parks.

Public officials are also aware that basic economic problems must be addressed if relief is to be found for widespread poverty and misery among the growing masses of unemployed. They realize that the shortage of jobs could worsen as the pressure on agricultural land increases and as the urban labor force swells with additions from the countryside and from the city's own natural increase. Political pressures to equalize the benefits of national economic growth have led to programs for rural development, which, if they should succeed, would relieve part of the population pressure on the cities. Because it has been a lagging sector of the CACM economy, agriculture has depressed the growth of other sectors. The United States AID program, among others, has been allocating increasing amounts of investment money to Central American agriculture. Agricultural development is a difficult problem, however, because of the technological backwardness and illiteracy of the farm population. Moreover, because modernizing the region's agriculture would reduce its labor intensity, this would add to the problem of rural unemployment and underemployment. One answer to this latter problem would be to introduce more resource-based manufacturing into the provincial towns—food processing, tobacco products, and so forth—which would increase the value of the local output and provide additional employment.

Even in the largest cities, however, the level of employment has been rising more slowly than it did in the 1960s. The stagnation of the industrialization program is a problem whose solution lies largely beyond the capabilities of individual nations, the domestic markets of which are much too small to permit the scale economies required by most industries. Despite the nationalistic feelings that have risen periodically in certain member countries, a cooperative approach to industrialization seems unavoidable. Even Honduras has belatedly recognized this and has initiated talks for resumption of a role in the CACM.

If their drive for industrialization is to regain its momentum, these countries will have to find a way to break through the present ceiling on industrial growth resulting from their having used up the opportunities for easy import substitution. One solution would be to find new kinds of manufacturing for export to world markets, to supplement the current group of industries whose sales are confined almost wholly to the five nations. Emulating such Asian nations as Taiwan, Korea, Hong Kong, and Singapore, the CACM countries might manufacture electronic products, small appliances, other labor-intensive items that would permit them to draw upon an important comparative advantage in the world economic system: their abundant and growing supply of cheap labor.

Initial steps in this direction have foundered because of artificial obstacles imposed by the political system. For example, one large North American electronics firm recently set up an assembly operation in Guatemala City, where conditions for production proved ideal. The Guatemalan management team was energetic and effective, and labor costs were low. Ultimately the venture had to be discontinued because of unreliable transport arrangements for bringing raw materials into the country and exporting the finished product and because of sluggish and unpredictable treatment of shipping documents by customs officials. These are problems that CACM members must quickly solve if they are to take advantage of a newly emerging pattern of international production that could employ much of the region's surplus labor.

Another underutilized opportunity for earning badly needed foreign exchange and generating employment is tourism. The comparative advantages in this case are the magnificent year-round climates, the spectacular scenery, the colorful and exotic Indian communities, and the many structures and artifacts remaining from the Mayan and Spanish colonial eras. At present the tourist industry is mainly confined to the capital cities and nearby areas. Although service trades of the capital regions gain much employment from this, most of the lesser centers in the hinterland rarely see tourists because they lack suitable accommodations. At the same time, some of the Western Hemisphere's best beaches go unused because of isolation. An investment in tourist facilities and better transport connections, together with better promotion, could solve the employment problems of many such outlying regions, thereby helping to relieve the capital cities of their present crush of job-seeking in-migrants.

Inflation, currently another grave problem for the CACM's urban population, shows some promise of subsiding as world recession slows the rise in prices of imported manufactured goods. The threat of further oil price rises remains, however, and national governments are still printing too much money. There is little that Central America can do about the cost of oil, a universal problem for less developed nationsl lacking domestic supplies; but the money supply can be controlled, given sufficient governmental restraint.

A prime goal of the CACM's secretariate at present is to promote a redistribution of incomes within member countries. As the disparities between the very rich and the very poor continue to widen, social and political unrest threatens the region's stability, which in turn tends to discourage both domestic and foreign investment. In the capital cities, where conspicuous consumption and human misery are in close proximity, the dangers are especially acute. Because politically powerful groups oppose much-needed tax reform, however, the best hope for the present is to redouble efforts to raise living standards of the low-income group through measures of the kinds just described.

Added to these problems generally shared throughout the Common Market are the special issues raised by disparities between CACM's capital cities. There is the plight of Tegucigalpa, which has difficulty in attracting modern industry. Here the only obvious remedy is closer cooperation among CACM members, which could funnel a larger share of the region's investment capital into this lagging center. A precondition for this, clearly, is Honduras' return to full active participation in the Common Market.

The most basic problem of all, however, and the one about which the least is being done, is the population issue—not just absolute numbers, but the rate at which those numbers are increasing. CACM's GNP is growing at a respectable rate, but the population is increasing so fast that many of the gains are lost. As Kingsley Davis (1965) has said, when speaking of the condition of cities in less developed lands, "The problem is not urbanization, not rural-urban migration, but human multiplication."

REFERENCES

ARRIAGA, E. (1968). "Components of city growth in selected Latin American countries." Milbank Memorial Fund Quarterly, 46:237-252.

––– (1969). "A new approach to the measurement of urbanization." Economic Development and Cultural Change, 18:206-218.

Banco Centroamericano de Integracion Económica (1971). Informe mensual de operaciones. Tegucigalpa, Honduras: Oficina de Administracion Financiera de Prestamos.

Battelle Memorial Institute (1969). Projections of supply and demand for selected agricultural products in Central America through 1980. Jerusalem: Publication Services Division of the Israel Program for Scientific Translations under contract with the U.S. Department of Agriculture.

DAVIS, K. (1965). "The urbanization of the human population." Pp. 10-12 in Cities. New York: Scientific American.

JOEL, C. (1972). "Growth trends in the Central American Common Market (CACM), 1960-71." Capto circular A-47. Guatemala City: U.S. Department of State, Regional Office for Central America and Panama.

––– (1975). "Economic assessment of the CACM region, 1974-75." Unpublished report. Guatemala City: U.S. Department of State, Regional Office for Central America and Panama.

KAMERSCHEN, D.R. (1968). "Further analysis of overurbanization." Economic Development and Cultural Change, 17:235-253.

THOMAS, R.N. (1968). "Internal migration to Guatemala City, Guatemala, C.A." Unpublished Ph.D. dissertation, Pennsylvania State University.

12

Southeast Asian Cities: Patterns of Growth and Transformation

YUE-MAN YEUNG

☐ IN THE GLOBAL PERSPECTIVE of population redistribution and spatial reorganization, the 19th century was noted for the remarkable urban transformation and attendant industrialization and rapid economic growth of the present developed countries rimming the North Atlantic (Weber, 1899; Lampard, 1955); the 20th century may be distinguished by the phenomenal urban growth in developing countries. With cities in the Third World growing at an average annual rate of 5% to 8%, both Nelson (1970:393) and Davis (1965:49) observe that urban populations double every 10 to 15 years. Between 1950 and 1970, many large Southeast Asian cities indeed assumed such growth patterns. For instance, Djakarta, Bandung, Kuala Lumpur, and Hanoi at least tripled their populations, while Surabaja, Medan, Rangoon, Manila, and Saigon more than doubled theirs (Davis, 1969:197-199).

Despite these almost runaway growth rates of individual large cities, Southeast Asia remains one of the least urbanized regions in the world. Of the 22 regions delineated by Davis (1972:191), only five regions had in 1970 a proportion of their urban population smaller than that of Southeast Asia. In 1970 one person in five was regularly domiciled in urban places in Southeast Asia, a rate of population concentration comparable to that of the Dutch Netherlands and Britain in 1800 (Lampard, 1969:3). At the same time, with 12.1% of the total population living in cities of 100,000 and over, the degree of urbanization was equivalent to that of Europe and North America between 1850

AUTHOR'S NOTE: *This chapter was written before the recent political upheavals in Cambodia and before the radically different policies of the new regime with respect to Cambodia's major cities.*

and 1900 (Lampard, 1965:548). However, given the signs and portents of present trends, urbanization will almost certainly continue apace. Goldstein (1973:85) anticipates that by 2000, one-third of the region's population will live in cities and towns, and the urban population will increase by more than threefold from 60 million in 1970 to 203 million.

The foregoing brief account of urbanization trends in the region implies that urban growth may well be the mainspring of progress in Southeast Asia in the decades ahead. Tentative futuristic statements will thus be offered toward the end of the essay. The primary focus of this paper, nevertheless, is to analyze the salient urban growth patterns since 1960, to evaluate development policies and strategies, and to examine aspects of transformation which urbanization, in the experience of developed economies, frequently entails.

URBAN GROWTH TRENDS

The survey begins here with regional trends (Davis, 1969:70). In the decade 1950-1960 the total population of Southeast Asia increased at an annual rate of 2.4%, as compared with a 5.7% increase in the urban population. The succeeding decade saw overall population growth accelerating at a rate of 2.8% per year, while the annual increase of the urban population was slightly moderated at 5.0%. Even at this reduced rate, the 1970 urban population will double itself in slightly over 14 years.

That urban populations have been growing more rapidly than total populations is better illustrated in Table 1, which shows that all countries individually exhibit this pattern. Laos and North and South Vietnam, especially, had their urban population increasing in the last decade at rates at least two times the population increase as a whole. Table 1 bears out, in addition, the low degree of urbanization of individual countries. With the exception of Brunei, a tiny oil-rich British protectorate, and the city-republic of Singapore, none of the other countries had its urban population close to 40% in 1970. It must be noted in this connection that the definition of "urban" widely varies, and in both Malaysia and the Philippines rather different census definitions of "urban" were applied in the 1970 census. Malaysia, for example, adopted in the 1970 census a threshold population of 10,000 for urban, instead of a threshold of 1,000 as in the 1947 and 1957 censuses (Ooi, 1975). Similarly, the Philippines used a more realistic but more complicated set of criteria to define its urban population in the latest census. According to the new definition, 11.68 million people lived in urban places in the Philippines in 1970, which is almost 400,000 short if the old definition had been used (Mijares and Nazaret, 1973:23). In applying the new urban definitions, particularly in Malaysia, the level of urbanization may thus appear to be lower than reported in previous censuses. It should also be noted in Table 1 that the increase in the proportion of urban populations (columns 2 and 4) has, to varying degrees, been rather modest over the last decade; only

Singapore recorded a reverse trend of a decreasing level of urban population, largely the result of a program of successful decentralization of population through public housing construction (Yeung, 1973). On the other hand, the increment in urban populations (columns 1 and 3) between 1960 and 1970 has been rather substantial in most countries. This suggests, in the light of the generally modest increase in urbanization levels, that urban growth has proceeded in sympathy with the immutable trend of overall population growth.

When the survey is extended to individual countries, the available information is extremely uneven on the urban situation. Although little is known about the extent of population redistribution in the Indochina states, incessant warfare over decades has certainly been responsible for considerable induced migration toward the cities. It is therefore no accident that the two Vietnam states, Laos, and Cambodia experienced the highest rates of urban growth in the last decade (Table 1). The high rate of urbanization in North Vietnam is rather surprising, since, at the height of the Vietnam conflict, American bombing was said to have caused massive evacuation from the cities. In South Vietnam the most dramatic population increases between 1960 and 1970 centered on small and middle-sized cities; cities in the 20,000 to 100,000 class increased from 15 to 28, with a growth of 161% in population, and cities in the 100,000 to 300,000 class jumped from 2 to 6, recording a population inflation of 307% (Goodman and Franks, 1974:26). Metropolitan Saigon escaped the major thrust of war-induced migration, its population having grown only by 26% in the stated period, and the city developed, politically and administratively, more by decree than by evolution (Goodman, 1973:3).

Malaysia is another country which has witnessed guerilla war-induced urbanization. During the period of Emergency (1948-1960), as part of the strategy to combat guerillas, more than half a million inhabitants were moved from rural settlements to towns and New Villages in peninsular Malaysia. More than 70 New Villages, each with a population of 2,000, were created (Lam, 1974; also Sandhu, 1964). This policy, while deployed with immense success against the insurgents, has had the unwitting effect of exacerbating the ethnic imbalance of the urban population further in favor of the Chinese, for most of the rural settlers involved in the program were Chinese. In the period of 1957-1970, however, some of the rural folk, as Saw (1972:115) has observed, returned to their farms in the former terrorist infested areas, and the previous rural-urban drift was partly reversed. Consequently, the rate of growth of urbanization slowed down appreciably in the period 1957-1970, as compared with the previous intercensal period 1947-1957. From 1957 to 1970 there is evidence of migration up the settlement hierarchy, with 13 settlements crossing the 10,000 threshold to become part of the 49 urban centers (Pryor, 1973; Ooi, 1975). However, the fastest growing urban places were the metropolitan areas of 75,000 and over, which experienced a 96.7% increase in population (Lam, 1974). Of these centers, Kuala Lumpur has grown most rapidly, specifically in the direction of the Klang Valley toward Port Swettenham. So speedily had

TABLE 1
URBAN POPULATIONS IN SOUTHEAST ASIA

Country	1960		1970		Annual Growth Rate 1960-1970		GNP per Capita (U.S. $)	Percentage of Labor Force in Manufacturing Industries
	Urban Population in Millions (1)	Percentage of Total (2)	Urban Population in Millions (3)	Percentage of Total (4)	Urban (5)	Total (6)	(7)	(8)
Brunei	.04	43.5	.05	44.1	3.6	3.5[a]	$1,793 (1971)[b]	4.3 (1971)
Burma	3.20	14.3	4.33	15.8	3.1	2.1	$85 (1973)	—
Cambodia	.55	10.1	.88	12.8	4.8	3.2	$130 (1970)	3.8 (1962)
Indonesia[c]	15.03	15.5	20.77	17.4	3.3	2.0	$112 (1972)	7.4 (1971)
Laos	.20	8.6	.40	13.4	7.2	2.4	$120 (1970)	—
Malaysia: West[d]	1.67 (1957)	26.5	2.53	28.7	3.3	2.6	$391 (1972)	15.5 (1967)
East[e]	.17 (1957)	14.0	.26	15.8	3.3	2.4		6.6 (1960)
Philippines[f]	8.17	30.2	11.68	31.8	3.6	3.0	$254 (1972)	14.5 (1967)
Singapore[g]	.91 (1957)	63.1	1.25	60.1	2.5	2.4	$1,780 (1973)	36.3 (1973)
Thailand	3.00	11.4	4.66	13.0	4.5	3.1	$193 (1972)	6.0 (1966)
North Vietnam	2.29	14.2	5.28	23.9	8.9	2.8	$100 (1970)	—
South Vietnam[h]	3.08	21.9	6.06	34.9	6.8	2.6	$174 (1971)	5.0 (1966)

SOURCES: Columns (1) to (5), Davis, 1969; Column (6), IBRD, 1973; Column (7), UN, 1973; IBRD, 1973; FEER, 1975; etc.; Column (8), ILO, 1973; You and Yeh, 1971; etc. unless otherwise indicated.

NOTES: a. Davis, 1969:151.
b. Chua, 1975:22.
c. Milone, 1966:3-4; Indonesia, 1972.
d. Chander, 1971; Ooi, 1975; Saw, 1973.
e. Chander, 1969, 1970.
f. Mijares and Nazaret, 1973:23.
g. Arumainathan, 1973:231.
h. Goodmann and Franks, 1974:26.

some of the large urban centers grown that by 1970, in eight of the nine largest cities, a substantial proportion of the population overspilled the urban boundaries; almost all these cities had become underbounded (Pryor, 1973:57). The expansion of the city area of Kuala Lumpur in 1974 from 37 square miles to 94 square miles was in part motivated by this phenomenon of conurbation and metropolitanization (Ooi, 1975).

In another mainland Southeast Asian country, Thailand, which has one of the fastest growing populations in the region, there have been significant changes in the differential patterns of growth by city size and a tendency for urban development to permeate other regions outside the central region. Since 1960, while the primacy of Greater Bangkok remains unchallenged, the larger cities have accounted for the bulk of the urban population increase. Goldstein (1973:90) shows that, since 1947, the cities in the smaller size categories have been losing consistently their share of the total urban population to the larger centers. In 1947, for instance, almost one-half of the urban population lived in places under 20,000, but by 1967 the figure had declined to one-fifth. In a similar vein, Romm (1972:19) draws attention to the fact that in the decade 1960-1970 urban places with more than 20,000 people have grown at average annual rates greater than 7%, whereas centers below that size category have grown more slowly than their natural rates of increase. Many smaller municipalities have in fact lost population by net outward migration. Romm further observes a trend toward equalization of the distribution of large urban centers among regions. In 1947, five of Thailand's 10 largest municipalities were located in the central region. By 1970, the share of the other regions has increased significantly, indicative of perhaps the result of a policy of fostering economic and urban development in the other three regions in the north, northeast, and south.

Within archipelagic Southeast Asia, the latest census results do suggest emergent trends of urban growth in Indonesia, by far the region's largest and most populous country. Between 1930 and 1971, the urban population grew at 4.7% per year, while the overall population increased annually at 1.7% (King, 1974:926). Table 1 indicates that the corresponding figures for the last decade are 3.3% and 2.0%. Like the Malaysian case, then, urban growth over the last decade has slowed down. More importantly, in comparing the growth rates of the 27 cities of over 100,000 inhabitants in 1971 (22 in 1961), it is readily apparent that the decrease in their annual growth rate in the period 1961-1971 from the growth rate between 1930 and 1971 has been sharp (McNicoll and Mamas, 1973:47). In many cases, the recent growth rates are but a fraction of what they were in the earlier period. Only three cities maintained a higher rate than before. However, Djakarta still grew at a booming annual rate of 4.6% over the last decase, which is a slight decline from the previous rate of 5.5%. The other high growth cities are all located in Sumatra and Kalimantan, their high growth rates explanable by their late start and the fact that the age-old development policy to decentralize congested Java has lately been gaining some momentum. In spite of the slackened tempo of growth in the larger cities, Fisher (1967:183) maintains that

> in Indonesia it is the high density metropolitan core which is poverty stricken and in direct economic difficulty, whereas the more sparsely peopled and relatively unsophisticated outlying regions have enjoyed the highest standard of living and the brightest prospects for the future.

In a country as diverse as Indonesia, national averages often mask significant regional variations. Therefore, King (1974:927) has shown that, by region, Djakarta, Kalimantan, Sumatra, and East Java have grown more rapidly in urban population in that descending order in the period 1961-1971.

Finally, in the Philippines, notwithstanding the small gain in the degree of urbanization between 1960 and 1970, the growth rate of the urban population has shown little signs of abatement (Mijares and Nazaret, 1973). In the period 1948-1960, the urban population grew at an average annual rate of 3.7%, which declined only marginally to 3.6% in the last intercensal period. Owing to the application of a unique set of definitional criteria for "urban" (including population density, size of settlement, and urban form and functions), the number of urban places in the Philippine urban system must rank among the highest in Southeast Asia. There was an increase of 807 urban places alone over the last decade, reaching a total number of 2,406 in 1970. The urban population is most unevenly distributed among the regions. In 1970, Manila and its suburbs, central Luzon, southern Luzon and adjacent islands, western Visayas, and eastern Visayas together contained 76.8% of the urban population. In other words, more than three-quarters of the urban population are concentrated in five of the 10 regions. Not surprisingly, these five regions also reveal the highest degrees of urbanization. Aside from Manila and its suburbs, which may be regarded as totally urban, the most rapidly urbanized region in the decade 1960-1970 was southern Luzon and adjacent islands, where the level of urbanization increased from 33.9% to 41.0%, with an absolute increase of almost one million inhabitants living in urban places.

A BRIEF URBAN PROFILE

After a survey of trends of recent urban growth, it seems appropriate to summarize some of the salient characteristics of the product of the urbanization process and the urban dynamics before attention is turned to urban problems and strategies.

Although with the exception of the Philippines most of the present Southeast Asian countries had some form of cities before the advent of the Europeans, these urban phenomena were for the most part inland, ephemeral, "orthogenetic" cult entities which have been eclipsed by subsequent urban expansion. The pattern of contemporary urban development dates from European

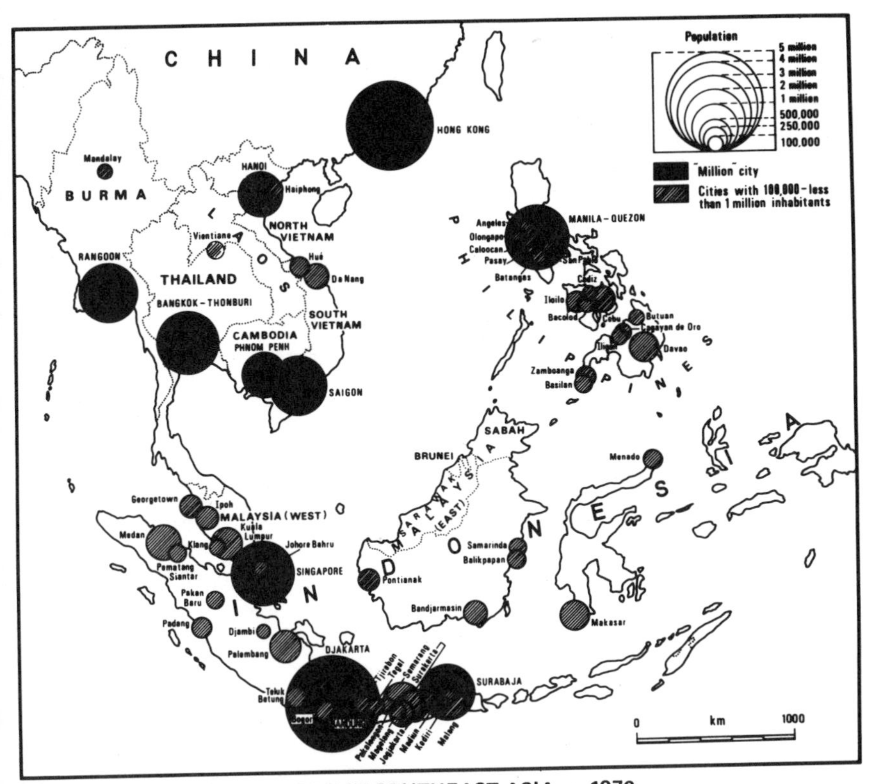

Figure 1: THE URBAN PATTERN IN SOUTHEAST ASIA, c. 1970 (Source: Yeung and Lo, 1975)

colonization of the region, beginning in the 16th century. Each of the Southeast Asian countries followed a pattern of development in which one city, almost invariably on a riverine or deltaic site for ease of external communications and transportation, was chosen to perform the main urban functions. These happened to be capital cities as well and soon developed to a size many times the size of the second largest city. Many of these urban centers have grown to "million" primate cities, and their dominance has often been accentuated in the postcolonial period (see Figure 1). In 1970 there were nine "million" cities in the region, which is an increase of four from the five studied by Fryer in the 1950s (Fryer, 1953). In step with the growth rate of the urban population, the growth of the primate cities was slightly slower in the period 1960-1970 than during the preceding decade. Djakarta's annual rate of growth declined from 10.8% in the decade 1950-1960 to 4.6% in the following decade, while the corresponding growth rates for Greater Bangkok decreased from 6.9% to 4.8% (Breese, 1969:78). However, even those growth rates are too large for the comfort of city administrators and policy-makers.

The pivotal role of the primate cities in the urban life of their countries can be easily gauged by the large proportion of the urban population they contain.

These range from a low of 21.7% for Djakarta to a high of 79.7% for Phnom Penh. Classic examples of primacy are provided by Greater Bangkok and Metropolitan Manila, which completely overshadow the respective second largest cities of Chiengmai and Cebu. To illustrate the overwhelming dominance of Greater Bangkok, Romm (1972:7) maintains that it "contains 77 per cent of the nation's telephones and about half its motor vehicles, consumes 82 per cent of its electricity, generates 82 per cent of its business taxes and 73 per cent of its personal income taxes, holds 72 per cent of all commercial bank deposits, and absorbs 65 per cent of the annual investment in construction." A similar set of statistics can be assembled for Metropolitan Manila (Viloria, 1972). The predominance of Bangkok and Manila is so complete that one begins to question whether diseconomies of scale, agglomeration, and centrality have set in.

Other important features of contemporary Southeast Asian cities pertain to the social and economic structures inherited from the colonial period. The urban population in most countries is ethnically diverse, with varying proportions of alien Asians, notably Chinese and Indians, who were recruited to help pioneer development under European colonialism. As later paragraphs will discuss, this multi-ethnic factor may pose a constraint to some development programs. Another inherited feature is the dualistic economy, differentiated into a modern and a traditional sector, or in Geertz's (1963) terminology, into a firm-type sector and a bazaar-type sector, with little functional interaction between them. Yet both coexist. As a consequence, retailing outlets, transportation modes, and industrial units are fragmented into a myriad of forms along two levels of operation. Because of their easy entry, labor intensiveness, and low visible productivity, bazaar-type functions are sometimes regarded as marginal and redundant. Hawkers, pedicab drivers, domestic servants, and shoeshine boys fall into this category. However, in view of burgeoning urban populations in the region, one must not underrate the role that these traditional occupations (with their high labor-absorption capacities) contribute in employment opportunities (Yeung, 1975a).

The dualistic urban structure throws into relief the nature of urbanization in Southeast Asia. In contradistinction to the Western experience, urbanization in the region, as in much of the Third World, has not been accompanied by industrialization and marked economic progress. Table 1 reaffirms that the general economic well-being of the countries, as denoted by a low per capita gross national product, is unsatisfactory and that the occupation structure has not transformed itself to a relatively large manufacturing work-force. Only Singapore and Malaysia are relatively advanced in economic terms. This condition of urban growth without commensurate expansion of economic opportunities has been variously termed "static expansion" (Wertheim, 1964:216-217), "pseudo urbanization" (McGee, 1967), and "urban involution" (Armstrong and McGee, 1968). As cities continue to grow in the demographic sense without sufficient qualitative change, dualism persists, symptomatic of the underdevelopment of the countries at large.

What is the nature of urban dynamics which have been shaping contemporary Southeast Asian cities and the urbanization process? At least five dynamic forces may be identified. Apart from the traditional sources of growth arising from population dynamics and migration, there have been significant changes in the urban population owing to warfare and insurgent activities, city boundary expansion, and census redefinition.

The demographic components of urbanization in Southeast Asia of today and those in developed countries at a comparable stage of development are not exactly identical (McGee, 1971:164). In the developed countries concurrent high rates of births and deaths in their cities permitted extensive rural-to-urban shift mediated by migration (Lampard, 1969:8). Migration was the principal mechanism to attain their urban transformation, for no evidence can be adduced to show that urban rates of natural increase were higher than rural rates. Only in a few instances did the contribution by natural increase outweigh that of migration (Lampard, 1973:13). On the other hand, Southeast Asia is trapped in what Davis (1971) depicts as a combination of preindustrial fertility and post industrial mortality. Crude birth rates are generally over 40 per 1,000, but crude death rates are as low as less than 10 per 1,000 in Singapore and Malaysia (You et al., 1974:460). Moreover, the data on differential annual growth rates between the urban population and the total population as presented in Table 1, appear to lend support to Davis's (1965) contention that the main factor in the rapid inflation of city population is derived from biological increase at an unprecedented rate rather than from rural-urban migration.

The prevailing patterns of population dynamics just described do not imply, nonetheless, that rural-urban migration does not contribute to overall urban growth. In fact, the ejective mechanism set in motion by lagging rural development and a progressively unfavorable man-land ratio has triggered a continual urbanward stream of population movement. The pressure felt by the primate cities is especially acute. In spite of the spotty evidence available for assessing the extent of rural in-migration to the primate cities, Sakornpan (1975:3; also Sternstein, 1974) indicates that of the 1.91 million native-born inhabitants in Greater Bangkok in 1960, one-quarter had been born in the provinces outside the metropolis. Similarly, almost half the residents in Manila had been born elsewhere (Hollnsteiner, 1975; also Dwyer, 1964), and about half of Djakarta's present annual growth results from migration (McNicoll and Mamas, 1973:33). In Malaysia, which has been pursuing the twin national goals of eradicating poverty and restructuring society–goals that underlie the Second Malaysia Plan (1971-1975)–indigenes have been encouraged to move into the cities. As a result, 14.9% of the Malays lived in cities in 1970, as opposed to 11.2% in 1957 (Ooi, 1975). As mentioned before, war-related factors have accounted for considerable migration of population from the rural areas to small and medium-sized urban centers in South Vietnam. However, the usual migration factors, including distance, are poor explanatory variables in the case of South Vietnam (Goodman, 1973).

Akin to the war-induced migration in Indochina was the reorganization of settlement forms stemming from counterinsurgency activities in the early stages of independence. These took place in Malaysia, the Philippines, and South Vietnam and led to particularized urban forms. The common strategy was to collect people from dispersed villages into centralized, easily controlled centers. Malaysia's New Villages during the Emergency were an outgrowth of this strategy and have since then been integrated into the urban hierarchy. The Huk rebellion in the Philippines in the early 1950s was likewise calmed down by a resettlement program which was geared to the rebels. Finally, in South Vietnam "Strategic Hamlets" were established and fortified out of existing settlements in order to counter insurgency activities. All of these are policies of population consolidation and have unwittingly induced urbanization.

The remaining factors affecting urban change are city boundary expansion and new urban definitions used in the latest census, as alluded to before. Kuala Lumpur, for instance, extended in 1974 its city area by more than 150%. When city-area redefinition of such a magnitude has taken place, wometimes the city takes on an entirely different name, such as Makasar in Sulawesi, which was recently renamed Ujung Pandang, to underscore its new entity.

URBAN PROBLEMS AND ISSUES

The galaxy of urban problems, such as housing shortages, inadequate jobs, insufficient infrastructural and social facilities, traffic jams, environmental deterioration, and a high incidence of crimes, is by no means unique to Southeast Asian cities. Yet these problems bedevil particularly the primate cities which, from the standpoint of the total urban environment, are demonstrably worse off than before. The urban situation is further aggravated in a generation of rising expectations, thanks to the success of the mass media and to the population and labor force explosions in the postwar period, the latter following on the heels of the former (Yeh and You, 1971). Needless to say, the primate cities differ greatly in their ability to cope with these problems. Singapore, perhaps more easily managed as a city-state, can be regarded as more successful in tackling its problems. The other larger cities are, unfortunately, hamstrung by a host of administrative, budgetary, and organizational difficulties.

In the provision of public housing for the low-income groups, Singapore is without doubt far ahead of the other cities. Public housing construction, typically high-rise and large-scale, has been undertaken since 1960, not only as a program to meet the initial housing shortage but as an agent for national development and nation-building as well (Yeung, 1973). With over 40% of the total population residing in public flats in 1974, the impact that this form of development has had on the urban structure, political participation, and the general way of life is indeed far-reaching (Yeh, 1975). Similar developments on a smaller scale have taken place in the larger urban places in Malaysia, such as

Kuala Lumpur, Georgetown, and Johore Bahru. Other cities, lacking adequate economic resources, have relied to a much smaller extent on high-rise residential development to meet housing shortages. Djakarta, Manila, and Bangkok have adopted a site-and-services approach, particularly in the more peripheral urban areas (Yeung, 1975b).

As a manifestation of prevailing housing shortages in Southeast Asian cities, at least one-quarter of the population in the primate cities live in squatter settlements (Hassan, 1972). Although most city governments are inclined to view squatter colonies as a passing stage of cities on the road to modernization, these urban elements figure prominently in the spatial patterning of contemporary Southeast Asian cities (Jackson, 1974). Laquian (1968), Hollnsteiner (1974), Poethig (1971), and Taylor (1973), among others, have recognized the positive contributions that squatters can make in nation-building. Laquian (1968, 1971) also demonstrates that slums and squatter areas are often stable and coherent economic and social entities bound by their own normative standards of social and functional relations.

Lack of employment opportunities is the compound of rapid population growth and sluggish economic development. The situation is particularly acute in the developing countries (Lewis, 1967), although Grant (1971) maintains that it is a global crisis. Singer (1970) has suggested that the international dualism in science and technology is in part responsible for the intensification of internal dualism in developing countries, where capital-intensive products and production methods are uncritically adopted, with attendant deleterious effects. Until the mid-1960s, unemployment even in relatively wealthy Singapore was higher than 10% (Oshima, 1967); and in 1970 unemployment in metropolitan Malaysian areas reached 10.1%, as compared with 5.4% in the rural areas (Ooi, 1975). In order to prevent unemployment from becoming an explosive social issue, there is the need in many Southeast Asian countries to explore and expand labor-intensive technologies and to develop the informal sector.

Among the primary factors leading to a worsening urban environment is the horrendous increase in the number of automobiles. Most of the Southeast Asian cities, originally built for foot and hoof, are ill prepared for such demands as generated by the sixfold increase of automobiles in Greater Bangkok between 1957 and 1969 (Romm, 1972:87) or the threefold increase in Singapore over the same period (Yeung, 1973:101). Many city governments are now watching with interest how effective are some of the draconian measures that Singapore has brought to bear on curbing car ownership, including high road taxes and restricted entry to the central city area at peak hours. Additionally, crime rates reach their peak in the large cities, as for example in Greater Bangkok, where the incidence of crimes is seven times the country rate (Watanavanich, 1975:21), and in Metropolitan Manila, where it is 60 times higher than in the rest of the country (Laquian, 1972b:10). Lately, too, the problem of drug addiction has begun to affect some of the youthful population in the cities (Punahitamond, 1974). Broadly speaking, these observations are symptomatic of the pitfalls of the process of modernization and rapid social change, and before conditions can improve a number of basic issues with respect to the city's functional base and social priorities need to be resolved.

The first general issue has to do with the ecological relations between the city and the countryside. Notwithstanding contrary viewpoints, Keyfitz (1965) and other researchers propose that cities exist as part of a process that includes instituting a food surplus from the countryside. However, the size of the population of a city must be proportional to its productive capacity. When the productivity of secondary industries in the city remains low and other spheres of economic activities similarly fail to provide growth dynamism, the city may be considered to have grown beyond its productive means. They become "parasitic" and a burden to the country, as Reissman (1968:131) states:

> The countries . . . where urban growth has progressed faster than industrial and middle class growth, are generally in a worse position than even the completely underdeveloped nations.

Under such conditions, frustrated individuals are likely to take on anomic and pathological attitudes and behavior.

Second, widespread urban poverty, resulting from the lethargic economic transition, prevails in Southeast Asia. Income distribution in the primate cities is highly skewed, with only a fraction of the work force really able to sustain a relatively high standard of living. Eames and Goode (1973:256) make the pessimistic projection that the extent of material deprivation, unlike the Western experience, is likely to grow in developing Third World cities in default of a demographic transition and because of the legacy of colonial exploitation. In any event, despite the lack of a growing middle class to provide leadership for social change as in Western cities of the past (Murphey, 1954), issues of equity and wealth-sharing will probably become sharpened when a heightened sense of justice is acquired by the proletariat and the urban poor.

Insofar as considerable urban poverty persists, the indigenes are too ready to blame immigrant groups who often hold the economic reins of power in the cities. The Chinese in particular have become the target of a variety of nationalist policies, which range from harsh outright nationalization in Burma to benign toleration in Thailand (Golay et al., 1969:452-459; Chang, 1973). In many countries concerned there was a clamor to extend the national revolution to the economy after the political battle with the European colonialists was over. According to Golay et al. (1969:455), the intensity of indigenism has been accelerated by urbanization, greater population mobility, and the "nationalization" of the traditionally cosmopolitan capital cities. Thus, Dwyer (1972c:30; also Osborn, 1974) forewarns the possibility of a future urban-centered confrontation of formidable dimensions in Kuala Lumpur, to which rural Malay in-migration has continued rapidly. Herein also looms the uncertainty, as in other Southeast Asian cities, whether urban change is effected through radicalism or incremental change, through revolution or evolution.

Finally, much as they are desired, evolutionary changes are difficult to attain given the fragmented administrative structures and overlapping responsibilities in many metropolitan governments. Metropolitan Manila, for instance, is composed of 5 cities and 23 towns in 4 provinces, with 28 separate and distinct administrative units (Hollnsteiner, 1974:105). With the present trend toward greater local autonomy, Laquian (1972a:643-644) also recognizes the obstacles in Metropolitan Manila to effective city governance and reform. Future accomplishments must await a restructuring of the governmental setup, with a view to establishing at least for some major urban services metropolitan-wide authorities.

DEVELOPMENT POLICIES AND STRATEGIES

If the present problems facing the large cities appear too daunting, it is hoped that an evaluation of development policies and strategies may provide some spots of optimism. It is fair to say that the will to develop is present in every Southeast Asian country, but the way to realize developmental goals in many cases is yet to be discovered. Unlike the struggle to win independence, which is readily targeted and terminable with victory, the struggle to achieve modernization and developmental change is a long and continuous process. It is also a complex process of mobilizing and organizing a country's total resources for the advancement of social and economic processes. In agrarian-based economies of the developing world, the key role that cities can play in the spatial diffusion of innovations and technologies is widely recognized by policy-makers and scholars (Goh, 1973; Mabogunji, 1973). It involves, in Friedmann's (1961) formulation, a social transformation in which there is a national integration of the social, political, and economic space. This may take place via the spatial diffusion mechanism down the urban hierarchy and from urban centers to interurban peripheries. In the kaleidoscopic shifts of rulers and regimes in the region after World War II, one gains the impression that, as far as national integration of the space economy and planned developmental changes are concerned, successive governments are more dynastic than dynamic. Crook's (1971:253) characterization of the planning process in the Third World is certainly apt for Southeast Asia as well:

> Sectoral emphasis, especially at the national level, concentrates on economic change and has ignored physical and social dimensions. Environmental planning has frequently received a low priority, and more often than not has been confined to the spatial arrangement of cities but not in context of regional urban systems or in the larger framework of the nation. Nations tend to view urbanization as a natural by-product of change and development rather than one of the components that ought to be manipulated as a means for achieving developmental goals.

However, the countries under review do provide a variety of developmental strategies which may vary in effectiveness but were born of the best intentions.

As indicated before, the preeminence of the primate cities is an outstanding feature of the urban systems in the region. Widely regarded as an unhealthy pattern of development, there has been no lack of alternative developmental strategies designed to alleviate primacy, to plan for decentralization, and to develop a more "balanced" urban system. Bradley's (1971:181-182) statement summarizes this line of thinking:

> The sort of planned change likely to succeed in this heterogenous region is one based on new urban centers to serve the large populations now remote from the cities. It requires devolution of decision making, education, business activity, and cultural control away from the primate cities. . . . Irrespective of whether existing states remain intact, their present major cities will remain, for they already serve vital functions.

The need for urban dispersion is nowhere more urgently felt than in Bangkok, the region's classic primate city. Romm (1972:130-132) has reported a noticeable decentralization of industries with attendant employment opportunities and new housing construction beyond suburban Greater Bangkok. Toward the same end, plans are in the making for five major universities and other institutions of learning to resite or expand to new locations as far away as 100 kilometers from the metropolis. Whether these policies will merely extend Bangkok's sphere of influence or be sufficient for secondary nodes of activities to sustain themselves remains to be seen. What is needed for Thailand is, in fact, not a Bangkok-related expansion of development, but rather policies aimed at a more "balanced" distribution of future urbanization and the formulation of a national urban strategy in which Bangkok is viewed in relation to the rest of the country and other urban centers are given enough opportunities to develop into important employment and cultural nuclei (Sakornpan, 1975:25).

In the Philippines, Laquian (1972b) also advocates the need for a national urban strategy. He recognizes that a national policy is embodied in the Philippine Four-Year Development Plan (1971-1974), but it has failed to spell out the way to overcome the spatial imbalance in which the agricultural sector has been lagging much behind the urban and industrial sectors. To translate development goals into reality, Laquian emphasizes, a national urban strategy is needed.

Other researchers have proposed for other countries more country-specific solutions. In a study of the development policy in Malaysia, for example, Osborn (1974:254) concludes that the most important problem for future development in that country is a reconciliation of the differing levels of income and opportunity between the rural rich and poor and between the urban rich and poor—most significantly between Malays and non-Malays. An alternative approach to current patterns of development is proposed, which calls for public capital investment and management in middle cities below the so-called "threshold of dynamism." Undynamic cities stagnate in the existing development policy which tend to be oriented toward Kuala Lumpur and the new towns. In a similar manner, were Indonesia to modernize itself effectively, Fisher (1967:185) prescribes the development of Java into a truly metropolitan region. This entails a three-pronged strategy of checking population growth in Java, of moving as many of the landless laborers from that island to useful construction work in the outer territories, and of expanding the industrial sector in Java itself.

In discussing the variety of urban strategies, it is necessary to stress the fact that population concentration is everywhere an *adaptive* process, and urban tradition is country-unique (Lampard, 1965:539). The number and size of the cities in an urban system is the outcome of the interplay of four variables: population, environment, technology, and organization (Lampard, 1965:522; 1969). The level of technological competence, the capacity for specialization and functional differentiation are crucial factors which will dictate (1) the fashion in which administrative, social, and economic functions are articulated through the space economy and, indirectly, (2) the number, size, and spacing of cities. So far, the theory and empiricism of the relationship between city-size distribution and economic development are at best conflicting (Berry, 1961, 1971; Hamzah, 1966; Böventer, 1973). Thus, city-size distribution per se must not be used as a policy tool in fashioning the course of economic development in a country. Despite the inordinate primacy of urban systems in most Southeast Asian countries, policies designed to create artificially or theoretically "balanced" urban hierarchies are certainly ill-founded (Ginsburg, 1972). While caution is needed to avoid planning and building a "balanced" city-size distribution for its own sake, the practical significance of national urban strategies, in conjunction with sectoral considerations, cannot be overemphasized. It should also be noted that in Southeast Asia, particularly in the archipelagic region, the extension of political-administrative authority from the capital city to the more distant regions is inhibited by distance and water bodies. As a result, the hold that Manila has upon Mindanao and that Kuala Lumpur has upon East Malaysia cannot realistically be said to be more than tenuous.

Apart from the urban development strategies reviewed above, regional development is often used as a policy instrument to improve urban-rural relations. At the present stage of development in most countries in the region, the underlying principles and rationale of regional development policies are more implicit than explicit. One way of inferring the implicit developmental strategies is to examine the pattern of allocation of public funds to different regions over periods of time, an approach which Osborn (1974) pioneered in Malaysia (also Unthavikul, 1970). With the general paucity of socioeconomic data in regional areal units in most of the countries, the approach is more often than not approximate and holds much scope for increasing sophistication and utility. Although growth centers may be identfied only on the basis of one or two variables rather than from a host of variables which would allow pinpointing input-output linkages, lead or growth sectors, and other critical levers of development, it is a step forward in reducing the dualistic pattern of development between the rural and the urban areas. Consequently, Poethig (1970) suggested that the rate of out-migration should be used as a criterion in the choice of growth centers, so that alternative points of attraction may be created to counteract the magnetic pull by Manila. Laquian (1973) also envisages growth poles to promote creeping urbanism which, used by Guyot (1969) originally, connotes a process of accelerating the spread of urban influence to the rural areas. Population migration is clearly linked to both of these strategies. More deliberate in dispersing the economic base is the suggestion by Tinker (1972:45; also Fryer, 1970:91) to build up small and medium-sized towns and industrial centers specializing in light and medium-scale industries.

These diverse approaches may suggest that while existing strategies in regional development are rudimentary, they do signify an awareness of the need to bridge the gap between the rural and the urban areas. Regional inequities in growth and resources complicate developmental priorities even in developed countries; and, where these are exacerbated by considerable urban-rural differences, it seems that some degree of efficiency will have to be compromised in order to narrow the gap between the rural and the urban sectors. This is a perennial debate of the dichotomy and trade-off between growth and equity. However, even a limited attack on the problem is an improvement on an otherwise helpless attitude to let development take its own course. As Onibokun (1973) has separately stated, the challenge of urbanization in developing countries is to evolve their own pragmatic comprehensive regional planning solutions. "Pragmatic" specifically implies selective actions, which are justified by scarce human and other resources at a country's disposal, but the decision-making must be preceded by a comprehensive endeavor to establish the interdependence of problems, with due regard to existing realities and contraints.

URBAN TRANSFORMATION CONSIDERED

The analysis so far of urbanization trends, problems, and strategies has highlighted a range of experiences in Southeast Asian countries. It remains to examine the nature and extent of the urban transformation with its corollary changes in demographic, economic, and social structures.

In the first place, a country may be said to be approaching an urban transformation when 40% to 50% of its population is concentrated in urban areas, or when from 25% to 30% live in large cities of 100,000 and over (Lampard, 1969:5-6). Judged by these yardsticks, no country except the city-state of Singapore in Southeast Asia is undergoing an urban transformation, as attested by the generally low proportion of urban populations (Table 1). On the basis of the second criterion of the percentage of the population living in large cities of 100,000 and over, the range is from no such population in Brunei to the highest of 17% in West Malaysia (Davis, 1969:126-127). Nevertheless, the rate of population concentration or incremental urban share exceeds, to varying degrees, the rate of total population increase (Table 1). This means that urbanization is occurring in all countries and most rapidly in the battle-torn Indochina states. As Davis (1965) postulates, urbanization is a definite process for which there is a beginning and an end. For Southeast Asia there is no question that it is in the middle of accelerating urbanization, but, given the fact that urbanization already occurs ahead of economic development, it can be argued (Murphey, 1966) whether the region should for its own good allow unbridled urban growth to reach a level customarily considered the threshold of the urban transformation. In this sense, it is fortunate that the sustained rural-urban migration as witnessed in developed countries in the past has not been repeated in its full dimensions.

With the urban transformation only in its incipient stages, the accompanying changes in the economic and social structures also are short of a true transformation. Of course, many of these changes are mutually reinforcing. The ecological relations between the city and the countryside can be taken as an illustration. Countries in Europe and North America underwent the urban transformation at a time when the farming sector went through remarkable growth in productivity and output brought about by technological innovations and increasing commercialization. As a consequence, there was less demand for farm laborers, who migrated to the cities, which were able to sustain the urban majority because of the enlarged agricultural surplus (Lampard, 1969:14). These basic changes in agriculture are still eluding the farmers in Southeast Asia, where the Green Revolution over the last decade has not produced the much vaunted cornucopia (Jacoby, 1972; Brown, 1968). This also explains why the economy and the functional organization of industries or occupations have not significantly been modified from traditional norms. Agriculture at present engages by far the largest proportion of the labor force, ranging from a low of about 55% in the Philippines and West Malaysia to a high of about 80% in East Malaysia, Cambodia, and Thailand, leaving Singapore out of this discussion (You and Yeh, 1971:526). In addition, little structural change of the economy is reflected in the continuing high share of agricultural earnings in the gross domestic product. Even in relatively prosperous West Malaysia, agriculture accounts for about 63% of the GDP (Wikkramatileke, 1972:484). The experience of the developed countries undergoing an urban transformation is such that as the gross national product grew, the proportion contributed by the agricultural output markedly and progressively decreased while nonagricultural sectors concomitantly gained their importance (Lampard, 1969:17). For lack of basic structural changes in the economy, Southeast Asian cities hence remain largely dualistic in character, marginal occupations abound, and their populations grow despite restricted economic opportunities provided by limited industrialization.

Finally, considering the question of social change in Southeast Asian cities, revolutionary changes in life styles, behavioral patterns, and social stratification have not yet affected the masses. McGee (1969:16) suggested that the persistence of pluralistic ethnic structures and a large bazaar sector have inhibited the kind of social change said to have occurred in Western cities. The normative standards of the Western "urban way of life" characterize a much smaller proportion of the urban population as compared with Western cities. Through the existence of squatter settlements and the bazaar sector, interlocking social and economic relations between city and rural folk continue. In fact, basic village forms of political and social organization prevail in these settlements (Laquian, 1971), and Hauser (1957) pointed to the persistence of "folk" conditions in many Southeast Asian cities. However, in view of the nature of urbanization in the region, Dwyer (1972c) has called attention to the city as a source of potential tensions, political conflict, and extremist behavior.

The far-reaching changes which went hand in hand with the urban transformation took more than a century to complete in Western cities. In the present period, when the aeroplane and the mule are put into use at the same time in the region, there are inevitably lags and overlaps between modernization and traditionalism. The period also provides an opportunity for these changes to be telescoped. Indeed, Hoan (1974:30) even argues for Southeast Asia that:

> The region cannot afford the luxury of waiting for the slow transformations that accompanied urbanization in the West. It faces a grim future unless modernizing influences usually associated with urbanization can be made to operate in the rural context as well.

Some of the measures adopted by the countries concerned do suggest that they do not have the patience to wait for slow evolutionary changes; the urban policies employed sometimes smack instead of a touch of radicalism. One example is the "closed door" policy applied since 1970 in Djakarta, in an attempt to stem the tide of rural-urban migration. In-migrants are required to show evidence of employment and housing accommodations before they are issued residency permits. They are also required to deposit with the city government for six months the equivalent of return fare to the point of origin (Berry, 1973:100-101). The policy is only partially successful, for, as Petersen (1966:26-27) has observed, many sets of rules have been applied in the game to block urbanization in the developing countries; the goalkeepers have never won. Similarly, to implement the New Economic Policy under the Second Malaysia Plan, the Urban Development Authority has since 1971 carried out a number of urban renewal and other projects which specifically encourage Malay participation. The impact that these will have on spatial reorganization, ethnic patterning, and relative economic roles is likely to be momentous and is being watched with care. Radical some of these changes may be, but change must come to effect progress. For Tinker (1969:116) has perspicaciously offered the comparison that, whereas in China, out of continuity will come change, out of change will emerge continuity in Southeast Asia.

FUTURE PROSPECTS

This essay should perhaps best be concluded by taking into account what future prospects should hold for Southeast Asian cities. The dramatic events that culminated in the ascendancy of Communist-led governments in Indochina in early 1975 have introduced new elements of uncertainty and political instability in the region, at least in the short run. Unquestionably, different development policies in which urban roles may be redefined will be adopted by the new governments. There is, however, little reason to believe that the Indochina states, guided by whatever political philosophies, would ignore the allomerative advantages inherent in the existing urban centers. In other words, Saigon, Phnom Penh, and Vientiane will most probably continue as primate cities of their respective countries, but how these and other cities are related to one another and to overall developmental strategies can only be determined by later events.

The political and economic impact that the recent events in Indochina on other Southeast Asian countries remains to be assessed. However, it is clear from the high rates of growth of the primate cities discussed before that, barring dramatic declines in fertility rates or success in migration and population control, the majority of the primate cities will become gigantic urban agglomerations of 10 million or more each by the year 2000 (McGee, 1972; Goldstein, 1973). Hamzah (1974) recognizes two opportunities in these huge urban regions: one, the integration of rural and urban occupations, thus effectively developing a viable domestic market, and the other, the reduction of ethnic identification with certain economic activities. More broadly, however, futurism involving a long-term projection of socio-cultural change should be employed if the kind of developmental change being sought is forward-looking and realistic. After all, "the most important social change of our time is the spread of awareness that we have the ability to strive and deliberately to contrive change itself" (Berry, 1972:42).

A striking feature which emerges from this paper is that Southeast Asian cities are of greater relative poverty compared with Western cities. This condition is characteristic of other Third World countries too. If they are to follow the traditional approach to development, they would need to attain unrealizable rates of economic growth at 9% to 11% per year in the gross domestic product to achieve reasonable employment objectives (Grant, 1971:115). These rates, moreover, would absorb only the increase in labor force in nonagricultural jobs and would far exceed corresponding rates in developed countries at comparable stages of development. In comparison with Western cities, it would thus appear that a larger majority of urban poor and a larger marginal class will be with Southeast Asian cities for a long time. To cope with this persistent condition, Dwyer (1972b:52) advocates that "Cities for the poor will have to be consciously designed, rather than the poor fitted into cities basically designed for the convenience of the moderately wealthy and the rich." There is the need, for instance, to reorient urban planning and revise developmental strategies. One example, as suggested before, may be the development of the informal sector by virtue of its high labor absorption capacities. Another illustration is the question of transference of technology. In the face of current levels of development, it is probable that only medium-level and labor-intensive types of technology are appropriate for most Southeast Asian countries. Greater emphasis could conceivably be put on men rather than on machines, as Turner (1970:262) has summarized:

> By facing the fact that they cannot substitute machines for men, the developing countries may well show us how to use our own sophisticated tools in humane and genuinely constructive ways. Their only chance for development and survival, after all, is to use the few tools that they can afford to stimulate and support the initiative to the mass of the common people.

Furthermore, it is necessary to revise the expectations of economic growth for its own sake. Imitation of Western aspirations in pursuit of economic growth and social life styles would seem to lead inevitably to the same nightmarish visions of ecological and social malaise as in developed countries. Above all, population growth must be brought under control if real economic progress in the urban and rural sectors can be attained. In short, the countries in question must "aim for a pattern of economic development that ensures employment and equity instead of growth and dualism, and built cities that can offer adequate opportunities for personal development" (Hoan, 1974:34).

All these revisions of developmental goals imply that the planning style in general will have to shift from what Berry (1973:174) terms ameliorative problem-solving to allocative trend-modifying, that is, from planning for the present toward planning for the future. Whether and how soon the shift comes about depends on the speed at which present governments can tackle their problems and how realistically development is envisaged. At any rate, Southeast Asian cities of today and tomorrow must be viewed as a frontier of opportunity as well as a challenge to which the creative energies of the people may be harnessed to serve purposeful developmental ends.

REFERENCES

ARMSTRONG, W.R., and McGEE, T.G. (1968). "Revolutionary change and the Third World city: A theory of urban involution." Civilisation, 18:353-378.

ARUMAINATHAN, P. (1973). Report on the census of population, 1970, Singapore (Vol. 1). Singapore: Department of Statistics.

BERRY, B.J.L. (1961). "City size distribution and economic development." Economic Development and Cultural Change, 9(July):573-587.

--- (1971). "City size and economic development: Conceptual synthesis and policy problems, with special reference to South and Southeast Asia." Pp. 111-115 in L. Jakobson and V. Prakash (eds.), Urbanization and National Development. Beverly Hills, Calif.: Sage.

--- (1972). "Deliberate change in spatial systems: Goals, strategies and their evaluation." South African Geographical Journal, 54(December):30-42.

--- (1973). The human consequences of urbanisation. London. Macmillan.

BOVENTER, E. von (1973). "City size systems: Theoretical issues, empirical regularities and planning goals." Urban Studies, 10(June):145-162.

BRADLEY, J. (1971). Asian development problems and programs. New York: Free Press.

BREESE, G. (ed., 1969). The city in newly developing countries: Readings on urbanism and urbanization. Englewood Cliffs, N.J.: Prentice-Hall.

BROWN, L.R. (1968). "The agricultural revolution in Asia." Foreign Affairs, 46(July): 688-698.

CHANDER, R. (1969). Annual bulletin of statistics, Sabah, 1968-1969. Kota Kinabalu: Sabah, Department of Statistics.

--- (1970). Annual bulletin of statistics, Sarawak, 1972. Kuching: Sarawak, Department of Statistics.

--- (1971). 1970 population and housing census of Malaysia: Field count summary. Kuala Lumpur: Malaysia, Department of Statistics.

CHANG, D.W. (1973). "Current status of Chinese minorities in Southeast Asia." Asian Survey, 8(June):587-603.

CHUA, W. (1975). "The petroleum industry of Brunei." Unpublished academic exercise, Department of Geography, University of Singapore.

CROOKS, R.J. (1971). "Planning for developing countries." Journal of Royal Town Planning Institute, 57(June):251-256.

DAVIS, D. (1965). "The urbanization of the human population." Scientific American, 213(September):41-53.

--- (1969). World urbanization, 1950-1970. Vol. 1: Population monograph series no. 4. Berkeley: University of California.

--- (1971). "The role of urbanization in the developing countries." Paper presented at Rehovot Conference.

--- (1972). World urbanization, 1950-1970. Vol. 2: Population monograph series no. 9. Berkeley: University of California.

DWYER, D.J. (1964). "The problems of in-migration and squatter settlement in Asian cities: Two case studies, Manila and Victoria-Kowloon." Asian Studies, 2(August): 145-169.

--- (ed., 1972a). The city as a centre of change in Asia. Hong Kong: Hong Kong University Press.

--- (1972b). "Future (urban) shock." Insight, (May):49-52.

--- (1972c). "Urbanization as a factor in the political development of Southeast Asia." Journal of Oriental Studies, 10:23-32.

EAMES, E., and GOODE, J.G. (1973). Urban poverty in a cross-cultural context. New York: Free Press.

Far Eastern Economic Review (1975). Asia 1975 yearbook. Hong Kong: Author.

FISHER, C.A. (1967). "Economic myth and geographical reality in Indonesia." Modern Asian Studies, 1:155-189.

FRIEDMANN, J. (1961). "Cities in social transformation." Comparative Studies in Society and History, 4(November):86-103.

FRYER, D.W. (1953). "The million city in Southeast Asia." Geographical Review, 43(October):474-494.

--- (1970). Emerging Southeast Asia. London: Philip and Sons.

GEERTZ, C. (1963). Peddlers and princes: Social change and economic modernization in two Indonesian towns. Chicago: University of Chicago Press.

GINSBURG, N. (1972). "Planning the future of the Asian city." Pp. 269-283 in D.J. Dwyer (ed.), The city as a centre of change in Asia. Hong Kong: Hong Kong University Press.

GOH, K.S. (1973). "Cities as modernizers." Insight, (August):46-50.

GOLAY, F.H. et al. (1969). Underdevelopment and economic nationalism in Southeast Asia. Ithaca, N.Y.: Cornell University Press.

GOLDSTEIN, S. (1973). "The demography of Bangkok: The case of a primate city." Pp. 81-119 in W.H. Wriggins and J.F. Guyot (eds.), Population, politics, and the future of Southeast Asia. New York: Columbia University Press.

GOODMAN, A.E. (1973). The causes and consequences of migration to Saigon, Vietnam: Final report to SEADAG. New York: Asia Society.

GOODMAN, A.E., and FRANKS, L.M. (1974). "Between war and peace: A profile of migrants to Saigon." SEADAG Papers. New York: Asia Society.

GUYOT, J. (1969). "Creeping urbanism and political development in Malaysia." Pp. 124-161 in R.T. Daland (ed.), Comparative urban research: The administration and politics of cities. Beverly Hills, Calif.: Sage.

GRANT, J.P. (1971). "Marginal man: The global unemployment crisis." Foreign Affairs, 50(October):112-124.

HAMZAH, S. (1966). "City size distribution of Southeast Asia." Asian Studies, 4(August): 268-280.

--- (1974). "Trends in urban development, with special reference to Southeast Asia." Speech read at the annual conference of the Asia Foundation, Penang.

HASSAN, R. (1972). "The problem of squatter relocation in Southeast Asia." Asia Research Bulletin, 11(April):756-757; 12(May):843-843.

HAUSER, P.M. (ed., 1957). Urbanization in Asia and the Far East. Calcutta: UNESCO Research Centre.

HAUSER, P.M., and SCHNORE, L.F. (eds., 1965). The study of urbanization. New York John Wiley.

HOAN, B. (1974). "Cities the poor build." Insight, (June):28-34.

HOLLNSTEINER, M. (1974). "The case of 'The people versus Mr. Urbano Planna Administrator,'" Pp. 84-111 in J. Hoyt (ed.), Development in the '70's. Manila.

--- (1975). "The urbanization of Metropolitan Manila." In Y.M. Yeung and C.P. L (eds.), Changing South-East Asian cities. Kuala Lumpur: Oxford University Press.

Indonesia, Republic of, Central Bureau of Statistics (1972). Indonesia, 1971 populatio census: Population by province and regency/municipality. Djakarta: Author.

International Bank for Reconstruction and Development (1973). "World Bank atlas." Finance and Development, 10(March):26-27.

International Labour Office (1973). Year book of labour statistics, 1973. Geneva: Author

JACKSON, J.C. (1974). "Urban squatters in Southeast Asia." Geography, 59(January) 24-30.

JACOBY, E.H. (1972). "Effects of the 'Green Revolution' in South and South-East Asia. Modern Asian Studies, 6:63-69.

JAKOBSON, L., and PRAKASH, V. (eds., 1971). Urbanization and national developmen Beverly Hills, Calif.: Sage.

KEYFITZ, N. (1965). "Political-economic aspects of urbanization in South and Southea Asia." Pp. 265-309 in P.M. Hauser and L.F. Schnore (eds.), The study of urbanizatio New York: John Wiley.

KING, D.Y. (1974). "Social development in Indonesia: A macro analysis." Asian Surve 14(October):918-935.

LAM, T.F. (1974). "Urban land use policy and development with reference to Malaysia. Paper presented at the International Seminar on Urban Land Policy, Singapore.

LAMPARD, E.E. (1955). "The history of cities in the economically advanced area Económic Development and Cultural Change, 3(January):81-102.

--- (1965). "Historical aspects of urbanization." Pp. 519-554 in P.M. Hauser and L. Schnore (eds.), The study of urbanization. New York: John Wiley.

--- (1969). "Historical contours of contemporary urban society: A comparative view Journal of Contemporary History, 4(July):3-25.

--- (1973). "The urbanizing world." Pp. 3-57 in H.J. Dyos and M. Wolf (eds.), Th Victorian city: Images and realities (vol. 1). London: Routledge and Kegan Paul.

LAQUIAN, A.A. (1968). Slums are for people. Manila: DM Press.
--- (1971). "Slums and squatters in South and Southeast Asia." Pp. 183-203 in L. Jakobson and V. Prakash (eds.), Urbanization and national development. Beverly Hills, Calif.: Sage.
--- (1972a). "Manila." Pp. 605-644 in W.A. Robson and D.C. Regan (eds.), Great cities of the world: Their government, politics, and planning (3rd ed., vol. 2). London: Allen and Unwin.
--- (1972b). "The need for a national urban strategy in the Philippines." SEADAG Papers. New York: Asia Society.
--- (1973). "Urban tensions in Southeast Asia in the 1970s." Pp. 120-146 in W.H. Wriggins and J.F. Guyot (eds.), Population, politics, and the future of Southeast Asia. New York: Columbia University Press.
LEWIS, W.A. (1967). "Unemployment in developing countries." World Today, 23(January):12-22.
MABOGUNJI, A.L. (1973). "Role of the city in the modernization of developing countries." Canadian Geographer, 17(spring):67-76.
McGEE, T.G. (1967). The Southeast Asian city. London: G. Bell.
--- (1969). "The urbanization process: Western theory and Southeast Asian experience." SEADAG Papers. New York: Asia Society.
--- (1971). "Catalysts of cancers? The role of cities in Asian society." Pp. 157-181 in L. Jakobson and V. Prakash (eds.), Urbanization and national development. Beverly Hills, Calif.: Sage.
--- (1972). "Man and environment in Southeast Asia: Ecology or ecocide?" Unpublished paper, partly published as "Urban ecocide." Insight, (June):29-31.
McNICOLL, G., and MAMAS, S.G.M. (1973). "The demographic situation in Indonesia." Papers of the East-West Population Institute, no. 28. Honolulu: East-West Population Institute.
MIJARES, T.A., and NAZARET, F.V. (1973). The growth of urban population in the Philippines and its perspective. (Technical paper no. 5). Manila: Republic of the Philippines, Bureau of the Census and Statistics.
MILONE, P.D. (1966). Urban areas in Indonesia: Administrative and census concepts (Research series no. 10). Berkeley: Institute of International Studies, University of California.
MUNSON, F.P. (1968). Area handbook for Cambodia. Washington, D.C.: U.S. Government Printing Office.
MURPHEY, R. (1954). "The city as the center of change: Western Europe and China." Annals of the Association of American Geographers, 44:349-362.
--- (1966). "Urbanization in Asia." Ekistics, 21(January):8-17.
NELSON, J. (1970). "The urban poor: Disruption or political integration in Third World cities?" World Politics, 22(April):393-414.
ONIBOKUN, A. (1973). "Urbanization in the emerging nations: A challenge for pragmatic comprehensive regional planning." Planning Outlook, 13(autumn):52-66.
OOI, J.B. (1975). "Urbanization and the urban population in peninsular Malaysia, 1970." Journal of Tropical Geography, 40(June):40-47.
OSBORN, J. (1974). Area, development policy, and the middle city in Malaysia (Research paper no. 153). Chicago: Department of Geography, University of Chicago.
OSHIMA, H.T. (1967). "Growth and unemployment in Singapore." Malayan Economic Review, 7(October):32-58.
PETERSEN, W. (1966). "Urban policies in Africa and Asia." Population Review, 10(January):24-35.
POETHIG, R.P. (1970). "Needed: Philippine urban growth centers." Ekistics, 180(November):384-386.
--- (1971). "The squatters of Southeast Asia." Ekistics, 183(February):121-125.

PRYOR, R.J. (1973). "The changing settlement system of West Malaysia." Journal of Tropical Geography, 37(December):53-67.

PUNAHITAMOND, S. (1974). A preliminary study of the problem of narcotic drugs and related issues in Thailand. Bangkok: Social Science Research Institute, Chulalongkorn University.

REISSMAN, L. (1968). "Urbanization: A typology of change." Pp. 126-144 in S.F. Fava (ed.), Urbanism in world perspective: A reader. New York: Crowell.

ROBERTS, T.D. et al. (1969). Area handbook for Laos. Washington, D.C.: U.S. Government Printing Office.

ROMM, J. (1972). Urbanization in Thailand. New York: Ford Foundation.

SAKORNPAN, C. (1975). "The social aspects of low-cost housing." IDRC-supported Low-Cost Housing, Thailand Monograph. Bangkok.

SANDHU, K.S. (1964). "Emergency resettlement in Malaya." Journal of Tropical Geography, 18(August):157-183.

SAW, S.H. (1972). "Patterns of urbanization in West Malaysia, 1911-1970." Malayan Economic Review, 17(October):114-120.

SINGER, H.W. (1970). "Dualism revisited: A new approach to the problems of the dual society in developing countries." Journal of Development Studies, 7(October):60-75.

STERNSTEIN, L. (1974). "Migration to and from Bangkok." Annals of the Association of American Geographers, 64(March):138-147.

TAYLOR, J.L. (1973). "The slums and squatter settlements of Southeast Asian cities." Pp. 175-189 in Kuala Lumpur Forum. Kuala Lumpur: Edinfo Press.

Thailand, Republic of, National Statistical Office (1973). 1970 population and housing census. Bangkok: Author.

TINKER, H. (1969). "Continuity and change in Asian societies." Modern Asian Studies, 3:97-116.

--- (1972). Race and the Third World today. New York: Ford Foundation.

TURNER, J.F.C. (1970). "Squatter settlements in developing countries." Pp. 250-263 in D.P. Moynihan (ed.), Toward a national urban policy. New York: Basic Books.

United Nations (1974). Yearbook of statistics, 1973. New York: Author.

UNTHAVIKUL, P. (1970). "Regional planning and development: Thailand." Ekistics, 180(November):416-423.

VILORIA, L.A. (1972). "The Manileños: Significant elites in urban development and nation-building in the Philippines." Pp. 16-28 in D.J. Dwyer (ed.), The city as a centre of change in Asia. Hong Kong: Hong Kong University Press.

WATANAVANICH, P. (1975). "Urbanization of the Bangkok region: A study of the relations of crime and other social factors." Unpublished paper, Thai University Research Associates, Bangkok.

WEBER, A.F. (1899). The growth of cities in the nineteenth century. New York: Macmillan.

WERTHEIM, W.F. (1964). East-West parallels: sociological approaches to modern Asia. Chicago: Quadrangle Books.

WIKKRAMATILEKE (1972). "The Jengka triangle, West Malaysia: A regional development project." Geographical Review, 62(October):479-500.

WRIGGINS, W.H., and GUYOT, J.F. (eds., 1973). Population, politics, and the future of Southeast Asia. New York: Columbia University Press.

YEH, S. (ed., 1975). Public housing in Singapore: A multi-disciplinary study. Singapore: Singapore University Press.

YEH, S., and YOU, P.S. (1971). "Labour force supply in Southeast Asia." Malayan Economic Review, 15(October):25-54.

YEUNG, Y.M. (1973). National development policy and urban transformation in Singapore: A study of public housing and the marketing system (Research paper no. 149). Chicago: Department of Geography, University of Chicago.

--- (1975a). "Hawkers and vendors: Dualism in Southeast Asian cities." Paper presented at the annual meeting of the Association of American Geographers, Milwaukee.

--- (1975b). "Locational aspects of low-cost housing development in Southeast Asia." Paper presented at the IDRC-supported low-cost housing conference, Tagaytay City, Philippines.

YEUNG, Y.M., and LO, C.P. (eds., 1975). Changing South-East Asian cities. Kuala Lumpur: Oxford University Press.

YOU, P.S., and YEH, S. (1971). "Aspects of population growth and population policy." Pp. 448-580 in ADB, Southeast Asia's economy in the 1970s. London: Longman.

YOU, P.S. et al. (1974). "Social policy and population growth in South-East Asia." International Labour Review, (May):459-470.

13

Chinese Urbanization Under Mao

RHOADS MURPHEY

THE HISTORICAL BASES OF MAOIST POLICIES

□ IT SHOULD BE MADE CLEAR at the start that the continuous Chinese experience with urbanization is perhaps longer than that of any other society. Far from being typical of the so-called Third or Developing World, China stands out as having been in advance of the rest of the world in the size, number, and complexity of its cities, as in other aspects of civilization, for some two thousand years. Although the level of urbanization remained relatively low, as it still is compared with most Western countries, and probably never exceeded 10% until the present century, cities large and small are nothing new to China. Sian, Peking, and Hangchow have each in its time been the largest city in the world. Although they owed their size primarily to their administrative and ceremonial functions as imperial capitals, there is also a long history in China of urban-based trade and of cities whose chief function was commerce. There have probably been more cities and more urbanites in China than in any other country; 10% or even 5% of China's immense population, over more than two thousand years, is still an impressive total.

Traditional cities were highly organized, managed, and planned in detail. They were centers of both local and long-distance trade, including both interprovincial and overseas flows. In absolute terms, this trade almost certainly exceeded trade totals in Europe or elsewhere until the 19th century (Murphey, 1972; Perkins, 1975). Western observers as late as the 1850s were uniformly impressed by the size and number of Chinese cities and by the hive of organized commercial activity centered in them, dominated by guilds but also taxed and regulated by, and often articulated with, the bureaucratic state. Cities were most importantly centers of imperial administration, laid out by the state according to a uniform pattern and managed by the bureaucracy. Urban forms and functions were in no sense haphazard but were part of a closely administered national system. Unlike other currently developing systems, where the urban dimension is either largely new or is being adapted from earlier Western models, China has its own urban experience as a base. This includes a reservoir of organizational and managerial skills which made the traditional cities, like the uniquely productive agricultural system, function so effectively as part of a sophisticated economy and imperium. In urban design and management, China can draw on this reservoir of familiarity with the problems and challenges posed by cities and is thus in a different position from that of most developing countries. The legacy of commercialism is important too. The Maoist design is not being imposed on a *tabula rasa;* the Chinese past helps to explain why current development has been so successful, but it may also help to keep the socialist ideals of the Maoist vision from being completely realized (as considered at the end of this essay).

There was not in imperial China the kind of split between urban and rural worlds which has been characteristic of at least the postmedieval experience in the West and of the experience in most developing countries. The city was seen as existing primarily in order to administer and serve the countryside. Agriculture was the chief source of wealth, as the land-and-grain tax was the chief source of state revenue. The countryside must be kept peaceful and productive, aided by state-directed and urban-based projects of water control, road building, and dissemination of improved techniques. Urban-based elites were drawn in large part from the countryside and retained close ties with it. There was not the denigration of rural circumstances and values common in the West, nor was there the same kind of dichotomy between urban and rural systems. Even in military terms, the city was not normally regarded as a bastion against rural-based rebellion but as being responsible for the defense and orderly management of the rural districts in its charge, roughly the size of U.S. counties.

Such a base of long urban experience and its management helps to make it possible for a revolutionary government since 1949 to plan and implement a new design for urban growth, new especially by contrast with the modern Western model. China did have some direct exposure to that model during its semicolonial period from 1842 to 1949 in the form of the Western-dominated treaty ports, and the new urban directions that China is now pursuing represent to some degree a nationalist reaction against this foreign effort to remake China, like the rest of the world, in the Western image. We have only recently become accustomed in the West to recognizing the city as presenting or generating dismaying problems. The commercial-industrial mode through which the modern West has achieved wealth and power has been largely created in and by the cities. We have tended to see this association as inevitable even while we recognize that cities have produced in many respects a deterioration in the quality of human life. Smoking chimneys and expanding metropolises have for too long been seen as symbols of prosperity and progress to make it easy either to undo their baneful effects or to envisage a different or better model for growth.

China has only recently been caught up in large-scale industrialization and the urban forms associated with it. Agriculture is still the principal sector of the economy, and the level of urbanization is still probably less than 25%. There is not the same association as in the West between prosperity and cities, nor the same assumption that industrialization, which is now seen by China, as by most other countries, as the chief engine of economic development, must necessarily take place only through the agency of huge, crowded, alienating, and polluted urban centers. The treaty port experience gave a revolutionary China a negative set against the Western industrial model of concentrated urban growth. Westerners established territorial bases on semipermanent leases adjacent to nearly all existing Chinese trade centers of major importance. With the stimulus of expanded foreign trade under Western direction, and the beginnings of Western-managed industrialization in their concession areas, these cities became the largest in the country, upsetting the traditional urban hierarchy and representing a new kind of city, radically different from traditional forms, where foreign trade and manufacturing were paramount. The treaty ports were integrated with the rural countryside only as they drew from it trade commodities for export or attempted to infiltrate it with new factory-made goods. They had no administrative responsibilities for the countryside, and their interaction was easily seen as purely exploitative, unlike traditional Chinese cities, and for foreign rather than Chinese benefit. As industrialization progressed in the treaty ports after about 1895, an economically depressed urban-industrial proletariat came into existence and was equally easily seen by nationalist-minded Chinese as demonstrating the exploitative and oppressive character of these Western-style cities.

It was also in Shanghai, Canton, Tientsin, Hankow, and the other major treaty ports that the Chinese collaborators lived, making a good thing out of the foreign presence and aping Western ways. The early struggles for power on the part of the Chinese Communist party were crushed in these same cities, despite Soviet and Marxist doctrine about the industrial proletariat, by the even greater power there of the new Western-style bourgeoisie and its support of the reactionary regime of the Kuo Min Tang government. In the end, it was a rural-based and peasant-manned movement, glorifying rural and peasant virtues as representing the real China in its struggle against these alien and evil cities, which carried the Communist party to victory, a Chinese cure for a foreign cancer. Cities are still seen as potential breeders of bourgeois counterrevolution and must be reshaped and carefully controlled to insure that their contribution is positive rather than negative. Under foreign domination, the treaty ports came more and more to resemble Western cities in physical appearance and in nature, clearly alien to China and reproducing many of the same problems of exploitation, overcrowding, human misery, and environmental degradation as the industrial revolution had already generated in European and American cities.

DISTINCTIVELY CHINESE POLICIES

It has been thus easier for a revolutionary China to attempt to break away from such models and to pursue its own path to industrialization and economic growth along different and more distinctively Chinese lines. Here the traditional symbiotic relationship between city and country is also relevant to the new design and its assumptions. Cities must now, as in the past, serve the countryside; they must, as the official position puts it, be transformed from consumer into producer cities: "it is only through the transformation of consuming cities into producing ones that the people's livelihood can be assured and also that urban construction can look forward to a bright future. This also represents the difference between socialist urban construction and imperialist urban construction" (Joint Publications Research Service, 1960). The country-side, where some three-quarters of China's people still live and work, is what counts; that is where the revolution began, as a Chinese answer to a Western threat, and that is where the peasant-socialist values on which the continuing success of the revolution depends are most strongly exemplified. There is no place in Mao's China for the Westerner Karl Marx's contempt for the "idiocy of rural life." It is the Western-tainted ex-treaty ports, still nearly all of China's largest cities, which are suspect, and the bourgeois behavior and values which such cities generate.

In the West, where commerce and manufacturing became proportionately more important, the symbiotic tie between city and countryside was, relatively speaking, obscured and forgotten. New classes arose in cities—which, of course is what the word bourgeoisie means. They and the new urban proletariat were divorced from the rural context, were indeed different kinds of people with different goals and values, and were consequently at chronic odds with the rural population and vice versa, as the lines drawn in the French revolution made classically clear, building on a history of "peasant wars" in early modern Europe. The treaty ports in China were cities of this sort, imposed on China by Westerners and concerned with draining exports out of the country rather than with serving its interests. Given also the Western concern for overseas exports, these cities, the country's largest, were heavily concentrated spatially along the east coast and the major inland waterway, the Yangtze river system. Their distribution was lopsided and did little or nothing for the development of most of the country. What relatively few railways were built served primarily the treaty ports; most of the country was without rail lines at all. The potentially promising beginnings of industrialization were limited almost entirely to the treaty ports, leaving the rest of the country in these terms almost wholly undeveloped.

It is understandable that a fully independent and anti-imperialist new government, let alone a Communist one, would see unacceptable flaws in such a pattern of urban and industrial growth and would strive to remedy it. Indeed the two major revolutionary goals, as identified by Chairman Mao, are the elimination of the twin and closely interrelated distinctions between mental and manual labor and between city and countryside. Cities are at least potentially the nests of elites—managerial, administrative, technical, professional, economic—and of mental workers. In the countryside, where manual work is uppermost, elites have less opportunity to grow or to become protected. But industrialization is recognized as essential to economic growth and to China's fervent drive to catch up with and overtake the West. Mass welfare, the aim of the revolution, cannot be served adequately without industrialization and new technology, nor can national strength be achieved. The Chinese solution is twofold: to transform existing cities so that they cease to breed selfish, bourgeois-minded elites and come to serve the people, most of whom are in the countryside; and to disperse new industrialization as widely as possible, into the former undeveloped inland provinces in new or newly industrialized cities and, on a smaller scale, onto rural communes in every area. No area, provincial or rural, should be primarily dependent on a distant city for the supply of what it needs but should be as much as possible self-reliant and self-sufficient. Every area, however small or remote, must take part in the experience of industrialization and modernization and must share equally in its benefits, rather than having this process so heavily confined to a few large cities, which are suspect in any case and which may continue to breed bourgeois elitist attitudes.

In the early stages of the revolution, there was talk of dismantling the former treaty ports, especially Shanghai, and relocating their populations and industrial plants among a number of smaller inland centers. This was soon replaced however by a policy of limiting the growth of the biggest cities (ex-treaty ports) and at the same time transforming their nature, from consumers to producers, from elitist bastions to servers of the people and of mass welfare. The existing industrial plant in the former treaty ports and in Manchuria was recognized as crucial to any national effort at further economic growth, as were their pools of skilled labor and technicians. Many technicians and workers were reassigned to take part in the industrialization of inland centers, but the existing bases were not allowed to atrophy. Since 1949, travel, housing, and employment have been all controlled by permit or assignment and are allocated on the basis of national need as perceived by the central planners but also in the light of the decision to restrict the growth of the largest cities and to disperse urbanization and industrialization more widely. Accurate urban population figures are very hard to come by, and many of those available are often both inconsistent and ambiguous in the sense that they may include, especially for the largest cities, the population of extensive areas which are purely rural-agricultural but administratively incorporated within the municipality. Shanghai, Peking, and Tientsin, the three largest cities, are clearly organized this way and administratively attached to the central government rather than to their respective provinces. The municipality of Shanghai covers extensive nonurban areas in the lower Yangtze delta, altogether over five times the urbanized area; Shanghai's population is given as 10 million (sometimes as 11), but with the distinction sometimes also provided that only 6 million of this refers to the urban areas, with the balance being represented by the rural areas within the municipality. Nevertheless, it does in general seem to be the case that the largest urban areas have been controlled in their growth, both through controls on movement of people and through planned allocation of investment, which favors new inland centers and developing areas.

But the big cities remain, over 20 of them with populations of a million or more, and nearly all of them are former treaty ports. Their nature must be transformed. In the light of traditional and now again prominent attitudes about the relative merits and roles of city and countryside, it is striking that one of the chief means used to prevent the emergence of selfish urban-based elites is to require virtually all white-collar or mental workers in cities to spend two months or more each year engaged in productive manual labor in the countryside. Only there, apparently, can potential elites be purged of the corrupting tendencies inherent in the city and reminded through the redeeming effects of physical labor among the peasants, in the still morally pure countryside, of their primary duty to serve the people (Murphey, 1972). While they remain in the cities, white-collar workers must also take part in manual labor wherever possible and break down the distinctions, sharply maintained especially in traditional China as in many developing societies now, between a professional elite whose status was emphasized by their refusal to work with their hands (symbolized by the long gown and fingernails of the Chinese gentry-literati) and the illiterate peasants and laborers. Engineers must get out on the shop floor to confront concrete problems of production and operation, technicians and theoreticians must get their hands dirty—and manual workers must also contribute to new designs and to plant management. The most talked-of achievements in industrialization are those won by manual workers who, from their own direct experience, have devised better tools or machines or have worked out improved production schemes.

The physical arrangement of cities has also been attacked, in an effort to combat alienation, waste, and environmental degradation. Unfortunately, we know relatively little as yet about the details of urban planning or of the internal structure of Chinese cities. The brief and managed visits to which foreign scholars are limited do not permit more than hurried glimpses and superficial discussions, nor have the Chinese published, at least not for outside circulation, more than the most general kinds of accounts. In the revolutionary fervor of the Great Leap Forward (1957-1958), when the commune system was first established, the cities too were organized into urban communes (Schurman, 1968), but this was essentially abandoned by the early 1960s in favor of an urban neighborhood model (which is discussed below). One of the other urban accompaniments of the Great Leap has however been retained, and further emphasized as a result of the Cultural Revolution from 1966 to 1968: the policy of removing large numbers of what are considered surplus people from the largest cities, notably Shanghai, and relocating them in the countryside, where their labor is considered more useful to the overall national goals of growth. Especially since about 1960, agriculture has been given priority in the drive for economic development—appropriately enough, given the present nature of the Chinese economy—and has been made the principal base for the generation of surpluses which can then be used to finance industrialization. As the Maoist formula now puts it, "Take agriculture as the base and industry as the leading sector." This has meant the somewhat belated, but rational, diversion of scarce investment funds from industry to agriculture and the replanning of the industrial sector to give more prominence to production of fertilizers, pumps, and agricultural equipment and machinery. It has also further underlined the chief role of cities, still the major industrial bases, as serving agriculture and the countryside.

EFFICACY OF MAOIST CONTROLS

Despite the tight controls on movement, employment, and housing, some people apparently still slip through and manage to move into the cities from the countryside, many of them presumably illegally and without ration cards or work permits, to be sheltered by friends or relatives. Others may actually find work of some sort, but all such extralegal migrants are regarded either as parasites or as supernumeraries who are potential sources of trouble and who, in any case, sabotage the national plan. At different periods, the Chinese press has carried accounts of the rounding up of many thousands of such people and their return to the countryside, so as both to use their labor where it is most needed and to prevent the big cities from growing bigger. It must nevertheless be discouraging that despite the persistent efforts to inculcate the opposite norms, many rural people still regard the cities as more attractive places to be. There are good reasons for this, not least among them the substantially higher wage rates in the cities. But the cities also, especially the largest ones, are in China as elsewhere attractive as centers of excitement, at least in a relative sense, and of amenities, if only in the form of parks, shop windows, and street crowds. Health and education services are on a far higher level in the cities; there are theaters and sports events; and there is the indefinable glitter of urban life, even in a Maoist China, where standards are puritanical and utilitarian, not to say drab. But there are department stores and cinemas and a general atmosphere which all Chinese recognize as profoundly different from the rural communes. Shanghai is the clearest case, not only as the largest city, but as the one probably still most clearly related to its pre-Liberation origins and nature. One suspects that despite Maoist goals, no part of the country, rural or urban, can ever become like Shanghai, and vice versa. The city-country distinction is also, in more general terms, probably permanent, at least to some degree, although it may become considerably less sharp in China than elsewhere. But even if wages, health care, and education can be equalized (which is very far from being realized so far), there really is no way to duplicate the manifold attractions of urban living in the rural communes. China will have to settle for reducing the sharp edges of the differences.

FUNCTIONAL ZONING OF THE CITIES

While the countryside is raising its economic, health, and educational levels and is also becoming involved to some extent in industrialization, the cities are being altered too. National planning decrees what kinds of industry and which plants shall be located in each city. Each city is in turn divided into functional zones, distinct areas of production, commerce, or transport with housing and service facilities associated with each. This is derived to some extent from earlier Soviet styles of urban development, as the Chinese themselves acknowledge, but seems to have been carried a good deal farther in China (Joint Publications Research Service, 1960, 1974, 1975). To some extent, this may approximate at least part of the purpose of the original urban commune, but the units are based largely on the presence of a large factory or other work unit or on a cluster of smaller factories or work units. Workers' housing is grouped in the immediate area, within easy walking or biking distance, and usually in large blocks, though sometimes in a cluster of smaller communities with buildings which we might describe as town houses. Most such areas are also supplied with a full range of services: schools, health clinics or small hospitals, shopping centers for vegetables, fresh fish and meat, dry, canned, preserved, and other foods, clothing, household equipment, and so on. There is often also a cinema and even a savings bank, run of course by the state. Urban wages are often high enough that many families are able to save, especially since all adult members of a family usually work, with aged parents (who normally live with their children) serving as supervisors of small children, shoppers, and house-tenders. Even in a Maoist China, savings bank deposits pay interest. A park and other recreational facilities complete the self-sustained complement of each area. It all seems familiar to a Western observer, perhaps especially the number of such "workers' villages" around the periphery of the still recognizable central business district of Shanghai. Foreign visitors to Shanghai are told that altogether 140 such work-housing complexes have been built there since 1949, approximately 40 of them encompassing over 50,000 square meters each, which is considered large.

A number of Western urban elements are missing however. Housing is strictly functional, and although certainly adequate (some observers have compared it favorably with the average quality of Soviet urban housing), the space alloted per person and per family seems spartan by Western, though not by Chinese, standards (Chao, 1968). Although houses may be and many are privately owned (unlike apartments in the large blocks of workers' housing), one does not see privately owned lawns and gardens, but communally maintained green spaces and plantings. In many such urban areas, one also sees fairly extensive plots or even fields of grain or vegetables; some are apparently run on the same system as the private plots on rural communes cultivated by each household, with the right to consume or sell the produce on an open market, but others are clearly managed on a collective basis with the produce considered as part of the total work output of the unit. Many houses and apartments, probably most, have a radio and at least one bicycle, many have a sewing machine and/or a camera, and many, at least in Shanghai, are provided with flush toilets and a one- or two-burner gas cooker, although on a national basis most cooking even in urban housing is probalby still done traditional style with wood, charcoal, or coal. But all of this represents an enormous improvement in living standards since 1949 and is certainly seen as such by nearly all Chinese, who also have almost no basis for comparison with living standards in other countries.

FUNCTIONAL ZONING AND COMMUTING

Perhaps the chief advantage of the functional zone system which forms the basis for urban planning in China, as compared with the West, is the very much lesser amount of commuting, at best a wasteful process and one which also contributes importantly to congestion, pollution, loss of work and recreational time, and dehumanization, all problems necessarily associated with most cities. There is little need for most urban Chinese to travel beyond the work-living complex in which they live, with its full complement of most services all in walking distance; most people therefore seldom cross the city or travel to the central business district, although there is some movement to cultural attractions, major parks, or sports events on holidays or days off. These tend to be staggered so that only a fraction of the work force is off on any one day, including Sunday. Public transport by bus (where possible electrified) seems still to run far behind demand, judging by the ubiquitous long queues, but there is an immense cloud of bicycles in all Chinese cities, and they surely account for the great bulk of intraurban movement beyond walking distance.

The private car is totally absent, giving China an enormous advantage over all other industrial or industrializing societies in a number of important ways. This is not only because, as a still relatively little industrialized country, it could not produce adequate numbers of private cars but also because, in its planning, it has decided to concentrate investment and production on essentials for economic growth and mass welfare and to deemphasize unnecessary luxuries for private consumption. As the manifold deleterious effects of the private car have become clearer in the West, the wisdom (and good luck) of the Chinese decision have become more striking. One hopes that as industrialization proceeds in China, the planners will be able to hold the line on this vital issue and to avoid the tragic mistake which both Westerners and now Japanese and Russians seem bent on perpetuating. Chinese truck and bus production is growing rapidly, and there is a very small output of cars, for official use only. Some foreign visitors have reported seeing official limousines with children peeking out from drawn curtains, bound for a family outing; this is disturbing, but probably only one of many reminders that the Chinese are as human as other people and that the bourgeois, corrupting nature of urban-based elitism is a perennial problem, which will continue to need perennial revolutionary correction.

NEIGHBORHOOD ORGANIZATION AND SOCIAL CONTROLS

Apart from the work-housing complexes, some of which are associated with newly constructed "satellite towns" arranged along the outer peripheries of the largest cities and usually specializing in a particular industry, such as the electronics district on the edge of Shanghai, urban populations are also grouped into defined neighborhoods. These are designed to implement self-government and local administration, but also to expedite mass mobilization and through it to further both national goals and the ideal of local self-reliance. Neighborhood and block committees are responsible for adjudicating disputes, maintaining houses and lanes, and keeping order. They form the basic units of all other forms of local government and planning, including the local organization links of the Communist party. They are responsible for controlling deviant behavior but depend more on highly effective group pressure, in an old Chinese pattern, rather than on specific sanctions. This structure is principally what remains of the original urban communes; in practice it functions, as the communes were intended to do, to shape individual and group behavior according to national revolutionary norms. No individual is free from its constraints, even in seemingly trivial matters such as the internal cleanliness of houses or the play activities of younger children. Chinese cities are well ordered, at least by contrast with cities elsewhere.

Many urban neighborhoods still retain some of the workshops that were part of the earlier urban commune scheme—small-scale industry designed to supplement and complement larger urban factories. Since about 1960, they seem to employ primarily former housewives, grandparents, and other previously unemployed or underemployed people, often on a part-time basis, although the number of such workshops has been growing. Larger projects of urban improvement, such as constructing new sewage facilities, cleaning up old stream beds, removing former city walls, constructing and maintaining new parks, and building new roads, are undertaken with mass labor drawn from neighborhood committees, factories, schools, and offices in the areas where the projects are located. Each such unit assigns a small part of its total labor force to the project for a set period, usually only a few days at a time, after which a new group of workers from other units takes over; this ensures that the work schedules of each production unit will not be disrupted. Schools everywhere operate as much as possible on the "half work, half study" system (Munro, 1967). Each school, even at the elementary level, maintains one or more workshops which are genuine producing units rather than simply training shops and which turn out goods either for sale or, more commonly, for delivery to a nearby factory as components. Pupils spend approximately half their time in such workshops, which thus serve both the national need and the training of the young in practical skills and attitudes.

The same system operates in the universities, although the proportion of the college-age population in such institutions or in technical schools is still extremely small. Most technical training is on the job, or in after-hours institutes attached to production units. School, university, and technical institute graduates are usually assigned to rural communes, where their help and skills are considered to be most needed. After a period in the countryside, some of them may be reassigned to an urban-based job. Very few students go directly from school to university but must first spend at least two years in productive labor in the countryside and then be recommended by their work units for university admission. In these ways as well, an effort is made to control the growth of urban populations and to disseminate into the rural areas the skills previously confined to the cities, at the same time educating a potential urban elite into the realities of life for the rural masses. More recently there has apparently been a trend toward assigning urban graduates to suburban areas within the municipality for their period of productive labor in agriculture (Peking Review, 1974). It is not clear whether this is simply a move toward greater efficiency and less travel or whether it suggests, as it surely does to some extent, that graduates can now begin to have higher expectations of ultimate reassignment to urban employment. It is, however, also true that there has been a consistent emphasis in recent years on creating a functional interdependent link between industry and agriculture even in the cities, stressing the need for self-sufficiency by increasing agricultural production within the municipality, and involving urban residents in the food production process through their work experience in their own immediate agricultural surroundings, which are perhaps thus less easily regarded or treated as a separate world or a lesser one, since its output is vital to the city.

ENVIRONMENTAL PROTECTION AND INDUSTRIAL DECISION-MAKING

Cities are also to be transformed by correcting their effects in degrading the environment. China has an enormous advantage here in being still only slightly industrialized and hence in being able to plan prevention rather than attempting cure. The absence so far of the private car is another priceless advantage in this effort. But Shanghai and other large industrial cities in the north and northeast (Manchuria) offer immediate examples of the environmental dangers of concentrated industrialization, and the earlier Western and Japanese experiences are also available as object lessons. The Western experience is in fact used to illustrate the tragic environmental consequences of the capitalist system, where the uncontrolled search for private profit leads to environmentally as well as socially destructive results (Peking Review, 1971). But the existing level of easily perceivable atmospheric and water pollution in Shanghai and other industrial cities is enough to sensitize the Chinese planners to the problem, fortunately before most of the country has gone far along the same road. Smog in Shanghai is as bad as in many American cities.

Environmental degradation is to be corrected through industrial dispersal (considered below) and through a combination of cure and prevention. Especially as a still relatively poor and developing country, China cannot afford to scrap its existing industrial plants. Factories of older design (roughly before about 1965, when pollution control began to be considered as a policy problem) are progressively being fitted with emission, waste disposal, and effluent controls as far as practicable and economically feasible. It is impossible to estimate how effective such controls have become or are likely to be in future, but there are frequent accounts in the Chinese press of successful efforts of this sort, and especially of those to make productive use of industrial wastes (New China News Agency, 1971). Many new enterprises have apparently been established using former industrial discharges as their basic raw materials; others reclaim wastes and return the resultant product to the factories. Electric buses are favored over internal combustion buses wherever possible. But the greatest progress will clearly have to be made in the design of new plants, to minimize their damaging environmental effects and to build into the plans devices or arrangements for the productive use of discharges, including the planned clustering of industrial enterprises, large and small, which can make use of each other's wastes. Given the long-standing importance of night soil in maintaining agricultural yields, this is not a new concept to Chinese, and indeed there are a number of new projects to channel urban biological wastes and nontoxic sewage more completely and effectively to the surrounding agricultural areas. All new factories have for several years now been required to meet rigid standards for nondestructive waste disposal and for minimal atmospheric and other forms of pollution. Foreign visitors are taken to see an early but advanced sample of a new model factory which has also been described in the Chinese press—a petrochemical works on the outskirts of Peking, where highly toxic discharges pass through successive treatment. Each stage removes materials which can then be reused, by the same or other enterprises, and at the end of the treatment line is a small lake of clear water in which ducks and fish are kept, for the supply of the factory canteen and for recreation during lunch hour.

As industrialization gathers more momentum, China has probably better prospects than most already industrialized systems to avoid or at least to minimize the destructive effects on the environment. Forewarned is forearmed, and in a planned society specific economic costs may be discounted or subordinated to what are regarded as higher priority requirements. But environmental degradation will also be eased by being in effect spread more thinkly, as part of the larger plan to disperse urbanization and industrialization, its potential problems and its benefits, as widely as possible over the country as a whole. As the so-called barefoot doctors, trained mainly in cities, carry their drugs and skills to the remotest villages, so industrial technology is spread to the countryside through the manufacturing which takes place on the rural communes. The comparison is appropriate in the sense that, in medicine as in industry, the big cities will remain centers of specialization, with fully trained physicians or surgeons and well-equipped hospitals and with large, sophisticated industrial plants. The rural areas are served through a hierarchy of paramedical personnel with only the equipment they can carry on their backs and with enough medical knowledge to distinguish cases which should be referred to the nearest commune hospital, where there is a higher level of equipment and skills, or to a provincial or even national medical center whose still more specialized facilities may be needed. Industrial production is not to be undertaken in remote villages, but each rural commune already does some manufacturing; for some components, or for some more sophisticated finished goods, it is, however, dependent on a provincial or national major industrial base (e.g., Shanghai).

But the ideal remains one of local self-reliance and self-sufficiency, especially in all agriculture-related industrial production such as tools, pumps, irrigation pipes, and fertilizers, but also in consumer goods for local use. Specialized pharmaceuticals, trucks, tractors, and other heavy equipment, for example, will probably continue to be made in one or a few large urban plants, but many rural communes now make their own iron and steel, using local ores and fuel, and from this they make their own electric-driven pumps, following a nationally distributed set of specifications and blueprints. Most generate their own power, and some build their own generators as well as hydro stations (Riskin, 1971; Sigurdson, 1973). This is a better planned effort than the overhasty campaigns of the Great Leap Forward in 1957-1958 and is really not to be compared with the unsuccessful backyard steel furnaces of that period. Rural communes average from 10,000 to 20,000 in population, with an average area of about 2,000 hectares (some are much larger), and most provide not only their own food but a large share of their own manufactured goods. Many produce a surplus in both categories, for sale to other communes or to urban industrial and consumer markets, or for delivery to the state. It is recognized that some areas have a better combination of advantages than others, and that some parts of the country are more productive. These areas remit most of their surpluses, directly or in the form of tax revenues, to the state for redistribution to less favored and poorer areas or communes. The municipality of Shanghai, for example, remits about 90% of its revenues to the central government for reallocation to less developed regions, which may keep all their revenues and receive a central subsidy in addition. Rural communes also attempt to maintain a set of what may be called urban services: schools, medical clinics and some small hospitals, modest recreation and sports facilities, and community buildings for meetings and for housing local or traveling theatrical performances or film showings. Some of the more prosperous communes include in their small hospitals even Western-style dental chairs and electric-driven drills, plus X-ray and other sophisticated equipment. What are generally missing are shops or a shopping center, plus of course an urban atmosphere of paved streets, crowds, and excitement.

GROWTH OF NEW CITIES

The other aspect of urban and industrial dispersal, so far and probably permanently more important at least in quantitative terms, is the growth of new cities of varying sizes in the inland areas, which before 1949 were very slightly urbanized and largely devoid of modern industry. Every province now has at least one major industrial city, most of them largely new, some transformed from former small towns in regional backwaters. These new developments have been keyed to recently discovered local raw materials (coal, iron, oil, natural gas) or to resources which had lain unused while industrial development centered so heavily in the treaty ports in their tiny dots along the coast. A national skeleton of railways and highways has been constructed to link these new centers with their hinterlands and with the still dominant major industrial bases in the former treaty ports, which have provided the original technical skills and sophisticated equipment, as they still do to some extent. But regional self-sufficiency is stressed as well as regional development per se, and the effort is to reduce any area's or city's dependence on distant sources of anything which it needs. Transport is expensive and wasteful, especially in a developing economy, but equally important is the need to mobilize and commit local and individual resources and energies, physical and human, to the national drive for economic growth. This is much easier to achieve if each area and its people can be made as much as possible dependent on their own efforts and resources and can work out their own problems without having to call in outside help, or outside experts; people learn better from their own experience and mistakes than from dependence on outsiders. This also has the effect of mobilizing local physical resources more quickly and more completely and of drawing the entire country into the process of development rather than having it depend on the slower and partial "trickle-down" process from a few large cities. In any case, change from the top down seldom works as well or as quickly as mass-based change from the bottom up, which was of course the basic strategy of the Maoist revolution from the beginning.

Some of these new cities, including a few which were in existence before but on a much smaller and very different basis, have been completely planned and are described in the Chinese press as models of urban development. This is especially the case with the wholly new oil drilling and refining city of Ta Ch'ing in the Sungari basin of Manchuria, not far from the older city of Harbin. Very few foreign visitors have been there, and the accounts in the Chinese press are frustratingly general or even vague, but Ta Ch'ing is extolled as a model which all cities should follow, especially in its allegedly successful effort to eliminate the distinctions between mental and manual labor and between city and countryside, industry and agriculture. It is described as being laid out around a smallish central town which provides the usual urban services, but surrounded by a ring of satellite communities of about 6,000 each in population which are involved in both manufacturing and agriculture. Between the satellite ring and the central town are more uniformly agricultural areas, and the whole complex is referred to as an agroville, using an originally Western term which Ta Ch'ing may well exemplify best, at least in concrete reality (Foreign Languages Press, 1972). Another model city is Chengchow, the capital of Honan province; before 1949 it was a small, backward, and unattractive town; now it is almost entirely replanned and rebuilt, with broad, tree-lined avenues and with housing, factories, and public buildings spread out over a very large urban area in functional clusters, separated by greenbelts which are used for productive agriculture as well as for recreation and pollution control.

Other cities are, like Ta Ch'ing, wholly new and based on newly discovered local resources, such as the iron and steel city at Paot'ou in Inner Mongolia. Paot'ou makes a good illustration of national planning policy in that when a rich field of iron ore was found in the vicinity, the decision was made to locate the iron and steel capacity there rather than hauling the ore to an existing industrial center or city closer to the center of the market. Paot'ou was picked in part precisely because it was remote, so that industrialization there could accelerate the development of this previously backward region as a whole and involve its people directly. The same approach underlies the boom industrial and urban development of distant Sinkiang, where the principal urban-industrial center at Urumchi is nearly 1,500 miles from Peking and considerably farther from the national center of population or production.

It is almost certainly more expensive to pursue industrialization on a dispersed as opposed to a concentrated basis. Unit costs of production may be particularly high for manufacturing on the rural communes by comparison with large urban plants which can enjoy the economies of scale, even if one allows for the costs of transport saved through not having to haul raw materials or finished products. Even for many of the new provincial urban centers of manufacturing, unit costs are probably often higher than in Shanghai or Manchuria. If rapid industrialization at least cost were the only goal, the former treaty ports would retain their quasi-monopoly, since by and large they remain the cheapest points of assembly and distribution and enjoy the lowest production costs, which were the chief reasons for their growth to urban dominance before 1949. They are still close to the center of the national market in population terms. But a Maoist China, while determined to industrialize, is concerned with more than strictly economic goals. As Mao himself put it, it is man that counts, or "people are the most precious things," not gross national product. Regional development is thus pursued as something axiomatically good whether or not it happens to be marginally or substantially more expensive. The experience and the fruits of industrialization, and of urban services, are to be brought as equally as possible to everyone and to all areas. Rural-urban distinctions are to be broken down, and the problems of urban overconcentration as well as urban corruption of values are to be corrected. Such goals are more important than least-coast solutions in purely economic terms. Indeed one of the cardinal sins, along with elitism, is identified as "economism," or concentration on personal material gain (especially in the cities) rather than on the selfless pursuit of national goals or group welfare. Nor is technology a god, an end in itself; it must serve the interests of the people, as must the cities themselves.

POLICY PROBLEMS

A great deal has been accomplished in all these respects to fundamentally alter China's urban patterns. This effort to develop a new approach to urbanism holds promise not only for China, but in some respects for the rest of the world, developed and developing, still grappling with the problems which Western-style urban-industrial concentration has brought with it. But if we are to examine the Chinese experience with profit, it is important not to be so carried away with its ideal goals; it is important to see its imperfections and the extent to which it has not achieved its own goals. The Chinese are more realistic and straightforward about this, in their own press, than are many starry-eyed but inadequately informed Western visitors, who are understandably impressed by what they see and hear and contrast it favorably with what they know in far more representative detail about their own cities at home. The Chinese freely acknowledge that there are still unresolved problems, still much to learn. And it is not difficult to detect conflicting goals, or at least conflicting trends, which seem likely to prevent the full or perhaps even partial realization of the most exciting aspects of the Maoist ideal, especially the transformation of the nature of cities and the elimination of rural-urban differences.

For all of the innovation and promise of rural-based commune manufacturing, it appears still to account for only about 10% of all industrial production; most production still centers in the former treaty ports (still by far the largest cities). Although their dominance is lessening as provincial and inland cities grow and industrialize, it seems unlikely to disappear. Cities over 100,000 have apparently continued to grow faster than small towns or rural areas. There were only nine cities at the time of the 1953 census with populations of a million or more; by the early 1970s, there were over 20, most of them former treaty ports. Shanghai remains by a large margin the country's biggest city. With all this new urban growth, it is difficult to prevent the consequent growth of new urban elites, and indeed the Chinese acknowledge this as a problem which must continue to be corrected. Even the wholly new worker-peasant agrovilles such as Ta Ch'ing are nevertheless cities, less different from the countryside and less elitist, no doubt, than most cities elsewhere, but still capable of generating both pollution and the moral degradation of bourgeois values. Living standards, wages, health care, education, recreation, and general stimulation are all on a higher level in virtually all Chinese cities—highest of all in the largest—than in the countryside, and seem certain to remain so. There is simply no way to transfer to the countryside the excitement and variety of urban living, nor the amenities or diversions represented by department stores, street crowds, museums, cinemas, and major sports events. It is primarily in the cities also that the most desirable jobs will continue to be found, not only in terms of pay rates but with the attractions of power, responsibility, creativity, leadership—in other words, elite status.

For all the inspiring quality of the Maoist vision, the old Adam has proved to be very resistant. Chinese are not different in these terms from any other national group. Many even of the cadres, supposedly the most selfless and responsible members of society, try to avoid rural and seek urban assignment, as do many of the university and technical graduates. Those with kin or other personal connections in the urban-based hierarchy try to use them to obtain favorable—i.e., urban—employment and higher status. The state still has to cope as well as it can with illegal migration to the cities from the countryside. Some outside visitors report the existence in the larger cities of a substantial subculture of such migrants, usually young people, living without official knowledge or approval as best they can, and some of them forming street gangs who live to some extent by pilfering, although crime even in the big cities is in general amazingly low.

There are some signs also of an equally disturbing problem which we would label consumerism. Urban stores display a surprising number of consumer goods, many of them far from necessities, although all are Chinese-made: cameras, wristwatches, transistor radios, a great variety of toys, flowered prints, and a wide range of household goods, plus others which may combine utilitarian with recreational or avocational appeal, such as bicycles, sewing machines, materials for drawing and painting, and assorted house furnishings and garden tools. One picks up complaints from time to time that some people even in the cities feel that the prices on many of these goods (all of which are set by the state) are too high, in relation to wage levels. As long as every individual or household does not have all these things, there will be a sense of relative deprivation, as well perhaps as the insidious pressures to keep up with the Joneses (Wangs). Wage and salary rates are far from equal even in given cities; the national spread is on the order of 1 to 10, and within individual cities (as in many rural areas) it may be as great as 1 to 4 or 5. This is, of course, a part of conscious policy, despite the egalitarian overtones of the Maoist system, and is related to the need seen to provide incentives (Howe, 1971). But if material incentives are thus granted some importance, as clearly they are, it is difficult to prevent the development of "economism" and hence also of consumerism. These sorts of conflicts with Maoist ideals are of course greatly sharpened in any comparison between wage and consumption patterns in the cities and in the rural areas, a contrast which at least so far has been only a little softened by the provision of goods and services, on a limited and largely utilitarian basis, to rural areas through the communes.

The basic conflict is between a China bent on increasing production and living standards and a China at the same time committed to austere and self-denying revolutionary ideals, with an inherited anti-urban component. Production and living standards are indeed increasing, for nearly everyone, at impressive rates.

But consumption levels are not increasing equally for everyone, and even if they were or could ultimately be made to do so, there is little reason to think that the Chinese, despite the inspiration of Maoist ideals, can remain immune to the blandishments of consumerism, any more than many of them so far seem to have been able to resist the attractions of the city and its disproportionate but selfishly bourgeois rewards. None of this should be taken to suggest that China and Chinese cities will soon or even ultimately become like the rest of the world, still less like the industrial West. The Maoist revolutionary vision has genuinely transformed China, its people and its cities. They will never again be the same, nor will they lose their distinctiveness by comparison with the rest of the world, even as the revolution eventually cools, as surely it must to some degree, as Mao himself and the early revolutionary struggles he directed recede into history. The legacy of his vision is permanent, and his innovative model for urban development will continue to be pursued. It will also go on being attractive to other developing countries, and many of its aspects will continue to appeal as well to developed countries which have come to see serious flaws in their own urban models and may find inspiration from some of the radical innovations which Maoist China has pioneered. But the sharpness and purity of the Maoist ideal, for all its inspirational quality, seems likely to become blurred as time passes and as the Chinese drive for industrialization and modernization gathers further momentum, based in large part and ineluctably in cities. They will not be like other cities, and there will be less of a split between them and their rural surroundings than in most of the rest of the world, but they will not conform either to the perfect model which the Maoist ideal has created in men's minds.

REFERENCES

CHAO, K. (1968). The construction industry in Communist China. Chicago: Aldine.

Foreign Languages Press (1972). Taching: Red banner on China's industrial front. Peking: Author.

HOWE, C. (1971). Employment and economic growth in urban China, 1949-1957. Cambridge: Cambridge University Press.

Joint Publications Research Service (1960). "Great achievements in urban construction during the last decade" (No. 18-29). Washington, D.C.: U.S. Government Printing Office.

--- (1974). "Sian: Ancient and new" (China Reconstructs, No. 28-35). Washington, D.C.: U.S. Government Printing Office.

--- (1975). "Shenyang: Socialist industrial city" (China Reconstructs, No. 17-72). Washington, D.C.: U.S. Government Printing Office.

MUNRO, D.J. (1967). "Marxism and realities in China's educational policy: The half-work half-study model." Asian Survey, 7:254-272.

MURPHEY, R. (1972). "City and countryside as ideological issues: India and China." Comparative Studies in Society and History, 14:250-267.

New China News Agency (1971). "Northwest China city makes multi-purpose use of industrial waste." Press release, May 15.

Peking Review (1971). "Multi-purpose use: Important policy for industrial production." 28:14.

--- (1974). "A good way to settle educated youth in the countryside." 31:13-14.

PERKINS, D. (1975). China's modern economy in historical perspective. Stanford, Calif.: Stanford University Press.

RISKIN, C. (1971). "Small industry and the Chinese model of development." China Quarterly, 46:269-273.

SCHURMAN, F. (1968). Ideology and organization in Communist China. Berkeley: University of California Press.

SIGURDSON, J. (1973). "Rural industry and the internal transfer of technology." In S.C. Schram (ed.), Authority, participation, and the cultural change in China. Cambridge: Cambridge University Press.

THE AUTHORS

JANET L. ABU-LUGHOD is Professor of Sociology and Urban Affairs and Director of the Comparative Urban Studies Program at Northwestern University. Her *Cairo: 1001 Years of the City Victorious* was published in 1971, and she is now working on a book comparing the major capitals of North Africa.

BRIAN J.L. BERRY is the Williams Professor of City and Regional Planning and Director of the Laboratory for Computer Graphics and Spatial Analysis in the Graduate School of Design at Harvard University. His most recent books include *The Geography of Economic Systems* (1976), *The Changing Shape of Metropolitan America* (1976), *The Social Burdens of Environmental Pollution* (1976), *Race and Housing* (1976), and *Essays on the Science of Geography* (1976).

L.S. BOURNE is Professor of Geography and Director of the Centre for Urban and Community Studies at the University of Toronto. His recent books include *Urban Futures for Central Canada* (1974) and *Urban Systems: Strategies for Regulation* (1976).

E.C. CONKLING is Chairman of the Department of Geography at the State University of New York at Buffalo. His most recent books are *The Geography of Economic Systems* (1976) and *Man's Economic Environment* (1976).

R.J. DAVIES is Professor and Head of the Department of Geography at the University of Cape Town. He is the author of papers on urban geography and coauthor with Leo Kuper and Hillstan Watts of a book, now being revised, on the *Racial Ecology of the City of Durban.*

ROY DREWETT is Lecturer in Geography at the London School of Economics and Political Science. His current research interests include the Urban Change Project and a cross-national comparative study of the Costs of Urban Growth in Europe coordinated by the Centre Europeen Sciences Sociales, Vienna, where he is project director. Recent publications include *Land Values and Urban Growth* and *The Containment of Urban England.*

T.J.D. FAIR is Director of the Urban and Regional Research Unit, University of the Witwatersrand, Johannesburg, South Africa. His most recent works are *Development in Swaziland* (with G. Murdoch and H. Jones, 1969) and *The Witwatersrand: A Study in Metropolitan Analysis* (with L.P. Green, 1974).

SPERIDIAO FAISSOL is a geographer and Superintendent of Geographic and Socio-Economic Studies at the Brazilian Institute of Geography and Statistics. He is also Professor at the Graduate Course in Geography of the Federal University of Rio de Janeiro-Brazil. He has published extensively, especially in the *Revista Brasileira de Geografia.*

JOHN GODDARD is Henry Daysh Professor of Regional Development Studies at the University of Newcastle upon Tyne, England. He is author of *Office Location in Urban dnd Regional Development* (1975), *Office Linkages and Location* (1974), and *The Communications Factor in Office Decentralisation* (1976). He prepared the report *National Settlement Strategies* (1973), which led to the commissioning of the study of the British urban system reported in this volume.

ROBERT G. JENSEN is Associate Professor and Chairman of the Department of Geography at Syracuse University. Since 1967 he has served on the Advisory Committee of *Soviet Geography: Review and Translation.* His articles in the field of Soviet geography have appeared in the *Annals of the AAG, Geographical Review,* and the *Journal of Regional Science,* and he has contributed to several books dealing with the Soviet Union, including *The Soviet Economy in Regional Perspective* (1973).

ELISABETH LICHTENBERGER is Chairman of the Department of Geography and Professor of Regional Research and Regional Planning at the University of Vienna, Austria, editor of *Mitteilungen der Osterreichischen Geographischen Gesellschaft,* and a member of the Advisory Board of Experts on Regional

Planning of the Austrian Federal Government. Her most recent publications include *The Eastern Alps* (1975) and a wide range of articles in the field of urban research.

M.I. LOGAN is Chairman of the Department of Geography and Chairman of the Board of Studies in Environmental Science at Monash University, Melbourne, Australia. From 1973 to 1975 he was an adviser to the Department of Urban and Regional Development in the Australian government. His most recent book is *Urban and Regional Australia: Analysis and Policy Issues* (with three others, 1975).

JAMES E. McCONNELL is an Associate Professor of Geography at the State University of New York at Buffalo. His most recent publication, *Exporting Business Decision Making, and Trade Policy* (1975), reflects a research emphasis in the areas of international trade and regional economics. He has also conducted field research in Central America.

M.L. McNULTY is Associate Professor of Geography and Program Director of the Center for the Study of Urban Growth in Developing Countries at the University of Iowa.

RHOADS MURPHEY is Professor of Geography and of Asian Studies at the University of Michigan, where he is also Director of the Center for Chinese Studies. He is the author of *An Introduction to Geography* (1961 and 1977), *Shanghai: Key to Modern China* (1953), *The China Treaty Ports* (1975), *The Outsiders: Westerners in India and China* (1976), and numerous other studies of China and of Asia.

EDWARD W. SOJA is a Professor of Regional Planning at the School of Architecture and Urban Planning, University of California, Los Angeles. He was formerly in the Department of Geography at Northwestern University and has also taught at the Universities of Nairobi and Ibadan and at the Interdisciplinary Institute for Spatial Planning in Vienna. Among his publications are *The Geography of Modernization in Kenya* (1968), *The African Experience* (3 vols. with John Paden, 1970), *The Political Organization of Space* (1971), and several articles on spatial development in Africa.

NIGEL SPENCE is a Lecturer in Geography at the London School of Economics and Political Science. He is interested in urban and regional planning problems with special emphasis on quantitative economic analysis. Currently he is codirecting research for the Department of the Environment on Urban Change in Britain and for the Social Science Research Council on Regional Unemployment Variations in Britain. He is coauthor of *British Cities. An Analysis of Population and Employment Trends 1951-1971* (1976).

CLYDE E. WEAVER is a Ph.D. candidate in the Urban Planning Program, School of Architecture and Urban Planning, University of California, Los Angeles.

YUE-MAN YEUNG is a Senior Program Officer of the Social Sciences and Human Resources Division of the International Development Research Centre, Canada. His recent publications include *National Development Policy and Urban Transportation in Singapore* (1973) and *Changing Southeast Asian Cities* (co-editor, 1976).